Green Tea Polyphenols

NUTRACEUTICALS OF MODERN LIFE

Green Tea Polyphenols

NUTRACEUTICALS OF MODERN LIFE

Edited by

Lekh R. Juneja • Mahendra P. Kapoor
Tsutomu Okubo • Theertham P. Rao

CRC Press
Taylor & Francis Group
Boca Raton London New York

CRC Press is an imprint of the
Taylor & Francis Group, an **informa** business

CRC Press
Taylor & Francis Group
6000 Broken Sound Parkway NW, Suite 300
Boca Raton, FL 33487-2742

First issued in paperback 2016

© 2013 by Taylor & Francis Group, LLC
CRC Press is an imprint of Taylor & Francis Group, an Informa business

No claim to original U.S. Government works

ISBN 13: 978-1-138-19937-8 (pbk)
ISBN 13: 978-1-4398-4788-6 (hbk)

Library of Congress Cataloging-in-Publication Data

Green tea polyphenols : nutraceuticals of modern life / editors, Lekh R. Juneja, Mahendra P. Kapoor, Tsutomu Okubo, Theertham Rao.
   pages cm
   "A CRC title."
   Includes bibliographical references and index.
   ISBN 978-1-4398-4788-6 (hardcover : alk. paper)
   1. Green tea--Therapeutic use. 2. Plant polyphenols--Health aspects. 3. Functional foods. I. Juneja, Lekh R., editor of compilation. II. Kapoor, Mahendra P., editor of compilation. III. Okubo, Tsutomu, editor of compilation. IV. Rao, Theertham, editor of compilation.

RM251.G74 2013
615.3'21--dc23   2013008834

Visit the Taylor & Francis Web site at
http://www.taylorandfrancis.com

and the CRC Press Web site at
http://www.crcpress.com

Contents

Preface

This book represents an extended and refreshed collection of global findings on the functional health benefits of green tea polyphenols. In the year 1211, the Japanese Buddhist priest Eisai Zenji proclaimed in the *Kissa Yojouki* that green tea is an excellent medicine and that it prolongs life. It is just in the past 50 years, however, that we have come to appreciate how true this visionary statement was. We have scientifically explored the health-promoting benefits of green tea and its various components including polyphenols, catechins, L-theanine, and caffeine; and to date there is a wealth of published research on its beneficial effects. This scientific research now supports Eisai Zenji's historic claim that green tea and its various components play an important role in supporting the health of our modern society and its unique lifestyle-related diseases. While the content presented in this book does not come close to covering all of the research to date, we have endeavored to accumulate the latest knowledge on the topic by inviting competent authorities in the field of green tea science to contribute. We have strived to make the book comprehensive and authoritative by adding extensive references to guide researchers, scientists, and regulatory bodies to make appropriate decisions on scientific and regulatory aspects. Each chapter includes a preview of specific themes and highlights the most recent research and development conducted in the field.

In the first three chapters we cover important topics on the processing, chemical composition, and properties of green tea polyphenols. The remaining chapters deal with the numerous health benefits associated with the consumption of green tea polyphenols related to cancer, cardiovascular disease, bone and muscle health, diabetes and weight management, protection of internal organs, allergies, oral care, inflammation, and gut health.

Standardized green tea components are of utmost importance to providing quality and efficacious commercial applications. Consumption of green tea polyphenols have been shown to enhance antioxidant activity, improve fat metabolism, increase energy expenditure, modulate appetite, and govern blood glucose management. Additionally, green tea polyphenols offer neuroprotection, cardiovascular support, reduced risk of cancer development and metastasis. Finally, the reduction of inflammation, skin aging and skin cancer protection, antibacterial and antiviral activity, testosterone metabolism, and enhancement of the glucuronidation detoxification pathways are effectively proven benefits of green tea polyphenols.

The natural botanicals market continues to mature and the broad application of green tea places it at the forefront of the industry. The prevalence of green tea polyphenol fortified foods, beverages, and dietary supplements can be attributed to the general consumer awareness of green tea and the need for natural antioxidants in their diet. The support of scientific research has further encouraged the food, beverage, supplement, and pharmaceutical industries worldwide to use green tea polyphenols in their products. Green tea polyphenol extracts are now being investigated for even further novel uses beyond food applications, including cosmetic, toiletry, and industrial applications.

We acknowledge all the researchers, scientists, and physicians who have contributed to this book by sharing their years of research and knowledge in this field related to the treatment and prevention of disease. We appreciate the support of Nagahiro Yamazaki, President of Taiyo Kagaku, who himself consumes a daily dose of green tea polyphenols. We would also like to thank our collaborators from industry and academia, who have played such an important role in this long journey for more than 25 years, striving to explore the health benefits of green tea. We also thank Scott Smith, Vice President of Taiyo International, and other colleagues at Taiyo Kagaku, Japan, for their endless support. The quest for health and longevity continues, and natural antioxidants such as green tea polyphenols will play an ever-increasing critical role.

About the Editors

Lekh R. Juneja, Ph.D., is internationally acclaimed for developing various nutraceuticals and functional foods and for his research into functional ingredients from green tea, eggs, water-soluble dietary fiber, and nutrient delivery systems of minerals, vitamins, and polyunsaturated fatty acids; and nanoporous materials. He has authored and coauthored more than 175 research papers as well as two books, including *Chemistry and Applications of Green Tea and Hen Eggs: Their Basic and Applied Science* (CRC Press), and holds more than 125 patents. Dr. Juneja is executive vice president of Taiyo Kagaku Co. Ltd., Japan, as well as CEO and chairman of Taiyo Lucid Pvt. Ltd., India, and Taiyo Green Power Co. Ltd., China. He is also president of Taiyo GmbH, Germany, and Taiyo-Labo Co. Ltd., Japan, a new company directly serving the health care business. Dr. Juneja has received many international awards and has been acknowledged for innovative research and commercial achievements in newspapers, magazines, and on television. He was director of the board and a member of the International Scientific Advisory Committee of the Advanced Food Materials and Nutraceuticals Network (AFMNet) Canada until 2011. He was president of the International College of Nutrition from 2010 to 2011 and has been a fellow since 2005. Other major positions include the advisory board of the IFT Nanoscience Panel, the Society for Novel Food, the Japan Society of Tea Catechinology, the Asian Food Community, and the Board of Foreign Investment, Mie, Japan. He is also affiliated with the New York Academy of Sciences, the American Association for the Advancement of Science, the American Oil Chemists' Society, and the Institute of Food Technologists. Dr. Juneja has given more than 100 lectures and keynote addresses at different conferences. He earned his Ph.D. at Nagoya University in Japan. (E-mail: juneja@taiyokagaku.co.jp)

Mahendra P. Kapoor, Ph.D., the lead contact for this book, is a senior manager at Taiyo Kagaku Co. Ltd., located in Yokkaichi, Mie, Japan. He represents the International Sales Division and provides marketing aids to develop, coordinate, and manage research programs in the areas of value-added novel products for food, beverages, nutraceuticals, cosmetics applications, and nanoporous materials. He served as the researcher at the Agency of Industrial Science and Technology, Japan, and Toyota Central R&D Labs., Inc., before joining as a senior scientist in the Nano-Function/Nutrition Division of the Taiyo Kagaku Co. Ltd. Dr. Kapoor currently serves as editor in chief of the *Open Catalysis Journal*, published by Bentham Science Publishers which focuses on various aspects of material science, catalysts, and reaction engineering. He also serves on the editorial board of the *E-Journal of Chemistry* of Hindawi Publishing Corporation. Dr. Kapoor has been active in professional societies, including the American (ACS) and Japanese (JCS) Chemical Societies, and has been frequently sought to speak on nanotechnology for energy, environment, and foods at professional conferences and symposia worldwide. (E-mail: mkapoor@taiyokagaku.co.jp)

Tsutomu Okubo, Ph.D., began his carrier in Taiyo Kagaku, Japan, about 25 years ago as an employed researcher working on the antimicrobial and antioxidative activities of plant extracts and their influence on human health. In the early days of his biotechnology studies, he researched fermentation technology, microbiology, and biological conversion of imminent food—such as green tea, eggs, and the beans of various medical herbs—into valuable commercial functional products primarily derived from natural resources. Later, he completed his Ph.D. in biotechnology and currently works as assistant general manager of R&D in the Nutrition Division of Taiyo Kagaku Co. Ltd. Dr. Okubo is responsible for several ongoing projects involving personalized nutrition and support initiative concepts on functional food development in the marketplace. Dr. Okubo's scientific activities have resulted in a number of peer-reviewed publications and various patents in the domestic and international arenas. (E-mail: tohkubo@taiyokagaku.co.jp)

Theertham P. Rao, Ph.D., is an assistant general manager at Taiyo Kagaku Co. Ltd. located in Yokkaichi, Mie, Japan. He is a technological advocate, new market developer, and key ally for international accounts. He provides marketing concepts and coordinates and collaborates international research programs in the areas of nutraceuticals and food ingredients. He served as a scientist at the International Crop Research Institute for Semi-Arid Tropics (ICRISAT), India, and researched the nutrition of crops of semiarid tropics. Later, Dr. Rao joined Taiyo Kagaku to focus research on the development of functional ingredients from Ayurveda. Dr. Rao qualified for his Ph.D. from Nagoya University, Japan, and for international food regulations from Michigan State University. Dr. Rao currently serves on the editorial board of NutraCos. (E-mail: tprao@taiyokagaku.co.jp)

Contributors

Vaqar M. Adhami
Department of Dermatology
University of Wisconsin–Madison
Medical Sciences Center
Madison, Wisconsin
E-mail: vmadhami@wisc.edu

Jack F. Bukowski
Division of Rheumatology, Allergy,
and Immunology
Department of Medicine
Brigham and Women's Hospital and
Harvard Medical School
Boston, Massachusetts
E-mail: jack.bukowski@Pfizer.com

Olivier M. Dorchies
Department of Pharmacology
University of Geneva
Geneva, Switzerland
E-mail: olivier.dorchies@unige.ch

Sanjay Gupta
Department of Urology
Case Western Reserve University
and University Hospitals Case
Medical Center
Cleveland, Ohio
E-mail: sanjay.gupta@case.edu

Tadashi Hase
Biological Science Research
Kao Corporation
Tochigi, Japan
E-mail: hase.tadashi@kao.co.jp

Masatomo Hirasawa
Department of Microbiology and
Immunology
School of Dentistry at Matsudo
Nihon University
Chiba, Japan
E-mail: hirasawa.masatomo@
nihon-u.ac.jp

Chi-Tang Ho
Department of Food Science
Rutgers University
New Brunswick, New Jersey
E-mail: ho@aesop.rutgers.edu

Hla-Hla Htay
Taiyo Kagaku Co. Ltd.
Mie, Japan
E-mail: hlahlahtay@taiyokagaku.co.jp

Santosh K. Katiyar
Department of Dermatology
University of Alabama at
Birmingham
Birmingham, Alabama
E-mail: skatiyar@uab.edu

Naghma Khan
Department of Dermatology
University of Wisconsin–Madison
Medical Sciences Center
Madison, Wisconsin
E-mail: nkhan2@wisc.edu

Isao Kouno
Department of Natural Product
    Chemistry
Graduate School of Biomedical
    Sciences
Nagasaki University
Nagasaki, Japan
E-mail: ikouno@nagasaki-u.ac.jp

Shiming Li
Department of Food Science
Rutgers University
New Brunswick, New Jersey
E-mail: shiming3702@gmail.com

Yosuke Matsuo
Department of Natural Product
    Chemistry
Graduate School of Biomedical
    Sciences
Nagasaki University
Nagasaki, Japan
E-mail: y-matsuo@nagasaki-u.ac.jp

Yoshinori Mine
Department of Food Science
University of Guelph
Guelph, Ontario, Canada
E-mail: ymine@uoguelph.ca

Hasan Mukhtar
Department of Dermatology
University of Wisconsin–Madison
Medical Sciences Center
Madison, Wisconsin
E-mail: hmukhtar@wisc.edu

Jeong Sook Noh
Institute of Natural Medicine
University of Toyama
Toyama, Japan
E-mail: noh@ms.toyama-mpu.ac.jp

Chan Hum Park
Institute of Natural Medicine
University of Toyama
Toyama, Japan
E-mail: parkchan@ms.toyama-mpu.
    ac.jp

Jong Cheol Park
Department of Oriental Medicine
    Resources
Sunchon National University
Jeonnam, Korea
E-mail: parkjong@ms.toyama-mpu.
    ac.jp

Molay K. Roy
Department of Food Science
University of Guelph
Guelph, Ontario, Canada
E-mail: ymine@uoguelph.ca

Urs T. Ruegg
Department of Pharmacology
University of Geneva
Geneva, Switzerland
E-mail: urs.ruegg@unige.ch

Tadashi Sakuma
Human Health Care Research
Kao Corporation
Tokyo, Japan
E-mail: sakuma.tadashi@kao.co.jp

Imtiaz A. Siddiqui
Department of Dermatology
University of Wisconsin–Madison
Medical Sciences Center
Madison, Wisconsin
E-mail: iasiddiqui@wisc.edu

Hirofumi Tachibana
Department of Bioscience and
    Biotechnology
Faculty of Agriculture
Kyushu University
Fukuoka, Japan
E-mail: tatibana@agr.kyushu-u.ac.jp

Kazuko Takada
Department of Microbiology and
    Immunology
School of Dentistry at Matsudo
Nihon University
Chiba, Japan
E-mail: takada.kazuko@nihon-u.ac.jp

Hideto Takase
Health Care Food Research
Kao Corporation
Tokyo, Japan
E-mail: takase.hideto@kao.co.jp

Takashi Tanaka
Department of Natural Product
    Chemistry
Graduate School of Biomedical
    Sciences
Nagasaki University
Nagasaki, Japan
E-mail: t-tanaka@nagasaki-u.ac.jp

Vijay S. Thakur
Department of Urology
Case Western Reserve University
    and University Hospitals Case
    Medical Center
Cleveland, Ohio
E-mail: vst2@case.edu

Ichiro Tokimitsu
R&D
Kao Corporation
Tokyo, Japan
E-mail: tokimitsu.ichirou@kao.co.jp

Takako Yokozawa
Institute of Natural Medicine
University of Toyama
Toyama, Japan
E-mail: yokozawa@inm.u-toyama.
    ac.jp

1

Green Tea

History, Processing Techniques, Principles, Traditions, Features, and Attractions

Mahendra P. Kapoor, Theertham P. Rao, Tsutomu Okubo, and Lekh R. Juneja

Contents

1.1 Introduction

The origins of tea are that it began as a medicine and only later grew into a beverage. The tea plant was known from very early times to Chinese medicine. The tracing of ancient tea history, however, is a tedious task and highly controversial. It is believed that tea was accidently discovered by a Chinese king, Shen Nong, in nearly 3000 B.C. (Lu, 1995; Shouyi, 1982). Another legend, originating from India, claims that tea was originally grown in India, and with the spread of Buddhism (through Prince Siddhartha, also known as Buddha) it was brought to China, Korea, and Japan. Despite the uncertainty surrounding the history of tea consumption, the first-ever encyclopedia description was summarized elsewhere (Ukers, 1935). According to the encyclopedia, currently both India and China are considered the birthplace of tea. Turkey was the first to introduce tea (nearly 600 A.D.) to the West. Later, China entered the tea market and during the Song dynasty (960 A.D.) began exporting tea. From around 1200 A.D., the Dutch and British established tea trade between Europe and China. Eventually, the Chinese domination of tea export ended around 1860 A.D., when the British started a planned cultivation of teas in Sri Lanka and India, specifically for export to European countries. Today, more than 25 countries are producing different kinds of teas not only for simple pleasure and a relaxing tea drink, but also for the health benefits due to growing knowledge and awareness of the scientific confirmation of their health benefits (Cabrera et al., 2006).

1.2 Botanical Classification of Tea Plants

Today the tea plant known as *Camellia sinensis* (L) O. Kuntze (named by the famous botanist C. Linne in 1753) is classified into two common species: var. *sinensis* and var. *assamica*.

These above two varieties are currently cultivated in more than 25 countries around the globe (Kitamura, 1950; Weatherstone, 1992). There are no botanical evidences of the wild-type sources of the aforementioned cultivars. Instead, semi-wild-type cultivars could be found near the areas surrounding Assam and Myanmar. The very first record of the botanical classification was summarized by Linnaeus in 1752, wherein tea cultivars were categorized as *Thea bohea* and *Thea sinensis*. Later, on the basis of classification, *Thea sinensis* became the designation for the small-leaved Chinese variety and *Thea assamica* the large-leaved Assam plant (Masters, 1844). A compendium of research on the genus of *Camellia*, dealing with more than 80 tea species for taxonomy, was summarized by Sealy in 1958 (Sealy, 1958, and references therein). In that work, *Thea* and *Camellia* were considered to be separate genera and cultivated tea plants included in the genus *Thea* and the nontea camellias in the genus *Camellia*. At present, both *Thea* and *Camellia* are considered to be synonymous, as the genus *Camellia* now also relates to the Thea family (Sharma and Venkataramani, 1974; Wight, 1962). Table 1.1 shows the different species of tea leaves categorized on the basis of their characteristic

TABLE 1.1
Different Tea Species Distributed in Relevant Category on the Basis of Their Characteristic Chemical Compositions

Subgenera/Species	Polyphenols	Theanine	Caffeine
Thea			
C. sinensis			
var. sinensis	13.52	1.21	2.78
var. assamica	17.26	1.43	2.44
C. taliensis	10.12	0.27	2.54
C. irrawadiensis	1.75	0.21	0
Camellia			
C. japonica			
var. japonica	5.06	0	0
var. decumbens	5.61	0	0
C. riticulata	0.37	0	0
C. saluensis	0.42	0	0
C. pitardii	6.89	0	0
C. sasanqua	0.02	0	0
C. oleifera	0.19	0	0
Hybrids			
C. sasanqua × C. sinensis	1.37	0.05	2.48
C. sinensis × C. japonica	2.81	0.15	0.25
C. japonica × C. kissi	0	0	0

chemical compositions. Growth habitat and leaf features continue to be used to distinguish between *C. sinensis* and *C. assamica*. Cultivars of var. *sinensis* are bush type, a small slow-growing shrub with small, serrate, narrow, dark green leaves. Cultivars of var. *assamica* are quick growing, tree type, with large, broad, horizontal, light green leaves (Harler, 1956; Sealy, 1958). The var. *sinensis* can survive in winter as cold as −12°C, whereas those of var. *assamica* perish at −4°C in a few weeks. Therefore, the var. *sinensis* have been cultivated in the lower temperature regions (Japan, eastern and southern parts of China) and var. *assamica* in the tropical and subtropical regions (Myanmar, Assam (India), and Yunnan (China)). A universally acceptable set of criteria for differentiating the three modern tea cultivars are provided by floral morphology (via disposition of the stylar arm), a variation in the number of styles and the degree of their fusion, and a globular or pubescent ovary is effectively considered. As an example, the styles are free for most of their length in *C. sinensis*,

fused for most of their length in *C. assamica*, and remain free for nearly half
their length in *C. assamica* ssp. lasiocalyx. Additionally, *C. irrawadiensis* and
C. taliensis cultivars are known for their potential contribution to the tea gene
pool, but are not grown commercially, as they only produce teas considered low
quality. A number of other *Camellia* species are historically grown as ornamen-
tal plants, wherein *C. japonica*, *C. reticulata*, and *C. sasanqua* are still popular
worldwide (Gao et al., 2005). In addition, seeds from species of *C. japonica*
and *C. oleifera* have been widely used for oil extraction in Japan and China,
respectively (Sealy, 1958). On the other hand, *C. japonica* and *C. reticulate*
(both camellias), and *C. sasanqua* (paracamellia), are known to contribute to
understanding the phylogenesis of *Camellia* species. Since the tea plant is easily cross-pollinated by fertilization of transferring pollens, the hybrid clines of
C. sasanqua × *C. sinensis*, *C. sinensis* × *C. japonica*, and *C. japonica* × *C. kissi* are
also known, since hybridization takes place promptly (Visser, 1969) (Table 1.1).
The classification has been continued with a number of revisions after careful
review and reexamination of their physical, chemical, and genetic properties.
In the 1980s, after examining 280 *Camellia* species, Chang (1981) classified
them into 4 subgenera and 14 sections. Later, Ming (2000) divided them into
2 subgenera and 14 sections on the basis of structural framework proposed by
Sealy (1958). The classification is still continuing using modern techniques of
genetic isolations and has been further grouped (Vijayan et al., 2009; Wachira
et al., 1997). When the classification based on the three major components
of tea, polyphenols, theanine, and caffeine, which are relevant for health
benefits of tea, was reviewed, all three components were present in varied
amounts in the species of subgenera *Thea*, mainly in *C. sinensis*, and its wild
relatives *C. taliensis* and *C. irrawadinesis*, or among the hydrides of *C. sinensis*
only. On the other hand, the species belonging to subgenera *Camellia* do not
contain even trace amounts of theanine and caffeine (Table 1.1), the major
contributors for the pleasure of drinking tea. It could be postulated that the
absence of these compounds in the species of subgenera of *Camellia* could
have led them to establish as a nontea plant species.

1.3 Traditional Teas: Cultivation and Processing

Ever since the initial discovery of tea, the tea plant has been highly valued
and considered preciously medicinal (Sato and Miyata, 2000). However,
with the advent of new harvesting and processing techniques, it spread all
over the world and became the world's most popular refreshing beverage

(Henderson et al., 2002). At present, tea as a beverage has been adopted in different cultures, and many countries have evolved their own tea traditions (Zhu et al., 2000). It has become generally accepted that, next to water, tea is the most consumed beverage in the world, with per capita consumption of <120 ml/day (Ahmad and Mukhtar, 1999; Balentine et al., 1998). Currently, tea is consumed as green, oolong, and black tea, and it is hard to believe that all teas originate essentially from the same plant, the *Camellia sinensis* (Cabrera et al., 2003; Reeves et al., 1987). Of the total amount of tea produced and consumed in the world, 78% is black, 20% is green, and less than 2% is oolong tea (McKay and Blumberg, 2002). Black tea is consumed primarily in Western countries and in some Asian countries like India, whereas green tea is consumed primarily in China, Korea, Japan, and a few countries in North Africa and the Middle East. Oolong tea production and consumption are particularly confined to southeastern China and Taiwan (Karori et al., 2007). The brewing and drinking of various kinds of tea followed with traditional tea ceremonies, wherein it is called *Gong Fu Cha* in China, for mainly the preparation of oolong tea, *Sado* in Japan for green tea, and *Samovor* in Russia for strong black tea.

The differences between individual types of tea result from variations in the processing of the tea leaves after they are harvested (Figure 1.1). Therefore, depending on the processing conditions, such as the extent of fermentation, the tea is classified into three basic categories: green tea (unfermented tea), oolong tea (midfermented tea), and black tea (fermented tea). Green teas are subject to minimal oxidation, whereas oolong and black teas are subjected to partial and extensive oxidation, respectively. The intent during the manufacturing of green tea is to preserve the healthy, natural, and active substances of the fresh tea leaves so they could be released into the consumer's cup upon infusion. After picking (by hand or via mechanical means), the green leaves are spread out in the hot air to wither (moisture elimination). Once they have become soft and pliable, they are traditionally panfried in traditional woks with a convex bottom specifically to inactivate enzymes (Wilson and Clifford, 1992). This process prevents the leaves from oxidizing (commonly known as fermenting), as occurs during the production of black tea. The subsequent "rolling" gives the leaves their shape and styles, such as twisted, curly, or balled, as well as increased durability. Rolling of the leaves also helps to regulate the release of natural substances and flavors during the steeping. Finally, the tea leaves are dried by firing, whereby the natural fragrances and flavors are stabilized, helping the tea leaves maintain their original texture and green color.

FIGURE 1.1
The types of teas depend on the processing conditions after they are harvested.

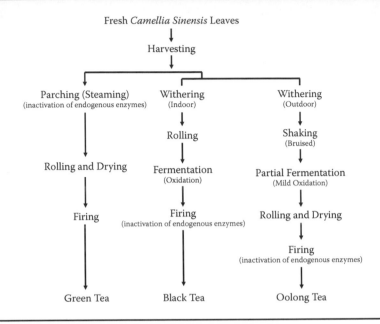

1.4 Green Tea—Elite Japanese Traditions

The culture of green tea in terms of production, processing, and ways of consumption had been perfected in Japan starting from the periods of Zen Buddhism (approximately 800 A.D.). In Japan, tea can be grown where the average temperature is 11.5 to 18°C and the average rainfall is 1500 to 2000 mm per year. The regions with commercial production of green tea in Japan are therefore in the southwest. The Tenryu region and the Kawane region located in Shizuoka prefecture are the best for green tea production. More than 40% of all Japanese highest-quality green tea (sencha and gyokuro) is produced in this area. Not only climate and soil quality, but also the knowledge, skill, and experience of the tea cultivator will affect the quality of his product—its taste, color, aroma, and appearance. To produce high-quality teas, the tea flushes are covered by shade prior to harvesting to protect them from direct sunlight, frost, or rain, which can damage the delicate tips of the shoots. All new shoots are plucked by hand (sencha is sometimes machine plucked). The highest quality of tea is made from the first few top new shoots.

In Japan, the tradition of tea drinking is not restricted to typical green tea but to a number of its variations that differ in taste and flavor. The variations are made through processing leaves as shown in the schematic illustration below:

Tea plant → 90% shading → fresh leaves → steaming → withering → drying → firing (mild) → *gyokuro* (high-quality, less bitter green tea with dominated *umami* taste)

Tea plant → no shading → fresh leaves → steaming → withering → drying → firing (mild) → *sencha* (regular green tea, with typical bitter stint)

Tea plant → no shading → fresh leaves → steaming → withering → drying → *bancha* → roasting → firing (strong) → *hojicha* (light to red brown colored with roasted flavor)

Tea plant → no shading → fresh leaves → roasting → withering → drying → firing (very strong) → *kamairicha* (slightly bitter with sweet, mildly roasted flavors)

Tea plant → 40 to 50% shading → fresh leaves → steaming → withering → drying → firing (mild) → *kabusecha*

Tea plant → 90% shading → fresh leaves → steaming → nonwithering → drying → *tencha* → grinding → firing (mild) → *matcha* (bright green colored with vegetative spell and combination of typical green tea bitter and umami taste)

Of the total annual cultivation of green tea in Japan, 78.6% of the teas are processed to produce sencha green tea, while only 0.4% gyokuro green tea is produced and the rest goes to others. Typical compositions of aforementioned traditional Japanese teas are listed in Table 1.2, along with oolong tea and black tea. In the good quality of sencha and gyokuro green teas the shape of each leaf is twisted, needlelike, and uniform in size. There are no stems or old leaves in these teas, and the color of infused tea is transparently clear, because the tea is graded lower when it contains more stems and old leaves, and when the end result is dark and cloudy. The aroma is fresh and the taste is umami (deep flavor, less bitter, more sweet) because the higher the quality, the less bitter the taste. In the case of sencha, the tea shoots are plucked and then spread out to dry, steamed, and rolled by four different rollers. After a primary drying and rolling, nearly 55% of the moisture can be removed. In a second roller the uniform moisture is obtained, and in a third roller the level is controlled to 30%. After the fourth rolling and drying, a moisture content of up to 13% is maintained. Tea leaves become a needle shape and are then

TABLE 1.2
Typical Compositions of Aforementioned
Traditional Japanese Teas

Type of Teas	Sencha	Matcha	Gyokuro	Bancha	Oolong	Black
Components						
Polyphenols (%)	14.7	6.5	13.4	12.45	16.03	
Caffeine (%)	2.87	3.85	3.1	2.02	2.34	
Theanine (mg/100 g)	1280	2260	2650	—	588	
Free amino acids (mg/100 g)	2700	5800	2730	770	993	
Total N (%)	5.48	6.36	5.48	3.83	3.43	
Vitamin B_1 (mg/100 g)	0.35	0.60	0.37	0.25	0.13	0.1
Vitamin B_2 (mg/100 g)	1.4	1.35	1.34	1.4	0.86	0.8
Vitamin C (mg/100 g)	250	60	257	150	43	32
Vitamin A (mg/100 g)	13	28.9	12.8	14	28.3	17.4
Niacin (mg/100 g)	4	4	4.1	5.4	5.7	10

placed in a dryer, followed by a refiner (mild firing to produce flavor), wherein the water content is controlled to ~6%, and finally, the processed tea leaves are packaged. Other relatively low qualities of green tea are bancha produced without any firing and hojicha performed by roasting prior to subjection to considerably strong firing. Kamairicha could be produced without steaming, but roasting prior to very strong firing, and kabusecha is another type of green tea usually produced through controlled shading of 40 to 50%, while the rest of the process is the same as that for gyokuro.

The production of matcha, an especially grinded tencha, is only 0.6% of the total tea leaves cultivation. Although its consumption is small compared to sencha, matcha historically enjoys a very important status in Japanese culture. For matcha production, the leaves are steamed, dried, and grinded without the rolling process.

1.5 General Features of Classical Teas

Tea polyphenols are known to contribute to the color and flavor of teas. Except in matcha, epicatechins are the main compounds in green tea, accounting for its characteristic color and flavor. In matcha, chlorophyll plays a major role in its distinct bright green color. In contrast, thearubigins and theaflavins in black tea, which are the oxidation products resulting from fermentation, contribute to the color and taste of black teas (Gardner et al., 2007; Reeves et al., 1987). Oolong tea possesses the intermediate characteristics of green and black teas, having both freshness and fragrance, produced by a moderate combination of withering, shaking, and cooling processes wherein their moisture content is reduced to less than 55% of the original leaf weight, resulting in the level of polyphenols in the tea leaves. The variety of fine teas available is huge, and all truly fine teas have in common that only the most aromatic, young, top two leaves and the unopened leaf bud are used. The activity of polyphenol oxidase, basically known as fermentability, and the typical content of polyphenols in young tea leaves greatly differ, depending on the cultivated varieties (cultivars). Thus, the composition of the tea leaves depends on multiple factors, including climate, season, adopted horticultural practices, and the variety type and physical age of the tea plant. For example, the preferable chemical composition of green tea beverages should be similar to that of the tea leaf in its derived from.

The common chemical composition of tea leaves is very complex, and like many other crude extracts from natural products, green tea contains several substances. In general, the main ingredients found in green tea are polyphenols (24 to 36%), protein (15%), lignin (7%), amino acids (3 to 4%), caffeine (2 to 4%), organic acids (2%), and chlorophyll (0.5%) (Stella, 2005; Lin et al., 2003). Green tea usually contains a mixture of polyphenolic compounds that generally include flavanols, flavandiols, flavonoids, and phenolic acids, mainly quercetin, kaempferol, myricetin, and their glycosides, and accounts for nearly 42% of the dry weight of green tea leaves (Graham, 1992). Flavanols are the most abundant fragment of the polyphenols in green teas, and are commonly known as green tea catechins (Peterson et al., 2005). The well-known major catechins in green tea are (–)-epicatechin (EC), (–)-epicatechin-3-gallate (ECG), (–)-epigallocatechin (EGC), and (–)-epigallocatechin-3-gallate (EGCG). Of the polyphenols, epigallocatechin (EGC) and epigallocatechin gallate (EGCG) are the most valuable antioxidant compounds (Sheila et al., 1997; Ohe et al., 2001), and it is estimated that a cup of green tea consists of 10 to 30 mg of EGCG (Balentine et al., 1997; Lakenbrink et al., 2000; Sajilata et al., 2008). Generally, the polyphenols are produced as secondary metabolites in higher plants and bushes and can be divided into two major categories: proanthocyanidins and the polyesters of gallic acid and hexahydroxydiphenic acid, including their derivatives (Fujiki et al., 1992; Pietta et al., 1998). The naturally occurring catechins are flavan-3-ols that are found abundantly in teas, and could be classified as C15 compounds. Their derivatives possess two phenolic nuclei (A-ring and B-ring) connected by three carbon units (C2, C3, and C4). The flavanol structure of catechins (3,3′,4′,5,7-pentahydroxyflavan) contains two asymmetric carbon atoms at C-2 and C-3. Catechins are generally synthesized in tea leaves through metabolism of malonic acid and shikimic acid, wherein the gallic acid is derived from an intermediary product produced in the shikimic acid metabolic pathway (Graham, 1992).

In black teas, the major polyphenols are theaflavin, theaflavinic acids, proanthocyanidin polymers, and thearubigin or theasinensis, which are the oxidation products of polyphenols during processing usually known as catechins and gallic acid complexes (Gardner et al., 2007; Lin and Liang, 2000). Aside from the above, black teas also contain caffeine, methylxanthines, and a small fraction of theophylline and theobromine (Robertson, 1992). Theaflavins are responsible for the astringency and brightness of black tea, whereas thearubigins contribute to its color and mouthfeel. The compositional details of tea components are discussed in subsequent chapters. The price of tea largely depends on the quality. There is acknowledgeable correlation between a tea's cost and its content of amino acids, particularly theanine and arginine. In general, tea contains a number of amino acids, but theanine, which is a specific characteristic of tea plants, is the most abundant amino acid and accounts for nearly 50% of the total amino acid content in tea. The established correlation coefficients between the total score of the sensory test of tea and its content of total nitrogen, theanine, and total free amino acids are estimated to be 0.91, 0.87, and 0.86, respectively (Oishi, 1983). In addition to polyphenols and amino acids, the teas also contain a significant level of carbohydrates and vitamins A, E, and K, including a reasonably low level of complex vitamins B (B_1, B_2, and niacin). It is noteworthy to mention that vitamin C is usually present only in green teas since it is readily decomposed during the fermentation process associated with manufacturing of oolong teas as well as black teas (Table 1.2). Teas also consist of appropriate contents of minerals like manganese, potassium, and a low level of fluoride ions (Friedman et al., 1984). The *Camellia sinensis* plant naturally extracts fluorides and minerals from the soil, which then accumulate in its leaves. Dry tea leaves may contain fluorides ranging from 1 to 2 ppm in green tea leaves, whereas the content in black tea is reported to be five times higher. The World Health Organization (WHO) and U.S. Food and Drug Administration (FDA) have already investigated that fluoride intake through teas is extremely safe (WHO, 1990).

1.6 World's Tea Production and Economy

Apart from India, China, and Sri Lanka, tea is also grown in several African countries, Malaysia, Argentina, Taiwan, Georgia and the southern part of

Russia, Iran, Turkey, and Brazil. Compared to many other commodities, tea consumption has been less affected by the economic downturn, although the demand for some premium-grade teas has declined. Tea prices were about $2.7/kg during 2009, nearly 13% higher than in 2008 and almost a dollar above their average in 2000 to 2007. After a tremendous sharp fall during the third quarter (Q3) of 2008, reflecting the credit crunch and the ensuing global economic downturn, tea prices climbed steeply to an all-time high of more than $3.0/kg in nominal terms during the fourth quarter (Q4) of 2009. Tea prices in 2010 averaged around $2.5/kg (Tea Market Updates, 2009). On the other hand, over the longer term, the prices are seen to be influenced by a continuous growing domestic demand from India, stimulus costs, ascending demand for higher-grade premium teas including organic teas, and concern of nature constraints, such as climate changes, drought or flood, and so forth. The Food and Agriculture Organization's (FAO) composite price for black tea reached $3.2/kg in September 2010, compared to an average price of $2.4/kg. At the end of 2010, the FAO predicted a drop of market prices (FAO, 2010), but the prices were higher in 2011, and the average price during the fourth quarter of 2011 was approximately $2.7/kg. The consistent increase in prices largely reflected the growing demand from growing economies like India, China, and Russia.

The world production of tea had been constantly growing at the pace of around 3% annually from 2005 (Table 1.3). China, followed by India, Kenya, and Sri Lanka, occupied the top ranking in production. India's *Business Standard* reported that India, Kenya, and Sri Lanka, which produce 80% of the world's total black tea, had a record strong output in the 2010 season. The Food and Agriculture Organization stated that world tea production reached an estimated 3.97 million tons in 2009, with markets such as Kenya, India, and Sri Lanka driving this robust growth. As expected, total world tea production maintained its healthy growth in 2009 and 2010 with greater possibility and touched the 4-million ton mark by 2010 (Figure 1.2). China, India, and the Russian Federation hold the top three positions as the largest consumers of tea in the world. The list of top 10 tea-consuming nations is presented in Table 1.4. An overall analysis of global tea production in 2011 revealed that Sri Lanka is +39.2 million kg, Kenya +84.8 million kg, South India +6.2 million kg, and North India −45.3 million kg. This must be viewed taking into consideration the shortfall of approximately 60 million kg in the first half of 2011, wherein India alone, the world's second-largest producer of tea, reported a shortfall of 85 million kg in 2011, as consumption of the beverage expands at a steady clip of 3 to 3.5% annually. The global tea output in

TABLE 1.3 World Tea Production (Descending Rank; ×1000 MT)				
Countries	2005	2006	2007	2008
China	935	1028	1140	1200
India	946	982	945	981
Kenya	324	311	370	346
Sri Lanka	317	311	305	319
Vietnam	133	143	148	166
Turkey	135	142	178	155
Indonesia	156	147	137	138
Japan	100	100	92	93
Argentina	80	88	87	72
Bangladesh	61	53	58	59
Uganda	38	37	45	43
Malawi	38	45	48	42
Tanzania	30	31	35	32
Myanmar	18	18	18	19
Iran, Islamic Republic	25	20	17	18
Taiwan, China	19	19	18	17
Rwanda	16	17	18	17
Nepal	13	14	15	16
Zimbabwe	15	16	14	8
PNG	7	7	7	7
Burundi	8	6	7	6
World Total	**3458**	**3580**	**3751**	**3804**

the first half of 2011 was down by more than 20.5 million kg. In Japan, the production and consumption of teas were somewhat constant for 2009 and 2010. Table 1.5 presents the overall figures of the tea economy in Japan.

Early estimates listed for 2009 export volumes showed an approximate 8% decline in Kenya and nearly 3% declines in both Sri Lanka and India. However, despite the lower export volumes, the global export revenues from tea have increased since 2009 due to its high prices. Still, Kenya was the top gross exporter of tea in 2008 with a record export of 0.38 million tons, followed by Sri Lanka and China, both with nearly 0.3 million tons of tea exports (Table 1.6). India produced 966 million tons of tea in 2010 and exported only 190 million tons, as domestic consumption is growing at a healthy pace by approximately 2.5% per year. This is because more people have entered the middle class and can afford a more diverse range of tea at various levels of quality, which in turn has pushed the tea exports downward.

FIGURE 1.2
Tea growth industry: Total world production and consumption of tea until year 2010.

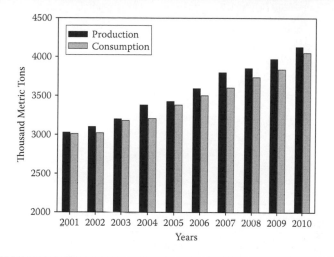

TABLE 1.4
Top 10 Tea-Consuming Countries in 2010

Rank	Country	Consumption (×1000 MT)
1	China	1000
2	India	817
3	Russian Federation	175
4	Turkey	153
5	Japan	134
6	United States	126
7	Pakistan	120
8	United Kingdom	119
9	Egypt	93
10	Iran	65

Every year, consumption in India grows by 30 to 35 million kg. The world export of tea by countries in 2010 is presented in Figure 1.3. It is noticeable that the top eight exporters of tea have retained their rankings, which did not change from 2005 to 2010, with Kenya topping the list with 26% of total world export. In contrast, for all time, Russia tops the list of net importers with 0.17 million tons of imported tea (Table 1.7).

TABLE 1.5
Production and Consumption of Tea in Japan

Volumes (×1000 MT)	2009	2010
Aracha volume	86	85
Export volume	5.87	5.91
Import	1.96	2.23
Domestic consumption	98.5	98.1

TABLE 1.6
Net Exports (Descending Rank; ×1000 MT)

Countries	2005	2006	2007	2008
Kenya	348	312	344	383
Sri Lanka	299	315	294	298
China	287	287	289	297
India	195	216	176	193
Vietnam	88	105	111	104
Indonesia	102	95	84	96
Argentina	66	71	75	77
Malawi	43	42	47	40
World Total	**1566**	**1579**	**1573**	**1638**

1.7 Conclusion

In summary, tea has a long history as a medicinal plant and is well known for its stimulating and relaxing effects on human physiology. Traditional beliefs in the health benefits of tea are continuously being confirmed on a scientific basis. Furthermore, green tea is a rich source of polyphenols called catechins, which are linked to the health benefits of green tea, ranging from a lower risk of certain cancers to weight loss and even protection against Alzheimer's disease, cardiovascular disease, some infectious diseases, and dental caries. Therefore, it is reasonable to conclude that drinking green tea and consuming green tea extracts is compatible with healthy dietary advice in helping to reduce the risk of chronic diseases and maintain overall health. The type of available commercial tea depends on the processing techniques, which involve withering, rolling, fermenting, firing, and drying as major steps. Specialized teas are made by different combinations of the aforementioned

FIGURE 1.3
Industry data on the world exports of tea by countries in year 2010.

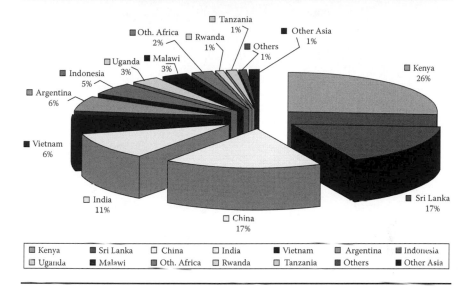

| Kenya | Sri Lanka | China | India | Vietnam | Argentina | Indonesia |
| Uganda | Malawi | Oth. Africa | Rwanda | Tanzania | Others | Other Asia |

TABLE 1.7
Gross Imports (Descending Rank; ×1000 MT)

Countries	2005	2006	2007	2008
Russian Federation	173	166	174	175
United Kingdom	128	135	131	130
United States	100	108	109	117
Pakistan	139	117	106	99
Egypt	74	79	69	94
Dubai	62	69	72	79
Other CIS	53	56	58	60
Iran, Islamic Republic	43	50	55	58
Morocco	50	51	53	48
Japan	51	48	47	43
Iraq	58	67	32	36
World Total	**1469**	**1487**	**1490**	**1532**

Note: CIS: Commonwealth of Independent States.

processes. The extraction of functional nutrients from green tea requires not only good quality teas, but also specific skills and practices to produce standard high-quality green tea extracts. The methods and practices of Japan, having perfected its green tea processing, are well recommended for the production of high-quality and standard green teas and extracts.

References

Ahmad, N., and Mukhtar, H. 1999. Green tea polyphenols and cancer: Biologic mechanisms and practical implications. *Nutr Rev*, 57: 78–83.

Balentine, D.A., Harbowy, M.E., and Graham, H.N. 1998. Tea: The plant and its manufacture; chemistry and consumption of the beverage. In *Caffeine*, ed. G.A. Spiller, 35–72. Boca Raton, FL: CRC Press.

Balentine, D.A., Wiseman, S.A., and Bouwens, L.C. 1997. The chemistry of tea flavonoids. *Crit Rev Food Sci Nutr*, 37: 693–704.

Cabrera, C., Gimenez, R., and Lopez, C.M. 2003. Determination of tea components with antioxidant activity. *J Agric Food Chem*, 51: 4427–4435.

Cabrera, C., Reyes, A., and Giménez, R. 2006. Beneficial effects of green tea—A review. *J Am College Nutr*, 25: 79–99.

Chang, H.T. 1981. A taxonomy of the genus *Camellia*. *Acta Sci Nat Univ Sunyatseni Monogr Ser*, 1: 1–180.

Friedman, M., Solouki, S., Gurevitz, S., Gedalia, I., and Onisi, M. 1984. Fluoride concentrations in tea: Its uptake by hydroxyapatite and effect on dissolution rate. *Clin Prev Dent*, 6: 20–22.

Food and Agriculture Organization of the United Nations (FAO). 2010. High tea prices. Media Center. http://www.fao.org/news/story/en/item/38315/icode/.

Fujiki, H., Yoshizawa, S., Horiuchi, T., Suganuma, M., Yatsunami, J., and Nishiwaki, S. 1992. Anticarcinogenic effects of (–)-epigallocatechin gallate. *Prev Med*, 21: 503–509.

Gao, J., Parks, C.R., and Du, Y.Q. 2005. *Collected species of the genus Camellia and illustrated outline*. Zhejiang, China: Zhejiang Science and Technology Press.

Gardner, E.J., Ruxton, C.H.S., and Leeds, A.R. 2007. Black tea—Helpful or harmful? A review of the evidence. *Eur J Clin Nutr*, 61: 3–18.

Graham, H.N. 1992. Green tea composition, consumption, and polyphenol chemistry. *Prev Med*, 21: 334–350.

Harler, C.R. 1956. *The culture and marketing of tea*. Oxford: Oxford Press.

Henderson, L., Gregory, J., and Swan, G. 2002. *National diet and nutrition survey: Adults aged 19 to 64 years*. London: FSA.

Karori, S.M., Wachira, F.N., Wanyoko, J.K., and Ngure, R.M. 2007. Antioxidant capacity of different types of tea products. *Afr J Biotechol*, 6: 2287–2296.

Kitamura, S. 1950. *Sinensis* var. *assamiea* (Masters). *Acta Phytotax Geobot* (Kyoto), 14: 59.

Lakenbrink, C., Lapczynski, S., Maiwald, B., and Engelhardt, U.H. 2000. Flavonoids and other polyphenols in consumer brews of tea and other caffeinated beverages. *J Agric Food Chem*, 48: 2848–2852.

Lin, J.K., and Liang, Y.C. 2000. Cancer chemoprevention by tea polyphenols. *Proc Nat Sci Counc Repub China B*, 24:1–13.

Lin, Y.S., Tsai, Y.J., Tsay, J.S., and Lin, J.K. 2003. Factors affecting the levels of tea polyphenols and caffeine in tea leaves. *J Agric Food Chem*, 51: 1864–1873.

Lu, Y.U. 1995. *The classic of tea: Origins and rituals*, trans. F.R. Carpenter. New York: Ecco Press.

Masters, J.W. 1844. The Assam tea plant compared with tea plant in China. *J Agri-Horticult Soc India*, 3: 61–69.

McKay, D.L., and Blumberg, J.B. 2002. The role of tea in human health: An update. *J Am Coll Nutr*, 21: 1–13.

Ming, T.L. 2000. *Monograph of the genus Camellia*. Kunming, China: Kunming Institute of Botany, Chinese Academy of Sciences, Yunnan Science and Technology Press.

Ohe, T., Marutani, K., and Nakase, S. 2001. Catechins are not major components responsible for anti-genotoxic effects of tea extracts against nitroarenes. *Mut Res*, 496: 75–81.

Oishi S. 1983. Development of tea manufacture in Japan (in Japanese). *Nousangyoson Bunka Kyokai Tokyo*, p. 38.

Peterson, J., Dwyer, J., Bhagwat, S., Haytowitz, D., Holden, J., Eldridg, A.L., Beecher, G., and Aladesanmi, J. 2005. Major flavonoids in dry tea. *J Food Comp Anal*, 18: 487–501.

Pietta, P.G., Simonetti, P., Gardana, C., Brusamolino, A., Morazzoni, P., and Bombardelli, E. 1998. Catechin metabolites after intake of green tea infusions. *Biofactors*, 8: 111–118.

Reeves, S.G., Owuor, P.O., and Othieno, C.O. 1987. Biochemistry of black tea manufacture. *Trop Sci*, 27: 121–133.

Robertson, A. 1992. The chemistry and biochemistry of black tea production: The non-volatiles. In *Tea: Cultivation to consumption*, ed. K.C. Wilson and M.N. Clifford, 53–86. London: Chapman & Hall.

Sajilata, M.G., Bajaj, P.R., and Singhal, R.S. 2008. Tea polyphenols as nutraceuticals. *Comp Rev Food Sci Food Safety*, 7: 229–254.

Sato, T., and Miyata, G. 2000. The nutraceutical benefit, part I: Green tea. *Nutrition*, 16: 315–317.

Sealy, J. 1958. *A revision of the genus Camellia*. London: Royal Horticultural Society.

Sharma, V.S., and Venkataramani, K.S. 1974. The tea complex. I. Taxonomy of tea clones. *Proc Indian Acad Sci*, 53B: 178–187.

Sheila, A.W., Douglas, A.B., and Balz, F. 1997. Antioxidants in tea. *Crit Rev Food Sci Nutr*, 37: 705–718.

Shouyi, B. 1982. *An outline history of China*, 55. Beijing: Foreign Languages Press.

Stella, A., Brochard, G., Beautheac, N., and Dozel, C. 2005. *The book of tea*. United Kingdom: Flammarion Publishing.

Tea Market Updates. 2009. *Publication of tea promotion division—Sri Lanka Tea Board*. Vol. 5(4). http://ceylontea.lk/Market/fourth%20quarter%202009.pdf.

Ukers, W.H. 1935. *All about tea*. Vol. I–II. New York: Tea and Coffee Trade Journal Company.

Vijayan, K., Zhang W.J., and Tsou, C.H. 2009. Molecular taxonomy of *Camellia (Theaceae)* inferred from nrITS sequences. *Am J Bot*, 96: 1348–1360.

Visser, T. 1969. Tea Camellia sinensis (L) O. Kuntz. In *Outlines of perennial crop breeding in the tropics*, ed. F.P. Ferwada and F. Wit, 459–493. Wageningen: H. Veenan and Zonen.

Wachira, F.N., Powell, W., and Waugh, R. 1997. An assessment of genetic diversity among *Camellia sinensis* L. (cultivated tea) and its wild relatives based on randomly amplified polymorphic DNA and organelle-specific STS. *Heredity*, 78: 603–611.

Weatherstone, J. 1992. *Historical introduction, in tea: Cultivation to consumption*, ed. K.C. Willson and M.N. Clifford, 1–23. New York: Chapman & Hall.

Wight, W. 1962. Tea classification revised. *Current Sci*, 31: 298–299.

Wilson, K.C., and Clifford, M.N. 1992. In *Tea, cultivation to consumption*, 603–647. London: Chapman & Hall.

World Health Organization (WHO). 1990. Toxicological valuation of certain foods additives and contaminants. *WHO Food Additives Ser*, 26: 156–160.

Zhu, B.T., Patel, U.K., Cia, M.X., and Conney, A.H. 2000. O-Methylation of tea polyphenols catalyzed by human placental cytosolic catechol-*O*-methyl transferase. *Drug Metab Dispos*, 28: 1024–1030.

2

Biochemical and Physicochemical Characteristics of Green Tea Polyphenols

Takashi Tanaka, Yosuke Matsuo, and Isao Kouno

Contents

2.1 Introduction

Polyphenols are plant metabolites that feature multiple phenolic hydroxyl groups, and which are widely distributed in the plant kingdom (Giannasi, 1988). Many angiosperms, gymnosperms, ferns, and even algae (Markham, 1988; Niemann, 1988; Targett and Arnold, 1998) contain polyphenols, and in some cases the polyphenols are accumulated in very high concentrations (from a few percent to 30% of the dried material) (Nutrient Data Laboratory Beltsville Human Nutrition Research Center Agricultural Research Service, 2004). Although there are other subtypes with different molecular skeletons and biosynthetic origins, catechins and proanthocyanidins are the most

important polyphenols found in foods (Figure 2.1), being widely distributed in fruits, vegetables, spices, and herbs. Catechins may also be called
flavan-3-ols, which represent a relatively small group of the flavonoid class
of natural products, while proanthocyanidins are defined as oligomers
and polymers of catechins, in which catechin units are connected through
carbon-to-carbon linkages. Catechin monomers and proanthocyanidin
oligomers are closely related biosynthetically and coexist in many plants,
such as grapes, apples, pine, and blueberries. Bananas (Tanaka et al., 2000)
and persimmons (Tanaka et al., 1994) are unusual fruits, as they contain
only polymeric proanthocyanidins. In contrast, the tea plant, *Camellia
sinensis* (L.) O. Kuntze (Theaceae), is also unusual, as it contains almost
exclusively catechin monomers (Hashimoto et al., 1992). This chapter first
introduces the chemical compositions of tea polyphenols. Then, extraction,
separation, and analytical methods are also briefly introduced. Finally, the
physicochemical characteristics of polyphenols as they relate to their biological functions are discussed.

2.2 Structure of Catechins and the Polyphenol Composition of Green Tea

Tea catechin molecules have two aromatic rings commonly called the A- and
B-rings (Figure 2.2). The A- and B-rings are biosynthetically derived from
acetic acid and shikimic acid, respectively, and they show significantly different chemical reactivities, as described later in this chapter. Most catechins
have hydroxyl groups at the C-5 and C-7 positions of the A-ring. In contrast, the B-ring usually possesses two or three vicinal hydroxyl groups at
C-3′, C-4′, and C-5′. Catechol-type B-rings having two hydroxyl groups at
the C-3′ and C-4′ positions are the most common in the plant kingdom.
However, tea leaves contain a particularly high proportion of catechins with
the pyrogallol-type B-rings (>70% of tea catechins), which have three vicinal
hydroxyl groups at the C-3′, C-4′, and C-5′ positions. This feature, along with
the aforementioned abundance of monomeric catechins, is characteristic of
tea polyphenols. Small quantities of catechins with 4-monohydroxyl-type
B-rings have also been observed in tea (Hashimoto et al., 1989). The central
part of the molecule is called the C-ring, and differences in the configurations at the C-2 and C-3 positions, which bear the B-ring and a hydroxyl
group, respectively, are important in tea chemistry. Catechins with a 2,3-*cis*
configuration are the major isomers found in fresh tea leaves and green tea.

FIGURE 2.1
General structures of flavan-3-ol (catechins) and proanthocyanidins.

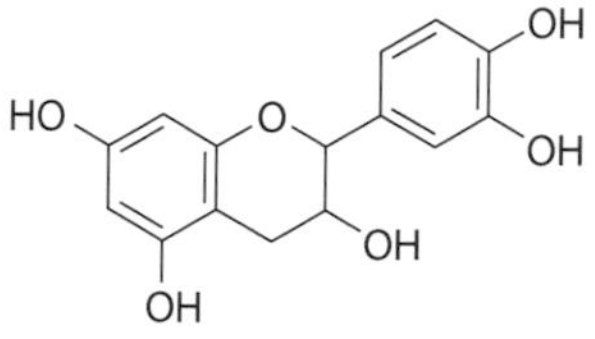

flavan-3-ol (catechins)

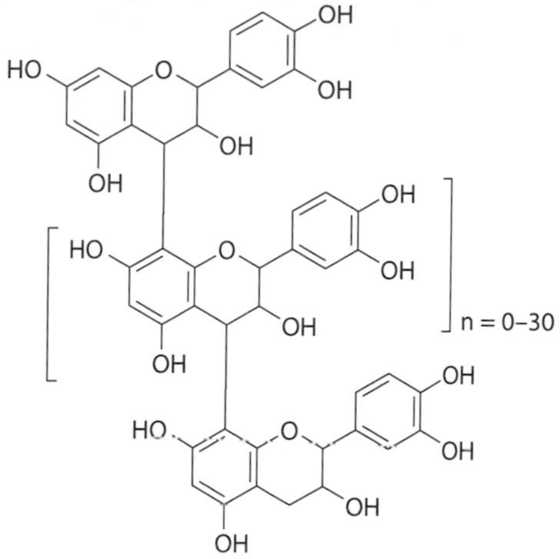

proanthocyanidins

n = 0–30

FIGURE 2.2
Structures of tea catechins.

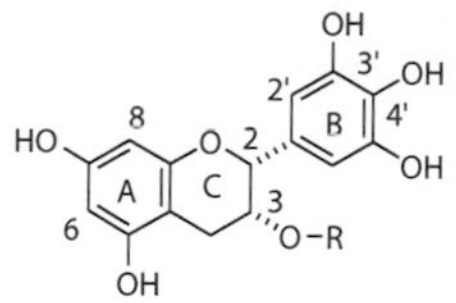

epigallocatechin R = H
epigallocatechin-3-O-gallate R = galloyl

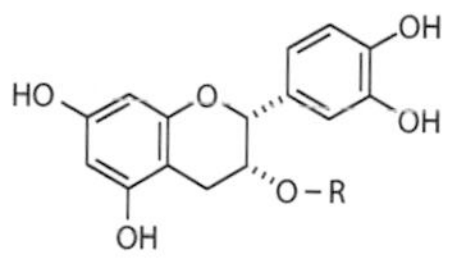

epicatechin R = H
epicatechin-3-O-gallate R = galloyl

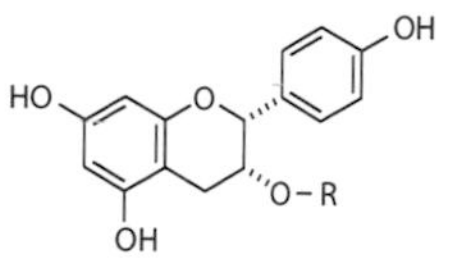

epiafzelechin R = H
epiafzelechin-3-O-gallate R = galloyl

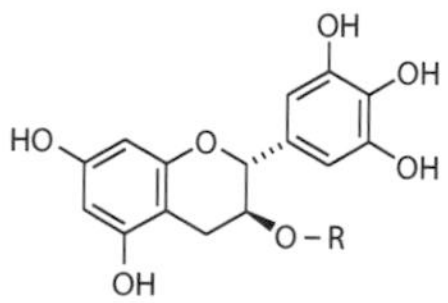

gallocatechin R = H
gallocatechin-3-O-gallate R = galloyl

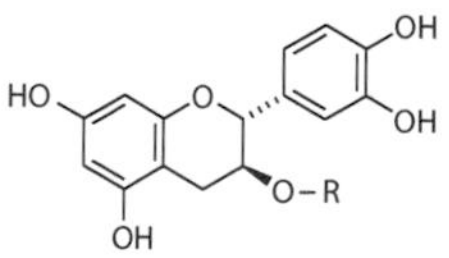

catechin R = H
catechin-3-O-gallate R = galloyl

galloyl =

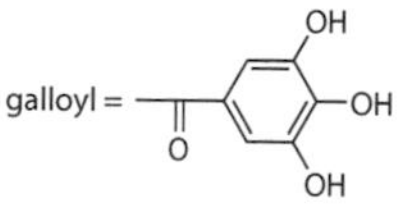

However, this configuration is converted to the 2,3-*trans* configurations on high-temperature treatment, such as during sterilization at temperatures of over 120°C, due to inversion of configuration at the C-ring C-2 position (Huang et al., 2004). Over 50% of tea catechins are derivatized with a galloyl ester at the C-3 hydroxyl group. The presence of the galloyl groups affects the physicochemical properties of tea catechins, in particular their oxidative degradation and interactions with coexisting biomolecules, such as proteins. Both from the point of view of biological activity and concentration in green tea leaves, the most important tea polyphenol is (–)-epigallocatechin-3-*O*-gallate, which has a pyrogallol-type B-ring and a galloyl ester at the C-3 hydroxyl group. This polyphenol is very characteristic of tea plants because there is no other plant that accumulates this compound in such high concentrations (3 to 9% of dried leaf).

The polyphenol composition of commercial green tea is similar to that of fresh tea leaves, because steaming or roasting at the initial stages of green tea manufacturing inactivates enzymes involved in the oxidation of tea polyphenols. The total polyphenol content is 13 to 30% of the dry weight, and the content depends on the cultivar of the tea plant studied (Astill et al., 2001). The concentration of each tea catechin measured varies considerably in different tea products and when different extraction conditions are used (Lin et al., 2003). Cultivars of tea plants commonly grown in the world belong to two major varieties: *C. sinensis* var. *sinensis*, which is usually cultivated for green tea production, and *C. sinensis* var. *assamica*, cultivated for black tea production. The Assam variety of tea contains particularly high quantities of polyphenols. The stage of harvesting is also important: the polyphenol content of leaves harvested in midsummer in Japan is higher than in tea harvested in May. The commercial value of green tea tends to be inversely proportional to polyphenol content because polyphenols taste bitter and astringent. The concentrations of polyphenols in a 60% ethanol extract of different representative commercial green tea products analyzed in the author's laboratory are listed in Table 2.1.

Besides the four major tea catechins, the presence of many minor polyphenols has been reported. Recently, the biological activities of methylated tea catechins have been reported and include the antiallergic activity of (–)-epigallocatechin-3-*O*-(3-*O*-methyl)gallate (Sano et al., 1999). The presence of hydrolysable tannins, proanthocyanidins (Nonaka et al., 1983, 1984), and flavonol glycosides (Finger et al., 1991) was also demonstrated (Figure 2.3). Tea catechins and caffeine usually have a colorless appearance, and flavonol glycosides, including *O*-glycosyl and *C*-glycosyl flavonoids, are responsible for the yellow color of green tea decoctions.

TABLE 2.1
Major Constituents of Green Tea (mg/g Dried Leaf, 60% EtOH Extract)

Compounds	Product A	Product B
Epicatechin	11.94 ± 0.76	9.95 ± 0.63
Epicatechin-3-O-gallate	16.67 ± 0.15	31.08 ± 0.77
Epigallocatechin	49.84 ± 0.53	21.70 ± 0.83
Epigallocatechin-3-O-gallate	76.93 ± 0.94	85.53 ± 1.80
Catechin	6.71 ± 9.56	2.56 ± 0.28
Gallocatechin	4.48 ± 0.21	2.69 ± 0.10
Gallocatechin-3-O-gallate	4.35 ± 0.09	8.83 ± 0.36
Gallic acid	0.14 ± 0.01	0.68 ± 0.04
Theogallin	1.06 ± 0.02	10.83 ± 0.40
Caffeine	27.70 ± 0.32	42.32 ± 1.28
Theobromine	0.91 ± 0.02	1.78 ± 0.04

FIGURE 2.3
Structures of minor polyphenols found in green tea.

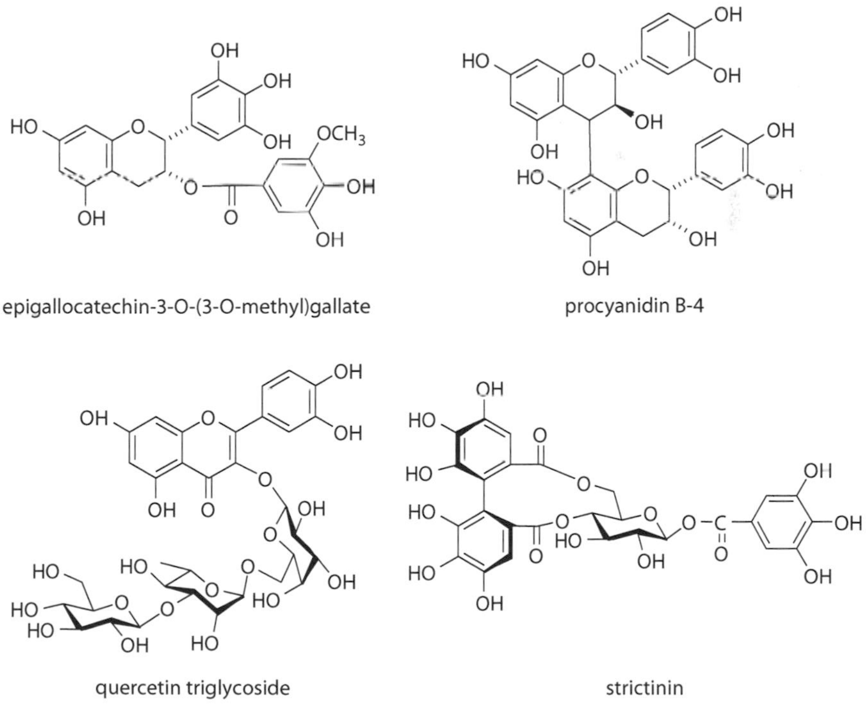

epigallocatechin-3-O-(3-O-methyl)gallate

procyanidin B-4

quercetin triglycoside

strictinin

2.3 Extraction, Separation, and Analysis of Tea Catechins

Polyphenols, especially those derivatized as galloyl esters, reversibly associate with proteins and other macromolecules in aqueous solutions by hydrophobic and ionic interactions, including hydrogen bonding. The extraction of tea polyphenols with cold water is therefore not successful. Considering that tea is usually consumed as a decoction with hot water, extraction should be performed with deionized hot water. The presence of heavy metals may decrease the stability of catechins because heavy metals, such as iron, interact strongly with phenolic hydroxyl groups by chelation, rendering them much more sensitive to oxidation. In addition, hypochlorous acid in tap water causes the oxidative decomposition of tea catechins. Usually, complete extraction of tea catechins with hot water is difficult, and some polyphenols remain in the debris. To cleave the interactions of catechins with macromolecules, mixtures of water and organic solvents, typically 60% aqueous acetone or 60% aqueous ethanol, are used for extraction in chemical studies (Nishizawa et al., 1984).

A typical scheme for the separation of tea catechins from green tea is shown in Figure 2.4. Fractionation of the extract using Sephadex LH-20 column chromatography with water containing increasing proportions of methanol is effective to obtain fractions containing exclusively tea catechins. Catechins with galloyl esters adsorb strongly to the gel compared to free catechins. Pyrogallol-type and catechol-type catechins can be separated using

FIGURE 2.4
A typical separation procedure of tea catechins in a laboratory scale.

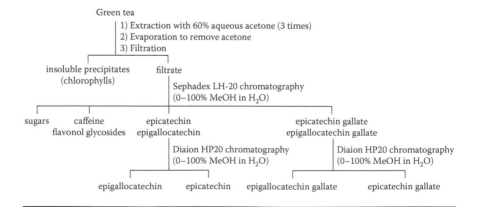

polystyrene gel, such as Diaion HP20, column chromatography. Further purification of each catechin may be achieved by crystallization from water or column chromatography on reversed-phase silica gel. Solvent partitioning between ethyl acetate and water is also effective, with most of the tea catechin content being extracted into the organic layer. However, a portion of epigallocatechin and gallocatechin often remains in the aqueous layer. Detailed separation procedures have been described in the literature (Hashimoto et al., 1987, 1989).

Tea polyphenols are analyzed using high-performance liquid chromatography (HPLC) using reversed-phase silica gel with acidic aqueous acetonitrile as the mobile phase. A gradient elution is necessary to detect polar and nonpolar constituents simultaneously (Hashimoto et al., 1992; Kusano et al., 2008). An example of the HPLC chromatogram obtained by analysis of a 60% ethanol extract of a commercial green tea product is shown in Figure 2.5.

2.4 Polyphenols of Oolong Tea, Postfermented Tea, and Black Tea

As mentioned previously, the polyphenol composition of green tea is similar to that of fresh tea leaves because steaming or roasting at the initial stages of green tea production inactivates enzymes involving the oxidation of chemical constituents of the leaves. In the case of oolong tea, though there are many varieties, the fresh leaves are first subjected to drying stress, and then partially rolled to produce a characteristic flavor. Four major tea catechins are found, similarly to green tea; however, polymeric polyphenols are also found in oolong tea (Figure 2.6a). The polymeric polyphenols are presumed to be a complex mixture of products derived from tea catechins and are detected as a broad hump on the HPLC baseline.

Postfermented tea is produced in South China by aerobic microbial fermentation of heat-processed green tea products. The catechin concentration is significantly lower than that in green tea, and polymeric substances are produced from the catechins (Figure 2.6c). The aging process, which can last for months or years, affects the quality and flavor of some postfermented teas, such as pu-erh tea. The HPLC chromatograms of postfermented tea products that have undergone aging in the open air for a few years only show peaks due to caffeine and polymeric substance (Figure 2.6d).

Green, oolong, and postfermented tea products are mainly consumed in East Asia, where the tea plant originated. However, in the rest of the world,

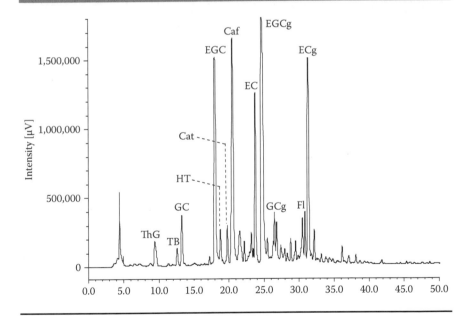

FIGURE 2.5
HPLC profile of a 60% ethanol extract of Japanese green tea. Conditions: Cosmosil 5C$_{18}$-AR II (Nacalai Tesque, Inc., Japan) column (4.6 mm i.d. × 250 mm) with gradient elution from 4 to 30% (39 min) and 30 to 75% (15 min) of CH$_3$CN in 50 mM H$_3$PO$_4$; flow rate, 0.8 ml/min; column temperature, 35°C; and detection, JASCO photodiode array detector MD-910. Abbreviations: ThG, theogallin; TB, theobromine; GC, gallocatechin; EGC, epigallocatechin; HT, hydrolysable tannin; Cat, catechin; Caf, caffeine; EC, epicatechin; EGCg, epigallocatechin-3-O-gallate; GCg, gallocatechin-3-O-gallate; Fl, flavonol glycoside; ECg, epicatechin-3-O-gallate.

black tea is very popular and accounts for almost 80% of world tea consumption. Black tea production begins with the mechanical crushing of partially withered leaves. At this stage, oxidation enzymes in the leaves are mixed with tea catechins to catalyze the oxidation of polyphenols to produce a characteristic color and flavor. The major oxidation products are theaflavins (Takino et al., 1964, 1965) and theasinensins (Hashimoto et al., 1988; Nonaka et al., 1983) (Figures 2.6b and 2.7). Theaflavins are the most important polyphenols in black tea and give black tea a reddish yellow color, while theasinensins are colorless, simple dimers of gallocatechin and its galloyl esters. In addition, chemically uncharacterized polymeric polyphenols, usually called thearubigins, are present as the major polyphenols (Roberts, 1962). Thearubigins are responsible for the dark reddish brown color of black tea and account for more than 60% of the polyphenols in the infusion (Haslam, 1998). Efforts to elucidate the chemical structure of thearubigins and related catechin oxidation products have been undertaken since the 1950s by a number of groups; however, the chemical structures of many of black tea polyphenols remain unknown (Haslam, 2003; Kusano et al., 2008; Ozawa et al., 1996; Tanaka and Kouno, 2003; Tanaka et al., 2010). The structures of typical oxidation products of epigallocatechin and its gallate are shown in Figure 2.7, and the polyphenol compositions of commercial oolong, postfermented, and black tea products are shown in Table 2.2 (Nonaka et al., 1986; Tanaka et al., 2005).

2.5 Chemical Reactivity of Tea Catechins

The chemical reactivity of catechins, especially that of the aromatic A- and B-rings, relates to their stability under physiological conditions, covalent bond formation with coexisting biomolecules, and the biological activities. These factors should be taken into consideration when evaluating biological activities of tea catechins.

The A-ring tends to carry out electrophilic substitution reactions with electrophiles such as aldehydes. Acetaldehyde and formaldehyde readily react at the C-8 or C-6 positions of the A-ring, affording condensation products having two catechin moieties (Figure 2.8). When an excess of aldehyde is added to an aqueous solution of tea catechins, further cross-coupling reactions occur, and finally the catechins are polymerized and insolubilized (Tanaka et al., 1999). Catechin-aldehyde condensation is a reversible reaction, and the products react further with amine and thiol groups. The reaction was applied to the absorption of toxic formaldehyde and also related to the polymerization and insolubilization of proanthocyanidins (Tanaka et al., 1994, 2008).

Recently, the electrophilic substitution reaction at the A-ring was applied to the development of lipid-soluble tea catechins. Derivatives can be prepared by reaction of epigallocatechin gallate with citronellal, a natural essential oil, and phytol, a component alcohol of chlorophyll (Fudouji et al., 2009). Although epigallocatechin gallate is insoluble in triglycerides, the newly prepared derivatives dissolve in the lipid and show strong radical scavenging activity (Figure 2.9).

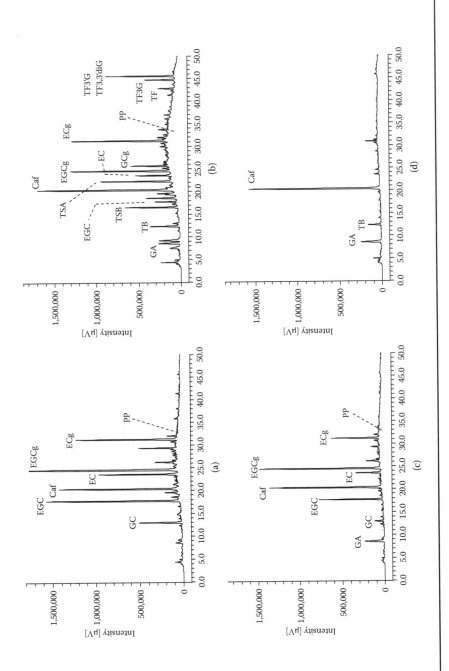

Figure 2.6

HPLC of 60% ethanol extracts of oolong, black, and postfermented teas: (a) oolong tea (Fujian, China), (b) black tea (India), (c) postfermented tea (Hunan, China), and (d) aged postfermented tea (Yunnan, China). Analytical conditions and abbreviations: see Figure 2.5; GA, gallic acid; TSB, theasinensin B; TSA, theasinensin A; TF, theaflavin; TF3G, theaflavin-3-O-gallate; TF3'G, theaflavin-3'-O-gallate; TF3,3'diG, theaflavin-3,3'-di-O-gallate; PP, polymeric polyphenols (PP of black tea: thearubigins).

FIGURE 2.7
Structures of theaflavins, theasinensins, dehydrotheasinensin AQ, and epitheaflagallin-3-O-gallate.

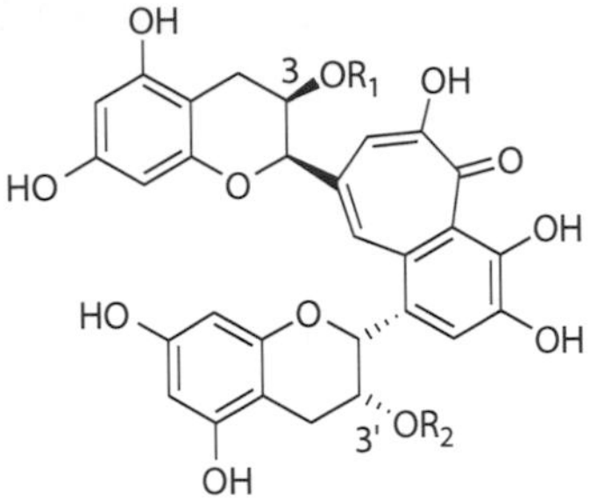

	R_1	R_2
theaflavin	: H	H
theaflavin-3-O-gallate	: galloyl	H
theaflavin-3'-O-gallate	: H	galloyl
theaflavin-3,3'-di-O-gallate	: galloyl	galloyl

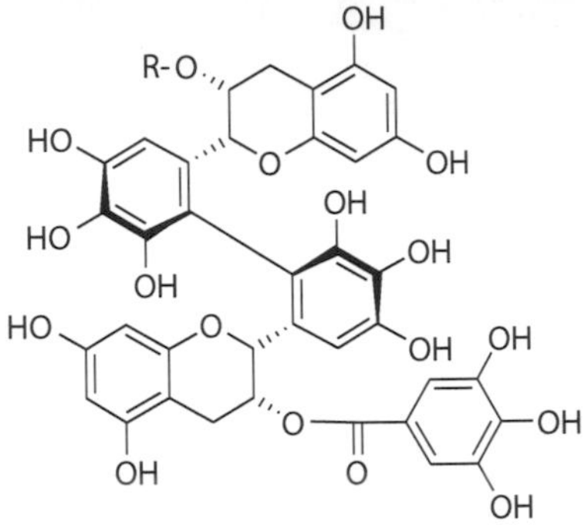

theasinensin A : R = galloyl
theasinensin B : R = H

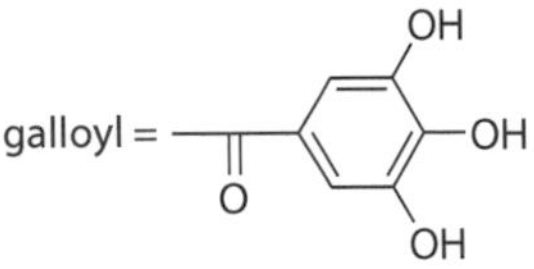

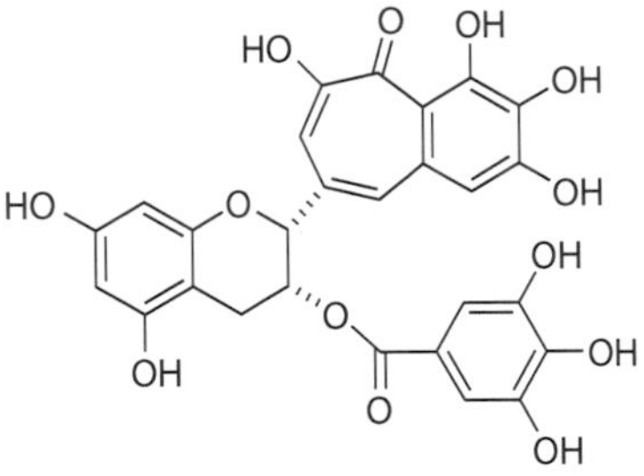

epitheaflagallin-3-O-gallate

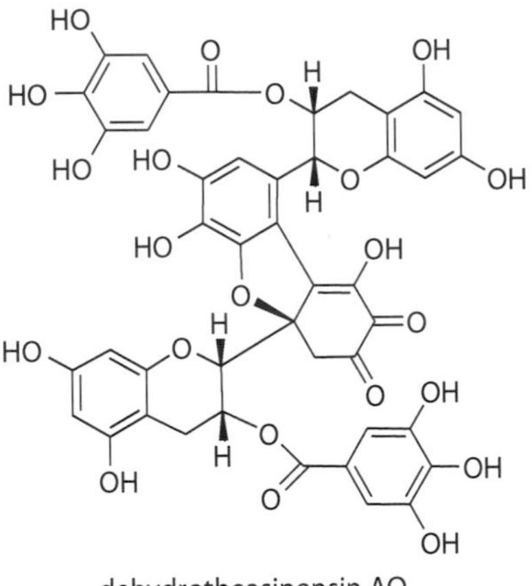

dehydrotheasinensin AQ

In contrast to the A-ring, B-rings of the major tea catechins have vicinal hydroxyl groups and tend to undergo oxidation. This chemical reactivity commonly accounts for the antioxidative activities of tea polyphenols. Oxidative stress caused by radicals and related electron deficient molecules in living organs is implicated in many diseases, such as diabetes and arteriosclerosis. Polyphenols are electron-rich compounds and react with the reactive oxidizing molecular species to reduce oxidative stress. The lower

TABLE 2.2
Polyphenol and Caffeine Content of Oolong, Black, and Postfermented Teas (mg/g Dried Leaf, 60% EtOH Extract)

Compounds	Oolong Tea[a]	Black Tea[b]	Postfermented Tea[c]	Aged Postfermented Tea[d]
Epicatechin	6.30 ± 0.95	1.81 ± 0.30	6.17 ± 0.13	0.55 ± 0.01
Epicatechin-3-O-gallate	11.67 ± 0.36	11.00 ± 0.28	13.00 ± 0.28	0.39 ± 0.04
Epigallocatechin	25.82 ± 0.59	7.65 ± 0.26	18.62 ± 0.40	—
Epigallocatechin-3-O-gallate	50.76 ± 0.55	12.69 ± 0.26	60.29 ± 0.66	0.77 ± 0.06
Catechin	1.28 ± 0.20	0.25 ± 0.49	0.56 ± 0.06	—
Gallocatechin	4.58 ± 0.05	—	3.30 ± 0.07	—
Gallocatechin-3-O-gallate	3.44 ± 0.33	1.12 ± 0.09	7.72 ± 0.02	—
Gallic acid	1.26 ± 0.11	3.53 ± 0.09	3.44 ± 0.02	3.44 ± 0.03
Theogallin	0.71 ± 0.02	3.58 ± 0.07	2.82 ± 0.08	—
Caffeine	26.38 ± 0.71	44.01 ± 0.76	42.16 ± 0.34	35.60 ± 0.32
Theobromine	0.40 ± 0.01	3.50 ± 0.07	1.11 ± 0.02	1.56 ± 0.02
Theasinensin A	1.14 ± 0.19	9.56 ± 0.25	0.80 ± 0.02	—
Theasinensin B	0.99 ± 0.09	6.31 ± 0.13	—	—
Total theaflavins[e]	—	8.90 ± 0.15	—	—

Note. —, not detected.
[a] Fujian, China.
[b] India.
[c] Hunan, China.
[d] Yunnan, China.
[e] Composition of theaflavins in black tea: TF, 13.9%; TF3G, 25.2%; TF3′G, 17.3%; TF3,3′diG, 43.5%.

oxidation-reduction potential of the pyrogallol-type B-ring, compared to that of the catechol-type B-ring, is reflected in the lower stability of epigallocatechin compared to that of epicatechin in neutral or weakly alkaline conditions. It is known that epigallocatechin gallate is oxidized by dissolved oxygen in neutral aqueous solution to give theasinensins, which are dimeric products originally identified as black tea polyphenols (Hatano et al., 2005; Tanaka et al., 2003) (Figure 2.10). The initial step of the oxidation generates *ortho*-quinones, which are electron-deficient molecules. The quinone tends to react with an electron-rich aromatic ring, which is usually another polyphenol molecule. The basic reaction mechanism is probably closely related to that of the enzymatic oxidation of epigallocatechin-3-O-gallate (Tanaka et al., 2010). Production of theasinensins also occurs on radical oxidation with the 2,2-diphenyl-1-picrylhydrazyl radical in organic solvents, though the mechanism may be different from that of enzymatic oxidation (Zhu et al., 2001). Another study on oxidation using a different radical initiator resulted in the production of a quinone dimer (Valcic et al., 1999), which was also

FIGURE 2.8
Reaction of epigallocatechin and acetaldehyde.

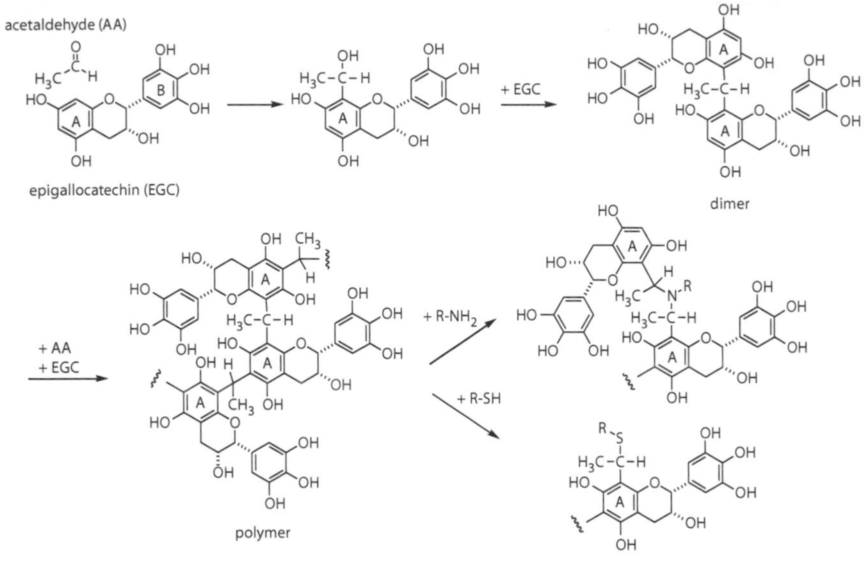

FIGURE 2.9
Lipid-soluble derivatives of epigallocatechin gallate prepared by reaction with natural aldehyde and alcohol.

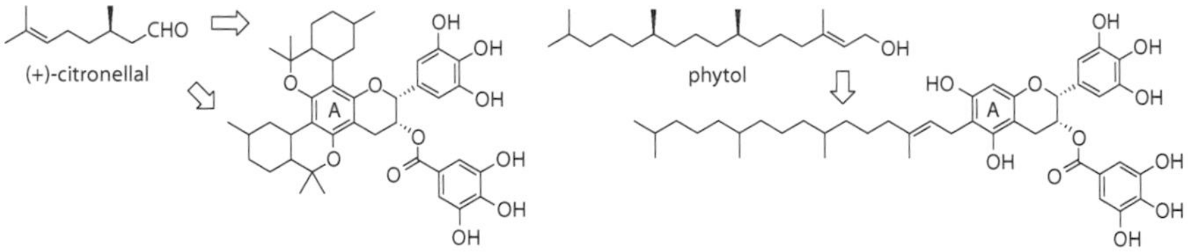

produced by enzymatic oxidation (Tanaka et al., 2003) (Figure 2.10). These results suggested that nonenzymatic reactions are somewhat good models for enzymatic oxidation. The reactive B-ring quinones also afford adducts with thiol groups, such as glutathione and cysteine residues of proteins (Tanaka et al., 2002). This may also have functional significance in biological activities of tea catechins.

When the catechins are oxidized by dissolving oxygen in an aqueous solution, oxygen is reduced to hydrogen peroxide (Akagawa et al., 2003).

FIGURE 2.10
Oxidation of epigallocatechin gallate with radicals or dissolved oxygen and production of catechin-protein conjugates.

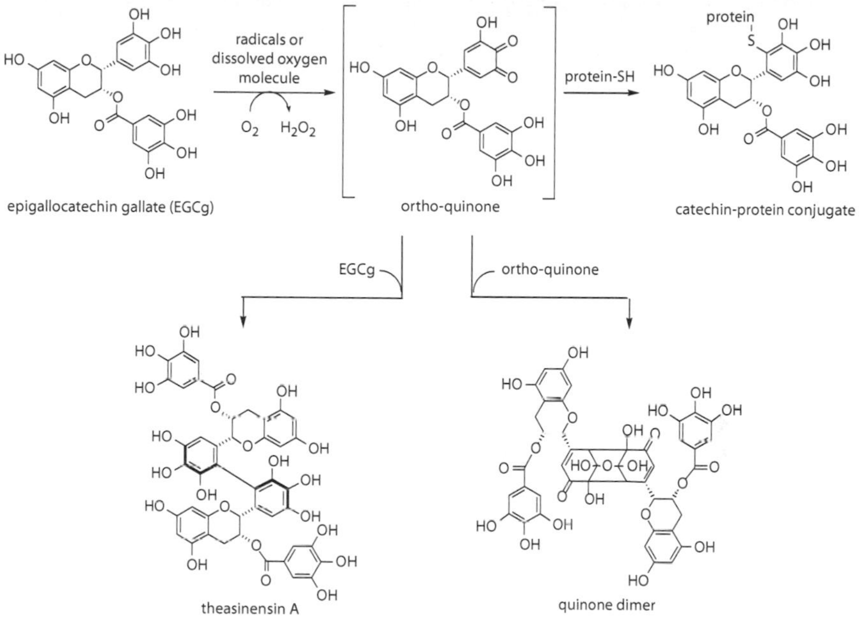

epigallocatechin gallate (EGCg)

ortho-quinone

catechin-protein conjugate

theasinensin A

quinone dimer

Recent studies suggest that the antibacterial activities of tea polyphenols are explained by the generation of hydrogen peroxide (Arakawa et al., 2004), which is a well-known antibacterial agent (Taguri et al., 2004, 2006). In addition to the above reactions, it should be noted that tea catechins are reductively degraded by human intestinal bacteria. The structures produced by this anaerobic metabolism have been recently clarified, and the reabsorption of these metabolites may be important to the *in vivo* biological activity of tea (Takagaki and Nanjo, 2010) (Figure 2.11).

2.6 Hydrophobic Interactions with Coexisting Compounds

The strongly astringent taste of epigallocatechin-3-O-gallate is related to the properties of tannins, which are a specific group of polyphenols that

FIGURE 2.11
Metabolites of epigallocatechin-3-*O*-gallate produced by mammalian intestinal bacteria.

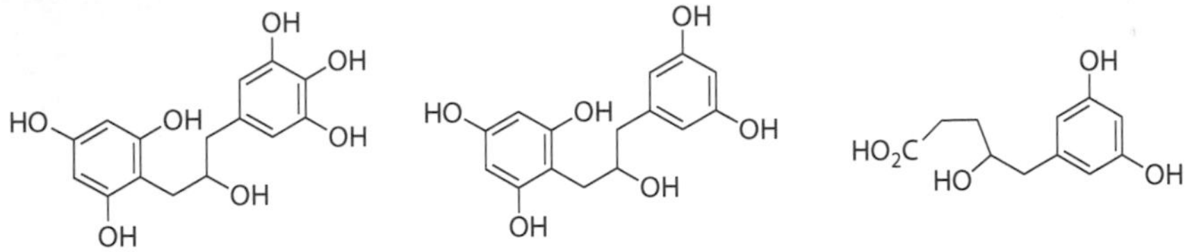

precipitate proteins, amines, and heavy metals from aqueous solutions. The astringent taste of tea is explained by the precipitation of salivary proteins (Luck et al., 1994; Murray et al., 1994). The scientific term *tannin* etymologically originated from leather tanning, where the collagen proteins of animal raw hides are precipitated by complexation with tannins to produce durable leather (Haslam, 1996). This property is related to the biological dysfunction of proteins. Inhibition of enzymes in animal digestive tracts functions as a feeding deterrent against herbivores. However, the property is of benefit to humans, who tend to absorb excess sugars and lipids, because inhibition of the digestive enzymes decreases sugar and fatty acid uptake (Hara and Honda, 1990; Nakai et al., 2005). The interaction between epigallocatechin-3-*O*-gallate and proteins are mainly accounted for by hydrophobic interactions, and the presence of the galloyl ester is important for this hydrophobicity. Disappearance of the C-3 hydroxyl group and addition of the aromatic ester provide a large hydrophobic site in the molecule. The hydrophobic interactions of polyphenols are significantly decreased by hydrolysis of the galloyl ester (Zhang et al., 2002). It has been reported that the interaction occurs in the vicinity of proline residues in peptide chains. The proline residue has a hydrophobic, flat structure and bends the peptide chain to form a β-turn structure, thereby providing a vacant space for the association with tea catechins. In addition to hydrophobic association, π-π interactions can occur with the aromatic rings of phenylalanine or tyrosine residues. Ionic interactions and hydrogen bonding are also involved in catechin-protein interactions. The interaction with a small, cyclic peptide, gramicidin S, is illustrated in Figure 2.12 (Zhang et al., 2002). Furthermore, interactions with the lipid bilayer also occur and are caused by ionic and electrostatic effects. These factors probably play important roles in determining the biological activities of polyphenols (Hashimoto et al., 1999; Ikigai et al., 1933; Ishii et al., 2008).

FIGURE 2.12
Hydrophobic interaction of epigallocatechin-3-O-gallate and a cyclic peptide, gramicidin S, in an aqueous solution.

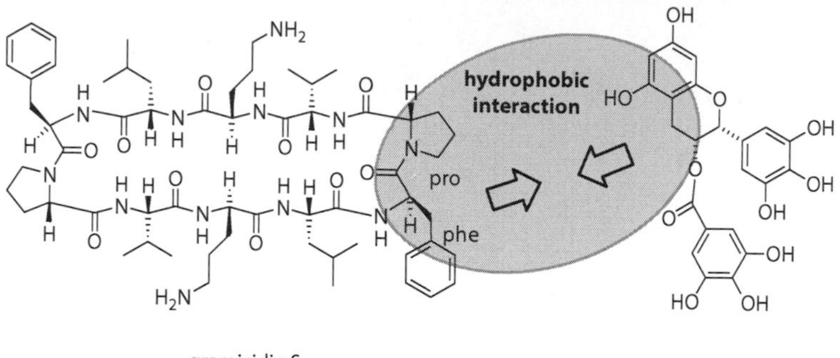

gramicidin S

2.7 Conclusion

Thousands of years ago in China, the tea plant was selected as a medicine probably because it contains caffeine, much as coffee was selected in Ethiopia, Africa, as far back as the 13th century. In addition to caffeine, modern scientific studies have revealed the health benefits of tea polyphenols. The characteristics of tea polyphenols are different from those of other polyphenol-rich plants, with low molecular weight polyphenols dominant over the more common proanthocyanidins, and the accumulation of polyphenols in particularly high concentrations. The chemical properties of epigallocatechin-3-O-gallate described above are apparently related to its biological activities, such as oxidative stress reduction and inhibition of various enzymes. However, the high reactivity of tea catechins also leads to their low stability, especially under physiological conditions in the presence of dissolved oxygen, proteins, and cell membranes at neutral pH. Thus, when *in vitro* activities are measured, the physicochemical properties of catechins should be carefully considered. Nevertheless, many *in vivo* biological and epidemiological studies are providing reliable evidence for the health benefits of tea, which justifies the selection of this plant by early herbalists from the many plants found in the East Asian jungle.

References

Akagawa, M., Shigemitsu, T., and Suyama, K. 2003. Production of hydrogen peroxide by polyphenols and polyphenol-rich beverages under quasi-physiological conditions. *Biosci Biotechnol Biochem*, 6: 2632–2640.

Arakawa, H., Maeda, M., Okubo, S., and Shimamura, T. 2004. Role of hydrogen peroxide in bactericidal action of catechin. *Biol Pharm Bull*, 27: 277–281.

Astill, C., Birch, M.R., Dacombe, C., Humphrey, P.G., and Martin, P.T. 2001. Factors affecting the caffeine and polyphenol contents of black and green tea infusions. *J Agric Food Chem*, 49: 5340–5347.

Finger, A., Engelhardt, U.H., and Wray, V. 1991. Flavonol triglycosides containing galactose in tea. *Phytochemistry*, 30: 2057–2060.

Fudouji, R., Tanaka, T., Taguri, T., Matsuo, Y., and Kouno, I. 2009. Coupling reactions of catechins with natural aldehydes and allyl alcohols and radical scavenging activities of the triglyceride-soluble products. *J Agric Food Chem*, 57: 6417–6424.

Giannasi, D.E. 1988. Flavonoids and evolution in the dicotyledons. In *Flavonoids*, ed. J.N. Harborne, 479–504. London: Chapman & Hall.

Hara, Y., and Honda, M. 1990. The inhibition of α-amylase by tea polyphenols. *Agric Biol Chem*, 54: 1939–1945.

Hashimoto, F., Nonaka, G., and Nishioka, I. 1987. Tannins and related compounds. LVI. Isolation of four new acylated flavan-3-ols from oolong tea. (1). *Chem Pharm Bull*, 35: 611–616.

Hashimoto, F., Nonaka, G., and Nishioka, I. 1988. Tannins and related compounds. LXIX. Isolation and structure elucidation of B, B′-linked bisflavanoids, theasinensins D-G and oolongtheanin from oolong tea. (2). *Chem Pharm Bull*, 36: 1676–1684.

Hashimoto, F., Nonaka, G., and Nishioka, I. 1989. Tannins and related compounds. LXXVII. Novel chalcan-flavan dimers, assamicains A, B and C, and a new flavan-3-ol and proanthocyanidins from the fresh leaves of *Camellia sinensis* L. var. *assamica* KITAMURA. *Chem Pharm Bull*, 37: 77–85.

Hashimoto, F., Nonaka, G., and Nishioka, I. 1992. Tannins and related compounds. CXIV. Structures of novel fermentation products, theogallinin, theaflavonin and desgalloyl theaflavonin from black tea, and changes of tea leaf polyphenols during fermentation. *Chem Pharm Bull*, 40: 1383–1389.

Hashimoto, T., Kumazawa, S., Nanjo, F., Hara, Y., and Nakayama, T. 1999. Interaction of tea catechins with lipid bilayers investigated with liposome systems. *Biosci Biotechnol Biochem*, 63: 2252–2255.

Haslam, E. 1996. Natural polyphenols (vegetable tannins) as drugs: Possible modes of action. *J Nat Prod*, 59: 205–215.

Haslam, E. 1998. Quinone tannin and oxidative polymerization. In *Practical polyphenolics from structure to molecular recognition and physiological action*, ed. E. Haslam, 335–373. Cambridge: Cambridge University Press.

Haslam, E. 2003. Thoughts on thearubigins. *Phytochemistry*, 64: 61–73.

Hatano, T., Ohyabu, T., and Yoshida, T. 2005. The structural variation in the incubation products of (–)-epigallocatechin gallate in neutral solution suggests its breakdown pathways. *Heterocycles*, 65: 303–310.

Huang, S., Inoue, K., Li, Y., Tanaka, T., and Ishimaru, K. 2004. Analysis of catechins in autoclaved tea leaves and drinks. *Jpn J Food Chem*, 11: 99–102.

Ikigai, H., Nakae, T., Hara, Y., and Shimamura, T. 1993. Bactericidal catechins damage the lipid bilayer. *Biochim Biophys Acta Biomembr*, 1147: 132–136.

Ishii, T., Mori, T., Tanaka, T., Mizuno, D., Yamaji, R., Kumazawa, S., Nakayama, T., and Akagawa, M. 2008. Covalent modification of proteins by green tea polyphenol (–)-epigallocatechin-3-gallate through autoxidation. *Free Radic Biol Med*, 45: 1384–1394.

Kusano, R., Andou, H., Fujieda, M., Tanaka, T., Matsuo, Y., and Kouno, I. 2008. Polymer-like polyphenols of black tea and their lipase and amylase inhibitory activities. *Chem Pharm Bull*, 56: 266–272.

Lin, Y.S., Tsai, J.Y., Tsay, J.S., and Lin, J.K. 2003. Factors affecting the levels of tea polyphenols and caffeine in tea leaves. *J Agric Food Chem*, 51: 1864–1873.

Luck, G., Liao, H., Murray, N.J., Grimmer, H.R., Warminski, E.E., Williamson, M.P., Lilleya, T.H., and Haslam, E. 1994. Polyphenols, astringency and proline-rich proteins. *Phytochemistry*, 37: 357–371.

Markham, K.R. 1988. Distribution of flavonoids in the lower plants and its evolutionary significance. In *Flavonoids*, ed. J.N. Harborne, 427–468. London: Chapman & Hall.

Murray, N.J., Williamson, M.P., Lilley, T.H., and Haslam, E. 1994. Study of the interaction between salivary proline-rich proteins and a polyphenol by ^{1}H-NMR spectroscopy. *Eur J Biochem*, 219: 923–935.

Nakai, M., Fukui, Y., Asami, A., Toyoda-Ono, Y., Iwashita, T., Shibata, H., Mitsunaga, T., Hashimoto, F., and Kiso, Y. 2005. Inhibitory effects of oolong tea polyphenols on pancreatic lipase *in vitro*. *J Agric Food Chem*, 53: 4593–4598.

Nakayama, T., Hashimoto, T., Kajiya, K., and Kumazawa, S. 2000. Affinity of polyphenols for lipid bilayers. *BioFactors*, 13: 147–151.

Niemann, G.J. 1988. Distribution and evolution of the flavonoids in gymnosperms. In *Flavonoids*, ed. J.N. Harborne, 469–478. London: Chapman & Hall.

Nishizawa, M., Yamagishi, T., Nonaka, G., and Nishioka, I. 1984. Quantitative determination of gallotannin in Paeoniae Radix (in Japanese). *Yakugaku Zasshi*, 104: 1244–1250.

Nonaka, G., Hashimoto, F., and Nishioka, I. 1986. Tannins and related compounds. XXXVI. Isolation and structures of theaflagallins, new red pigments from black tea. *Chem Pharm Bull*, 34: 61–65.

Nonaka, G., Kawahara, O., and Nishioka, I. 1983. Tannins and related compounds. XV. A new class of dimeric flavan-3-ol gallates, theasinensins A and B, and proanthocyanidin gallates from green tea leaf. (1). *Chem Pharm Bull*, 31: 3906–3914.

Nonaka, G., Sakai, R., and Nishioka, I. 1984. Hydrolysable tannins and proanthocyanidins from green tea. *Phytochemistry*, 23: 1753–1755.

Nutrient Data Laboratory Beltsville Human Nutrition Research Center Agricultural Research Service. 2004. USDA database for the proanthocyanidin content of selected foods. U.S. Department of Agriculture. http://www.nal.usda.gov/fnic/foodcomp.

Ozawa, T., Kataoka, M., Morikawa, K., and Negishi, O. 1996. Elucidation of the partial structure of polymeric thearubigins from black tea by chemical degradation. *Biosci Biotechnol Biochem*, 60: 2023–2027.

Roberts, E.A.H. 1962. Economic importance of flavonoid substances: Tea fermentation. In *The chemistry of flavonoid compounds*, 468–512. Oxford: Pergamon Press.

Sano, M., Suzuki, M., Miyase, T., Yoshino, K., and Maeda-Yamamoto, M. 1999. Novel antiallergic catechin derivatives isolated from oolong tea. *J Agric Food Chem*, 47: 1906–1910.

Taguri, T., Tanaka, T., and Kouno, I. 2004. Antimicrobial activity of 10 different plant polyphenols against bacteria causing food-borne disease. *Biol Pharm Bull*, 27: 1965–1969.

Taguri, T., Tanaka, T., and Kouno, I. 2006. Antibacterial spectrum of plant polyphenols and extracts depending upon hydroxyphenyl structure. *Biol Pharm Bull*, 29: 2226–2235.

Takagaki, A., and Nanjo, F. 2010. Metabolism of (–)-epigallocatechin gallate by rat intestinal flora. *J Agric Food Chem*, 58: 1313–1321.

Takino, Y., Ferretti, A., Flanagan, V., Gianturco, M., and Vogel, M. 1965. The structure of theaflavin, a polyphenol of black tea. *Tetrahedron Lett*, 4019–4025.

Takino, Y., Imagawa, H., Horikawa, H., and Tanaka, A. 1964. Studies on the mechanism of the oxidation of tea leaf catechins. Part III. Formation of a reddish orange pigment and its spectral relationship to some benzotropolone derivatives. *Agric Biol Chem*, 28: 64–71.

Tanaka, T., Jiang, Z.H., Nonaka, G., and Kouno, I. 1999. Modification of the solubility of tannins: Biological significance and synthesis of lipid-soluble polyphenols. In *Plant polyphenols 2: Chemistry, biology, pharmacology, ecology*, ed. G.G. Gross, R.W. Hemingway, T. Yoshida, and S.J. Branham, 761–778. New York: Kluwer Academic/Plenum Publishers.

Tanaka, T., Kondou, K., and Kouno, I. 2000. Oxidation and epimerization of epigallocatechin in banana fruits. *Phytochemistry*, 53: 311–316.

Tanaka, T., and Kouno, I. 2003. Oxidation of tea catechins: Chemical structures and reaction mechanism. *Food Sci Technol Res*, 9: 128–133.

Tanaka, T., Matsuo, Y., and Kouno, I. 2005. A novel black tea pigment and two new oxidation products of epigallocatechin-3-O-gallate. *J Agric Food Chem*, 53: 7571–7578.

Tanaka, T., Matsuo, Y., and Kouno, I. 2010. Chemistry of secondary polyphenols produced during processing of tea and selected foods. *Int J Mol Sci*, 11: 14–40.

Tanaka, T., Matsuo, Y., Yamada, Y., and Kouno, I. 2008. Structure of polymeric polyphenols of cinnamon bark deduced from condensation products of cinnamaldehyde with catechin and procyanidin. *J Agric Food Chem*, 56: 5864–5870.

Tanaka, T., Mine, C., Inoue, K., Matsuda, M., and Kouno, I. 2002. Synthesis of theaflavin from epicatechin and epigallocatechin by plant homogenates and role of epicatechin quinone in the synthesis and degradation of theaflavin. *J Agric Food Chem*, 50: 2142–2148.

Tanaka, T., Takahashi, R., Kouno, I., and Nonaka, G. 1994. Chemical evidence for the de-astringency (insolubilization of tannins) of persimmon fruit. *J Chem Soc Perkin Trans*, 1: 3013–3022.

Tanaka, T., Watarumi, S., Matsuo, Y., Kamei, M., and Kouno, I. 2003. Production of theasinensins A and D, epigallocatechin gallate dimers of black tea, by oxidation-reduction dismutation of dehydrotheasinensin A. *Tetrahedron*, 59: 7939–7947.

Targett, N.M., and Arnold, T.M. 1998. Predicting the effects of brown algal phlorotannins on marine herbivores in tropical and temperate oceans. *J Phycol*, 34: 195–205.

Valcic, S., Muders, A., Jacobsen, N.E., Liebler, D.C., and Timmermann, B.N. 1999. Antioxidant chemistry of green tea catechins. Identification of products of the reaction of (–)-epigallocatechin gallate with peroxyradicals. *Chem Res Toxicol*, 12: 382–386.

Zhang, Y.J., Tanaka, T., Betsumiya, Y., Kusano, R., Matsuo, A., Ueda, T., and Kouno, I. 2002. Association of tannins and related polyphenols with the cyclic peptide gramicidin S. *Chem Pharm Bull*, 50: 258–262.

Zhu, N., Wang, G., Wei, M.J., Lin, J.K., Yang, C.S., and Ho, C.T. 2001. Identification of reaction products of (–)-epigallocatechin, (–)-epigallocatechin gallate and pyrogallol with 2,2-diphenyl-1-picrylhydrazyl radical. *Food Chem*, 73: 345–349.

3

Metabolism, Bioavailability, and Safety Features of Green Tea Polyphenols

Shiming Li and Chi-Tang Ho

Contents

3.1 Introduction

Tea (*Camellia sinensis*) originated in the southern part of Yunna province in southwest China and has since spread worldwide. There are two major varieties of tea in the Theaceae family: *Camellia sinensis* var. *sinensis*, which is characterized by its small leaves and bushlike plants, originating in China and growing in some countries in Southeast Asia with a mild cold climate, and

Camellia sinensis var. *assamica*, which is a large-leaved tree discovered in the region of southwest China and India and immigrated to several other countries with a semitropical climate. Owing to its unique flavor and taste, the *sinensis* tea dominates green tea production, whereas the *assamica* tea is mainly used for black tea production because of its high contents of catechins and tannins. Original tea consumption was mainly for its central nerve-stimulating and soothing effects, and tea drinking has been linked to health for centuries.

Tea varieties are closely associated with the growing conditions and regions of tea plants, and the pluck and process of tea leaves. Currently, there are five varieties readily available: (1) white tea, harvested leave buds with white trichomes; (2) green tea, harvested buds and top two adjacent leaves; (3) black tea, fully fermented green or white tea; (4) oolong tea, partially fermented green or white tea; and (5) pu-erh tea, partially or completely fermented green or white tea.

Regarding the existence of tea polyphenols, white tea and green tea contain only catechins (compounds I to IV in Table 3.1), whereas black tea and oolong tea have both catechins and theaflavins (compounds V to VIII in Table 3.1). The tea catechin contents of white tea are approximately one-third less than that of green tea. The existence of green tea polyphenols is ubiquitous among varieties of tea, even though they are always perceived to be found only in green tea. Table 3.2 shows a clear picture of the predominance of catechins in tea across all varieties.

Epidemiological evidence shows that the intake of green tea polyphenols has a myriad of health beneficial effects, including antioxidant, anticancer, anti-inflammation, antidiabetes, antiatherosclerosis, antihyperlipidemia, antibacteria, and antiviral activities (Arab and Il'yasova, 2003; Crespy and Williamson, 2004; Mejia et al., 2009; Yang et al., 2009).

Green tea polyphenols possess multiple phenolic hydroxyl groups on the A-ring and B-ring. Owing to the nature of the aromatic system, the phenolic hydroxy group acts as an electron donor to the benzene ring, which contributes to the electron-rich property of the aromatic system; that is, the polyphenols are a very electron-rich system, readily losing electrons and easily being oxidized. Therefore, the polyphenols can catch free radicals and react with reactive carbonyl species (RCS), such as superoxide, peroxyl, and hydroxy radicals, nitric oxide, nitrogen dioxide, peroxyl nitrite, and singlet oxygen. Thus, green tea polyphenols can protect DNA, RNA, and cells from oxidative stress caused by free radicals and RCS. *In vivo* studies have demonstrated that the intake of green tea polyphenols (GTPs) has increased total plasma antioxidant activity, as well as increased the superoxide dismutase (SOD) activity in serum and glutathione-*S*-transferase (GST) activity in the liver,

TABLE 3.1
Nomenclatures of Tea Polyphenols

Tea Polyphenol	Structure	No.	Name	R	R'	Acronym
Catechins		I	Epicatechin	H	H	EC
		II	Epigallocatechin	H	OH	EGC
		III	Epicatechin gallate	Galloyl	H	ECG
		IV	Epigallocatechin gallate	Galloyl	OH	EGCG
Theaflavins		V	Theaflavin	H	H	TF1
		VI	Theaflavin-3-monogallate	galloyl	H	TF2a
		VII	Theaflavin-3'-monogallate	H	galloyl	TF2b
		VIII	Theaflavin-3,3'-digallate	galloyl	galloyl	TF3

TABLE 3.2 Tea and Its Polyphenols		
Tea		*Content of Polyphenols*
Green tea		Catechins
White tea		Catechins
Black tea (red tea)		Catechins, theaflavins, thearubigins
Oolong tea		Catechins, theaflavins
Pu-erh tea	Partially pu-erh tea	Catechins, minor theaflavins
	Fully fermented pu-erh tea	Gallic acid

and decreased the concentration of nitric oxide (NO) in plasma and malondialdehyde, a biomarker of oxidative stress, in the liver (Negashi et al., 2004; Skrzydlewska et al., 2002; Yokozawa et al., 2002).

Numerous animal and human studies have tested green tea or green tea extract (GTE) for the prevention and treatment of cancer. These studies have revealed the prevention, delay, and treatment properties of green tea catechins for cancers in various organs, including the lungs, liver, breast, esophagus, intestine, stomach, pancreas, colon, skin, bladder, and prostate (Crespy and Williamson, 2004; Mejia et al., 2009; Khan and Mukhtar, 2007; Yang et al., 2009). Mechanism studies of anticarcinogenic activity of green tea polyphenols have revealed that the GTPs play significant roles in the inhibition of tumor cell proliferation and induction of tumor cell apoptosis.

In an nicotine-derived nitrosamine ketone (NNK)-induced lung cancer study, ingestion of 2% green tea in the diet resulted in a decrease in the number of lung tumors within 3 months compared with a placebo-controlled group (Xu et al., 1992). The number of lung tumors was also decreased in diethylnitrosamine-induced mouse lung tumors at all dosages tested (Cao et al., 1996).

Oral administration of green tea extract (1.5%, wt/v) to 7,12-dimethylbenz[α]anthracene (DMBA)-induced oral tumors in hamsters reduced the incidence of dysplasia and carcinoma in 3 to 4 months (Li et al., 1999, 2002). Both green tea extract and EGCG reduced rat duodenal and colon tumors to show the tea polyphenols as active components (Yamane et al., 1996).

In diethylnitrosamine-induced mouse liver tumors, GTE decreased both the number and diameter of liver tumors, and reduced the number and volume of liver foci at a dosage of 1.25% wt/v (Cao et al., 1996). Weight reduction of a rat liver primary tumor induced by diethylnitrosamine was observed at a GTE dosage of 2% wt/wt (Zhang et al., 2002). In DMBA-induced rat mammary carcinogenesis, GTE ingestion dramatically prolonged the tumor latency and reduced the number of invasive tumors. The administration of

GTE to mice at the time of transplantation showed similar effects on transplanted mammary cells (Kavanagh et al., 2001). Green tea catechins have been found to be associated with prevention, delay, or inhibition of prostate cancer development. Intake of 3 cups or more of green tea has demonstrated the declined risk of prostate cancer (Jian et al., 2004), whereas another clinical phase II trial with green tea intervention resulted in a significant decrease in urinary 8-hydroxydeoxyguanosine after drinking decaffeinated green tea (4 cups/day) among smokers over a 4-month period (Hakim et al., 2003). A questionnaire survey of 40- to 69-year-old men (n = 49,920) concluded that consumption of green tea showed a dose-dependent decrease in the risk of advanced prostate cancer (Kurahashi et al., 2008).

An immune response to bacterial, viral, mechanic, or chemical damages on tissues causes inflammation, which involves activation and directed migration of leukocytes, such as neutrophils, monocytes, and eosinophils, from the venous system to damaged sites. An indication of chronic inflammation is the persistent accumulation and activation of leukocytes (Coussens and Werb, 2002). Chronic inflammation is directly related to other diseases, such as cancer, cardiovascular diseases (CVDs), diabetes, and Alzheimer's disease. Hence, chemical agents that have the ability to reduce the inflammation or downregulate the inflammation process not only decrease the inflammatory disorders, but also reduce the risk of other diseases. Research data have shown that GTPs are directly associated with the inhibition of enzymes involved in inflammation, and anti-inflammatory activity of GTPs has been confirmed by animal and human studies. The anti-inflammatory activities of GTP include the suppression of chemokine production at the inflammatory site by EGCG (Takano et al., 2004) and inhibition of the transcription factor (NFκB) activity by ECG and EGCG, two dominant GTPs in green tea (Abboud et al., 2008; Kawai et al., 2005). Also, in a study of GTE-administered rats, GTE showed effective reduction of nitrotyrosine formation and tumor necrosis factor α (TNFα) production (Muià et al., 2005).

Hypertension and hyperlipidemia are well-known risk factors for CVD. Studies have been shown that GTPs interfere with lipid metabolism by various pathways and have antiatherosclerotic effects. Hence, consumption of GTPs or GTE plays significant roles in reducing or decreasing the appearance of CVDs. It is observed that in studies with rats, mice, and rabbits, intake of GTE at various dosages decreased blood pressure, liver and plasma triglycerides, plasma total cholesterol, and low-density lipoprotein (LDL), as well as LDL peroxidation (Loest et al., 2002; Raederstorff et al., 2003; Tijburg et al., 1997). A recent randomized, double-blind, and placebo-controlled study with a dosage of 400 mg GTE (>320 mg GTP) and 200 mg of theanine/person/day demonstrated the decrease of hypertension, LDL cholesterol, oxidative stress, and serum amyloid-α (SAA), a marker of chronic inflammation, which are all independent risk factors for CVDs (Nantz et al., 2009).

Type II diabetes mellitus, characterized by insulin resistance, involves resistance of glucose and lipid metabolism in peripheral tissues and inadequate insulin secretion by pancreatic β-cells. A study in rats treated with alloxan concluded that green tea lowered serum glucose (Sabu et al., 2002), suggesting the interaction between GTP and glucose. Lipid metabolism is usually modified in type II diabetes to elevated levels of plasma and liver triglyceride and plasma cholesterol. However, these elevated values were reduced with the GTE intake in both Zucker rats and rats administered with a sucrose-rich diet (Hasegawa et al., 2003; Yang et al., 2001). At a dose of 130 mg/kg, GTE effectively reduced the body weight, liver weight, and adipose tissue weight of Zucker rats (Hasegawa et al., 2003). Hence, GTP has effects of both antidiabetes and antiobesity (Kao et al., 2006). Tea catechins, that is, GTPs, have been drawing more and more attention recently from both scientific research and new product development in the food and dietary supplement industry.

3.2 Metabolism

Due to the structural nature of multihydroxy groups and electron-rich aromatic rings of green tea polyphenols, they are subjected to extensive biotransformation, including methylation, glucuronidation, sulfation, and ring fission metabolism, of which the metabolism study of GTPs has been a focus of Yang and several other research groups (Auger et al., 2008; Kohri et al., 2001; Lambert and Yang, 2003a, 2003b; Li et al., 2000, 2001).

3.2.1 Methylation

Catechins are readily methylated by catechol-O-methyltransferase to form mono- or dimethylated catechins. Therefore, methylation is a major metabolic pathway of GTPs and forms the following metabolites: (1) from EC, 3′- and 4′-O-methyl-(–)-EC and O-methyl-(–)-EC-glucuronide (Kuhnle et al., 2000); (2) from ECG, 4″-O-methyl-ECG; (3) from EGC, 4′-O-methyl-EGC; and (4) from EGCG, 4′-O-methyl-EGCG, 4″-O-methyl-EGCG, 4′,4″-O,O-dimethyl-EGCG, 4″-O-methyl-EGCG-3′-O-glucuronide, and 3′,4′- or 3′,5′-O,O-dimethyl-EGCG-4″-O-glucuronide (Feng, 2006). The GTP methylation metabolites were identified and characterized by metabolite standards and various analytical instrumental techniques such as liquid

FIGURE 3.1
Metabolism of EGCG and EGC.

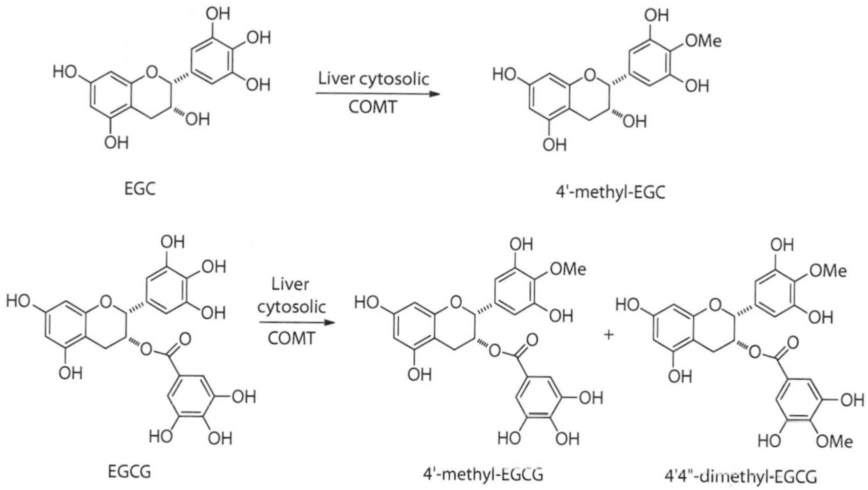

chromatography and tandem mass spectroscopy (LC/MS/MS) and nuclear magnetic resonance (NMR). In a study on the enzymology of cytosolic catechol-*O*-methyltransferase (COMT)-catalyzed methylation of EGCG and EGC in humans, mice, and rabbits, Lu et al. (2003b) identified that the methylation products of EGCG were mainly 4'-methyl-EGCG at a high concentration of EGCG and 4',4"-*O,O*-dimethyl-EGCG at a low EGCG concentration, whereas 4'-*O*-methyl-EGC was the major methylation product of EGC. Figure 3.1 shows the above methylation reaction scheme.

It is also found that the methylation on the B-ring or the D-ring of EGCG was greatly inhibited by the glucuronidation on the same ring, but the methylation was not affected by the glucuronidation on the A-ring of EGCG or EGC. The methylation product of EGCG-3'-*O*-glucoronide was identified as 4"-*O*-methyl-EGCG-3'-*O*-glucuronide and the methylation metabolites of EGCG-4"-*O*-glucuronide were identified as 3',4'-*O,O*-dimethyl-EGCG-4"-*O*-glucuronide and 3',5'-*O,O*-dimethyl-EGCG-4"-*O*-glucuronide.

3.2.2 Glucuronidation and Sulfation

It has been reported that EGCG and EGC glucuronidation was one of the biotransformation pathways, and EGCG-4"-*O*-glucuronide and EGC-4"-*O*-glucuronide were found to be the major metabolite formed by human, mouse, and rat microsomes (Lambert and Yang, 2003a, 2003b). The most efficient catalysis for glucuronidation occurs in mouse intestinal microsomes, followed by the liver of mice, human, and rats, and the rat small intestine, in decreasing order. Vaidyanathan and Walle (2002) studied sulfation and glucuronidation of (–)-EC using human liver and intestine microsomes, cytosol, and recombinant SULT isozymes. They found that in EC sulfation catalyzed by human and rat intestinal and liver cytosol, the human liver is the most efficient catalyst. (–)-EC was efficiently sulfated mainly through SULT1A1 in human liver cytosol. In the intestine, both SULT1A1 and SULT1A3 also contributed to the sulfation. Other SULT isozymes contributed little to sulfation of (–)-EC. Sulfation of (–)-EC was considerably more efficient in humans than in rats. Surprisingly, (–)-EC was not glucuronidated by human and small intestine microsomes, which indicated that sulfation of (–)-EC was a major pathway in the human liver and intestine with no glucuronidation metabolic pathway.

A systematic study of glucuronidation of EGCG and EGC was conducted by Lu et al. (2003a, 2003b) in human, mouse and rat liver microsomes and in nine different human UGT isozymes expressed in insect cells. From EGC, 4'-O-methylated EGC and four glucuronides were identified from EGC: EGC-3'-O-glucurondide, EGC-7-O-glucuronide, 4'-O-methyl-EGC-3'-O-glucuronide, and 4'-O-methyl-EGC-7-O-glucuronide. From EGCG, methylated EGCG and six glucuronides, 4'-O-methyl-EGCG-glucuronide, 4',4''-O,O-dimethyl-EGCG-glucuronide, EGCG-7-O-glucuronide, EGCG-3'-O-glucuronide, EGCG-3''-O-glucuronide, and EGCG-4''-O-glucuronide, were biosynthesized. For EGCG, EGCG-4''-O-glucuronide was the main metabolite in all incubations. The catalytic efficiency for the formation of EGCG-4''-O-glucuronide was in the following decreasing order:

Mouse intestine > mouse liver > human liver > rat liver

Glucuronidation of EGC was much lower than that of EGCG. The catalytic efficiency for EGC-3'-O-glucuronide production followed the order of:

Mouse liver > human liver > rat liver

In 4''-O-glucuronidation of (–)-EGCG and 3'-O-glucuronidation of EGC, it was observed that mouse was more similar to human than rat. It was also observed that human UGT1A1, UGT1A8, and UGT1A9 had high activities with EGCG (Feng, 2006).

3.2.3 Ring Fission

Bacteria in the small intestine and colon catalyze many reactions, such as hydrolysis, hydrogenation, dehydroxylation, decarboxylation, deconjugation, oxidation, and ring fission. Unlike enzymes in animal and human tissues, these bacterial enzymes catalyze the breakdown of polyphenols to small molecules such as phenolic acids.

(–)-EC and (+)-C were found to have similar microbial metabolism profiles (Scheline, 1991) both *in vitro* and *in vivo*. Thirteen metabolites of (–)-EC and (+)-C were found *in vitro* and in animal and human systems:

Two 1,3-diphenylpropanols; 1-(3′,4′dihydroxyphenyl)-3-(2″,4″,6″-trihydroxyphenyl)-propan-2-ol (I) and 1-(3′-hydroxyphenyl)-3-(2″,4″,6″-trihydroxyphenyl)-propan-2-ol (II)

Three γ-valerolactones: δ-(3′,4′-dihydroxyphenyl)-γ-valerolactone (III), δ-(3′-hydroxyphenyl)-γ-valerolactone (IV), and δ-(4′-hydroxy-3′-methoxyphenyl)-γ-valerolactone (V)

Eight acids: 3-hydroxybenzoic acid (XII), 3,4-dihydroxybenzoic acid (VII), vanillic acid (VI), 5-(3′,4′-dihydroxyphenyl)-γ-valeric acid (VIII), 5-(3′-hydroxyphenyl)-γ-valeric acid (X), 3-(3′-hydroxyphenyl)-propionic acid (IX), 3-(3′-hydroxyphenyl)-hydracrylic acid (XIII), and 3-hydroxyhippuric acid (XI)

The metabolic pathway is illustrated in Figure 3.2.

Hydrolysis is the first step of ECG microbial metabolism, yielding (–)-EC and gallic acid. (–)-EC followed the metabolic pathway discussed above, and gallic acid underwent decarboxylation to produce pyrogallol and pyrogallol derivatives (Takizawa et al., 2003). Following oral administration in rats, EGCG was microbially metabolized to EGC, gallic acid, 5-(5′-hydroxyphenyl)-γ-valerolactone-3′-O-β-glucuronide, (–)-5-(3′,4′,5′-trihydroxyphenyl)-γ valerolactone, (–)-5-(3′,5′-dihydroxyphenyl)-γ-valerolactone, and (–)-5-(3′,4′-dihydroxyphenyl)-γ-valerolactone (Meng et al., 2002). Similar to EGCG, it was detected that EGC was also microbially metabolized to the above three metabolites: (–)-5-(3′,4′,5′-trihydroxyphenyl)-γ-valerolactone, (–)-5-(3′,5′-dihydroxyphenyl)-γ-valerolactone, and (–)-5-(3′,4′-dihydroxyphenyl)-γ-valerolactone (Meng et al., 2002).

Green tea extracts or green tea infusion, a mixture of green tea catechins (GTPs), namely, EC, ECG, EGC, and EGCG, was also evaluated for its microbial metabolism. Owing to the complex nature of its components,

FIGURE 3.2
Ring fission metabolites of (–)-EC and (+)-C.

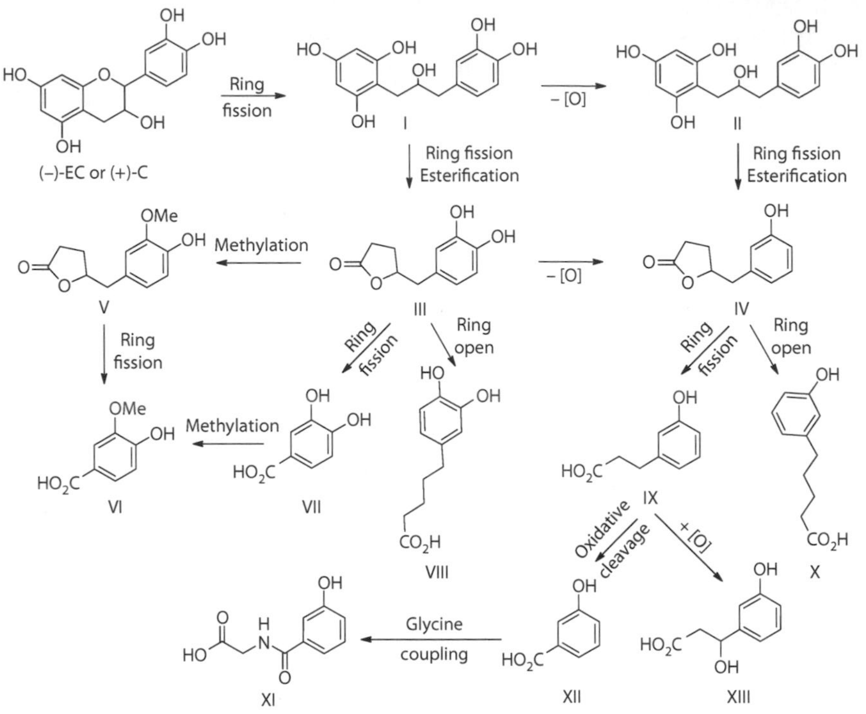

the microbial metabolites cannot be readily assigned to their parent GTP. Figure 3.3 summarizes the microbial metabolite profile of green tea (Feng, 2006). The metabolites include 1,3-diphenylpropanols (direct ring fission products from GTP), γ-valerolactones (ring fission products from 1,3-diphenylpropanols), glucuronides and sulfates of the γ-valerolactones, acids (benzoic acids from phenyl lactone cleavage), and hippuric acids (coupling product between benzoic acids and glycine).

Therefore, the GTP microbial metabolic pathway can be generalized as (1) hydrolysis of gallate ester for ECG and EGCG; (2) ring fission at the C-ring, cleavage between O1 and C2 of the GTP bond by enzyme-catalyzed reduction, and proton transfer to the O1 and C2; (3) formation of lactones and partially de-oxygenated B-ring lactones; (4) formation of phenyl-substituted aliphatic acids from hydrolysis, enzymatic oxidation, and coupling with glycine; and (5) formation of benzoic acids and their derivatives by enzyme-catalyzed oxidative cleavage, methylation or de-oxygenation of phenolic benzoic acids, and coupling with glycine.

FIGURE 3.3
Microbial metabolites of GTPs.

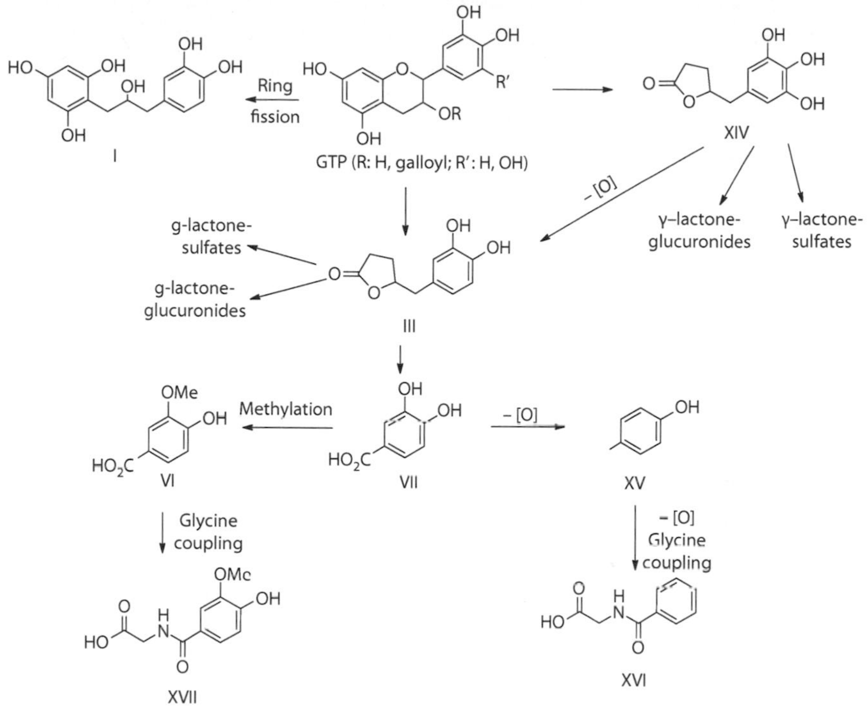

3.2.4 Metabolism in Animals

Metabolism in animals and humans is a complicated process and an overall effect of many enzymatic catalyzed reactions; that is, the GTP metabolites are the overall effect of methylation, glucuronidation, sulfation, and ring fission. For example, from mouse urine the following GTP metabolites were found:

10 EC metabolites: 2 EC-sulfates, 2 EC-glucuronides, 3 methyl-EC-glucuronides, and 3 methyl-EC-sulfates

13 EGC metabolites: 3 EGC-glucuronides, 2 EGC-sulfates, 4 methyl-EGC-glucuronides, and 4 methyl-EGC-sulfates (Li et al., 2001)

From mouse plasma, urine, feces, liver, kidney, and small intestine, many EGCG metabolites were identified, including 3'-O-methyl-EGCG, 3"-O-methyl-EGCG, 3',3"-O,O-dimethyl-EGCG, 4"-O-methyl-EGCG, and 4',4"-O,O-dimethyl-EGCG (Lambert and Yang, 2003a; Meng et al.,

2002). Additionally, ring fission products from EC, EGC, and EGCG in mice were detected as (−)-5-(3′,4′,5′-trihydroxyphenyl)-γ-valerolactone, (−)-5-(3′,4′-dihydroxyphenyl)-γ-valerolactone, and also their sulfates and glucuronides (Li et al., 2001; Meng et al., 2002). GTP metabolites are closely associated with animal species. For instance, from a rat urine sample, two EC-glucuronides and two methyl-EC-glucuronides were identified, but no EC-sulfates were detected, whereas two EC-sulfates were found from mouse urine (Li et al., 2001).

Following the intravenous (i.v.) injection of ECG in rats, metabolites of methylation, glucuronidation, sulfation, ester cleavage, and ring fission were found in plasma, urine, and bile, including EC, 3′-O-methyl-ECG, 4′-O-methyl-ECG, 4″-O-methyl-ECG, 3′,4′-O,O-dimethyl-ECG, pyrogallol, (−)-5-(3′,4′-dihydroxyphenyl)-γ-valerolactone, (−)-5-(3′,4′,5′-trihydroxyphenyl)-γ-valerolactone, (−)-5-(4′-hydroxy-3′-methoxyphenyl)-γ-valerolactone, m-coumaric acid, 3-(3-hydroxyphenyl)-propionic acid, and (−)-5-(3′,4′-dihydroxyphenyl)-4-hydroxy-valeric acid, as well as their sulfate and glucuronide derivatives (Kohri et al., 2001; Takizawa et al., 2003). Metabolism studies in rats with other single GTPs (EC, EGC, and EGCG) were also conducted and found the metabolites following the same pattern as that of ECG—methylation, glucuronidation, sulfation, ester cleavage for EGCG, and ring fission—and generating similar metabolites as O-methyl-GTPs, GTP-glucuronides, GTP-sulfates, O-methyl-GTP-glucuronides, O-methyl-GTP-sulfates, γ-valerolactones, and their glucuronides and sulfates. For EGCG, additional 4′,4″-O,O-dimethyl-EGCG was found in rat urine, plasma, feces, liver, kidney, and small intestine (Kohri et al., 2001; Li et al., 2001; Meng et al., 2002).

3.2.5 Metabolism in Humans

Significant progress has been made in human metabolism of GTPs over the last decade. In human urine, several EC metabolites were identified as (−)-EC-3′-O-glucuronide, 4′-O-methyl-(−)-EC-3′-O-glucuronide, 4′-O-methyl-(−)-EC-5 (or 7)-O-glucuronide, three EC-sulfates, two O-methyl-EC-sulfates, (−)-5-(3′,4′-dihydroxyphenyl)-γ-valerolactone, and other microbial metabolites and their glucuronides. Also from human urine, the metabolites of EGC detected include 4′-O-methyl-EGC, two EGC-sulfates, three O-methyl-EGC-sulfates, three O-methyl-EGC-glucuronides, and microbial biotransformation products (−)-5-(3′,4′-dihydroxyphenyl)-γ-valerolactone, (−)-5-(3′,5′-dihydroxyphenyl)-γ-valerolactone, (−)-5-(3′,4′,5′-trihydroxyphenyl)-γ-valerolactone, and their sulfates and glucuronides. From human plasma and urine, the EGCG metabolites were characterized as 4′,4″-O,O-dimethyl-EGCG,

(–)-5-(3′,4′-dihydroxyphenyl)-γ-valerolactone, (–)-5-(3′,5′-dihydroxyphenyl)-γ-valerolactone, (–)-5-(3′,4′,5′-trihydroxyphenyl)-γ-valerolactone, and their sulfates and glucuronides (Li et al., 2000, 2001; Meng et al., 2001, 2002).

In summary, GTPs underwent extensive methylation, glucuronidation, and sulfation metabolic pathways. To date, at least 18 of EC metabolites, 6 of ECG metabolites, 19 of EGC metabolites, and 22 of EGCG metabolites have been recorded (Feng, 2006).

3.3 Bioavailability

Bioavailability of a drug is defined as the amount of the drug that reaches the blood circulation system and to the tissue. Its determination is a combined rate of absorption, distribution, metabolism, and excretion (ADME). Absorption describes how well a drug passes to the systemic circulation after oral administration. Metabolism is how fast a drug is eliminated from the systemic circulation. Distribution describes how well a drug reaches the tissues. Excretion is the rate of a drug being secreted from the systemic circulation. In general, after the ADME process, only the free form of a drug or nutrient can reach the tissue to interact with the molecular target (Van de Waterbeemd et al., 2003).

3.3.1 GTP Bioavailability in Animals

The general observation of GTP bioavailability is low. The plasma bioavailability of rats was determined as 1 to 3.3% ECG following an oral administration, and 1.6% for EGCG after the intragastric administration. After an oral administration in rats, the plasma concentration was 1.2 μM of EC at a dose of 50 mg/kg, 1.05 μM of ECG at the same dose, and 12.3 μM of EGCG at a dose of 500 mg/kg. The plasma concentration of EGCG was 0.04 μM at an intragastric dose of 75 mg/kg and 10.3 μM at an intravenous (i.v.) dose of 10 mg/kg. The plasma protein binding in rats was 81% at 13.8 μM for EC, 100% at 20.3 μM for ECG, and 97% at 78.5 μM for EGCG (Feng, 2006).

It is observed that except for EC, the bioavailability of GTPs in mixed form was very low. The bioavailability of GTP in rats following oral administration was 39% for EC, 6 to 31% for ECG, and 8 to 14% for EGCG, and following intragastric administration was 31% for EC, 13.7% for EGC, and 0.1% (12.4% in mice) for EGCG (Feng, 2006). The plasma concentration in rats after i.v. injection was found to be 138 μM at a dose of 15 mg/kg of EC,

145.8 µM at a dose of 39 mg/kg of ECG, 29.4 µM at a dose of 1.7 mg/kg of EGC, and 514.8 µM at a dose of 150 mg/kg of EGCG. The plasma exposure level after a single i.v. injection was associated with the dose increase: EC from 2.5 to 15 mg/kg, ECG from 6.5 to 39 mg/kg, and EGCG from 25 to 150 mg/kg (Zhu et al., 2001).

The distribution of GTPs in tissues was reported. After feeding decaffeinated green tea to mice, EGCG was detected in the small and large intestine, colon, liver, and other organs. Following intragastric ingestion, EGCG showed high area under the curve (AUC) levels in the small intestine (126.2 nmol.h/g) and colon (38.8 nmol.h/g), with half-lives of 2.2 and 3.8 h, respectively (Feng, 2006). After i.v. injection of green tea and decaffeinated green tea in rats, the distribution of EC was mainly in the intestine (21.4 nmol.h/g) and kidney (19.3 nmol.h/g). The major distribution tissues of EGC in rats after i.v. administration were bladder, kidney, lung, and intestine with AUC levels in the kidney, lung, and intestine being 77.7, 65.8, and 47.3 nmol.h/g, and half-lives of 0.5, 0.7, and 0.7 h, respectively. After oral administration in rats, EGCG was mainly in the small intestine, and the major distribution site after i.v. administration of EGCG in rats was also the intestine (62.1 nmol.h/g). In mice, it was also found that EGCG was present in its free form in the lung, prostate, and other tissues at levels of 0.31 to 3.56 nmol/g after i.v. injection (Feng, 2006).

3.3.2 GTP Bioavailability in Humans

Studies have demonstrated that the plasma bioavailability of GTPs in humans is very low. After the administration of 697 mg of green tea to healthy volunteers, the plasma EGC and EC content was 0.26 to 0.75% compared with EGCG and ECG with 0.07 to 0.20%, and the same trend was observed in urine (Henning et al., 2008). When an individual green tea catechin was used, plasma concentration was reported as high as 7.8 µM at a dose of 2000 mg for C, 1.53 µM at a dose of 1050 mg for EC, 3.1 µM at a dose of 664 mg for ECG, 5 µM at a dose of 459 mg for EGC, and 6.35 µM at a dose of 1600 mg for EGCG. T_{max} and $T_{1/2}$ were ranged between 0.5 and 4 h and 1 and 6.9 h, respectively (Feng, 2006). It has been observed that methylated metabolites of EGC and EGCG tended to have longer T_{max} and $T_{1/2}$ than their parent compounds. For instance, EGC had a T_{max} value of 0.5 h and a $T_{1/2}$ value of 1 h, whereas the T_{max} and $T_{1/2}$ of its metabolite 4'-O-methyl-EGC were 2 and 4.4 h, respectively (Meng et al., 2001).

It has been speculated that several presystematic processes contribute to the low oral bioavailability of GTPs, including chemical degradation, intestinal metabolism, microbial metabolism, hepatic metabolism, poor membrane permeability, and efflux transporter-mediated intestinal excretion.

3.4 Safety Study of Green Tea Polyphenols

Because of the popularity of tea drinking, it is of great significance to elucidate the safety level of GTPs. Adverse events associated with high flavanol intake are rarely observed, which may be explained by their relatively low absorption, rapid metabolism, and fast elimination. A single dose of EGCG up to 1600 mg or a 10-day repeated oral dose of EGCG up to 800 mg per day was found to be safe and very well tolerated (Feng, 2006). Cancer patients, consuming 6 g/day of a caffeinated green tea extract in three to six equal doses, reported mild to moderate nausea, abdominal pain, and diarrhea, along with agitation, insomnia, dizziness, and tremors, likely due to the caffeine content (Mejia et al., 2009). All of the above suggest that relatively large amounts of flavanols are well tolerated. At present, it would seem that a safe upper limit of chronic intake is on the order of 1 g or more of GTPs per day. In typical brewed tea, that is, 3 g dry weight of tea in 240 ml of water, the GTP content is 50 to 150 mg, depending on the tea variety, water temperature, and length of brewing.

3.5 Conclusion and Perspectives

Not only is tea a pleasant, popular, socially accepted, economical, and safe drink that is enjoyed daily by hundreds of millions of people across all continents, but it also provides a dietary source of biologically active GTPs that help prevent a broad spectrum of diseases. Research in the past three decades has identified tea as nature's gift for promoting human health.

The knowledge of pharmacokinetics (PK) and pharmacodynamics (PD) and ADME of green tea polyphenols is essential to determine their potential bioactivities and overall significance in disease prevention and treatment. Progress has been made in understanding biotransformation and bioavailability of the GTPs and their fate in animals and humans. Studies have demonstrated that the bioavailability of GTPs in humans is relatively low.

Therefore, a number of unresolved questions need to be addressed in further development of GTP products for clinical use. Questions exist particularly concerning dosage, specificity, potency, feasibility, and short- and long-term side effects in humans. It is extremely critical to realize that most of the literature-described molecular effects of GTPs in cell culture systems are obtained with rather high doses of GTPs. Future research needs to define *in vivo* effective dosage to have health benefits, establish the safe range of GTPs or tea consumption associated with these beneficial health effects, and elucidate potential mechanisms of biological action. Furthermore, definitive conclusions concerning the disease prevention and treatment of tea or GTPs have to come from well-designed observational epidemiological studies and intervention trials.

References

Abboud, P.A., Hake, P.W., Burroughs, T.J., Odoms, K., O'Connor, M., Mangeshkar, P., Wong, H.R., and Zingarelli, B. 2008. Therapeutic effect of epigallocatechin-3-gallate in a mouse model of colitis. *Eur J Pharmacol*, 579: 411–417.

Arab, L., and Il'yasova, D. 2003. The epidemiology of tea consumption and colorectal cancer. *J Nutr*, 133: 3310S–3318S.

Auger, C., Mullen, W., Hara, Y., and Crozier, A. 2008. Bioavailability of polyphenon E flavan-3-ols in humans with an ileostomy. *J Nutr*, 138:1535S–1542S.

Cao, J., Xu, Y., Chen, J., and Klaunig, J.E. 1996. Chemopreventive effects of green and black tea on pulmonary and hepatic carcinogenesis. *Fundam Appl Toxicol*, 29: 244–250.

Coussens, L.M., and Werb, Z. 2002. Inflammation and cancer. *Nature*, 420: 860–867.

Crespy, V., and Williamson, G. 2004. A review of the health effects of green tea catechins in *in vivo* animal models. *J Nutr*, 134: 3431S–3440S.

Feng, Y.W. 2006. Metabolism of green tea catechins: An overview. *Curr Drug Metab*, 7: 755–809.

Hakim, I.A., Harris, R.B., Brown, S., Chow, H.H., Wiseman, S., Agarwal, S., and Talbot, W. 2003. Effect of increased tea consumption on oxidative DNA damage among smokers: A randomized controlled study. *J Nutr*, 133: 3303S–3309S.

Hasegawa, N., Yamda, N., and Mori, M. 2003. Powdered green tea has antilipogenic effect on Zucker rats fed a high-fat diet. *Phytother Res*, 17: 477–480.

Henning, S.M., Choo, J.J., and Heber, D. 2008. Nongallated compared with gallated flavan-3-ols in green and black tea are more bioavailable. *J Nutr*, 138: 1529S–1534S.

Jian, L., Xie, L.P., Lee, A.H., and Binns, C.W. 2004. Protective effect of green tea against prostate cancer: A case-control study in southeast China. *Int J Cancer*, 108: 130–135.

Kao, Y.H., Chang, H.H., Lee, M.J., and Chen, C.L. 2006. Tea, obesity, and diabetes. *Mol Nutr Food Res*, 50: 188–210.

Kavanagh, K.T., Hafer, L.J., Kim, D.W., Mann, K.K., Sherr, D.H., Rogers, A.E., and Sonenshein, G.E. 2001. Green tea extracts decrease carcinogen-induced mammary tumor burden in rats and rate of breast cancer cell proliferation in culture. *J Cell Biochem*, 82: 387–398.

Kawai, K., Tsuno, N.H., Kitayama, J., Okaji, Y., Yazawa, K., and Asakage, M. 2005. Epigallocatechin gallate induces apoptosis of monocytes. *J Allergy Clin Immunol*, 115: 186–191.

Khan, N., and Mukhtar, H. 2007. Tea polyphenols for health promotion. *Life Sci*, 81: 519–533.

Kohri, T., Nanjo, F., Suzuki, M., Seto, R., Matsumoto, N., Yamakawa, M., Hojo, H., Hara, Y., Desai, D., Amin, S., Conaway, C.C., and Chung, F.L. 2001. Synthesis of (–)-[4–3H]epigallocatechin gallate and its metabolic fate in rats after intravenous administration. *J Agric Food Chem*, 49:1042–1048.

Kuhnle, G., Spencer, J.P., Schoeter, H., Shenoy, B., Debnam, E.S., Srai, S.K., Rice-Evans, C., and Hahn, U. 2000. Epicatechin and catechin are O-methylated and glucuronidated in the small intestine. *Biochem Biophys Res Commun*, 277: 507–551.

Kurahashi, N., Sasazuki, S., Iwasaki, M., Inoue, M., and Tsugane, S. 2008. Green tea consumption and prostate cancer risk in Japanese men: A prospective study. *Am J Epidemiol*, 167: 71–77.

Lambert, J.D., and Yang, C.S. 2003a. Cancer chemopreventive activity and bioavailability of tea and tea polyphenols. *Mutat Res*, 523–524: 201–208.

Lambert, J.D., and Yang, C.S. 2003b. Mechanisms of cancer prevention by tea constituents. *J Nutr*, 133: 3262S–3267S.

Li, C., Lee, M.J., Sheng, S., Meng, X., Prabhu, S., Winnik, B., Huang, B., Chung, J.Y., Yan, S., Ho, C.T., and Yang, C.S. 2000. Structural identification of two metabolites of catechins and their kinetics in human urine and blood after tea ingestion. *Chem Res Toxicol*, 13: 177–184.

Li, C., Meng, X., Winnik, B., Lee, M.J., Lu, H., Sheng, S., Buckley, B., and Yang, C.S. 2001. Analysis of urinary metabolites of tea catechins by liquid chromatography/electrospray ionization mass spectrometry. *Chem Res Toxicol*, 14:702–707.

Li, N., Chen, X., Liao, J., Yang, G., Wang, S., Josephson, Y., Han, C., Chen, J., Huang, M.T., and Yang, C.S. 2002. Inhibition of 7,12-dimethylbenz[a] anthracene (DMBA)-induced oral carcinogenesis in hamsters by tea and curcumin. *Carcinogenesis*, 23: 1307–1313.

Li, N., Han, C., and Chen, J. 1999. Tea preparations protect against DMBA-induced oral carcinogenesis in hamsters. *Nutr Cancer*, 35: 73–79.

Loest, H.B., Noh, S.K., and Koo, S.I. 2002. Green tea extract inhibits the lymphatic absorption of cholesterol and alpha-tocopherol in ovariectomized rats. *J Nutr*, 132: 1282–1288.

Lu, H., Meng, X., Li, C., Sang, S., Patten, C., Sheng, S., Hong, J., Bai, N., Winnik, B., Ho, C.T., and Yang, C.S. 2003a. Glucuronides of tea catechins: Enzymology of biosynthesis and biological activities. *Drug Metab Dispos*, 31: 452–461.

Lu, H., Meng, X., and Yang, C.S. 2003b. Enzymology of methylation of tea catechins and inhibition of catechol-O-methyltransferase by (–)- epigallocatechin gallate. *Drug Metab Dispos*, 31: 572–579.

Mejia, E.G., de Ramirez-Mares, M.V., and Puangpraphant, S. 2009. Bioactive components of tea: Cancer, inflammation and behavior. *Brain Behav Immun*, 23: 721–731.

Meng, X., Lee, M.J., Li, C., Sheng, S., Zhu, N., Sang, S., Ho, C.T., and Yang, C.S. 2001. Formation and identification of 4′-O-methyl-(-)-epigallocatechin in humans. *Drug Metab Dispos*, 29: 789–731.

Meng, X., Shang, S., Zhu, N., Lu, H., Sheng, S., Lee, M.J., Ho, C.T., and Yang, C.S. 2002. Identification and characterization of methylated and ring-fission metabolites of tea catechins formed in humans, mice and rats. *Chem Res Toxicol*, 15: 1042–1050.

Muià, C., Mazzon, E., Paola, R.D., Genovese, T., Menegazzi, M., Caputi, A.P., Suzuki, H., and Cuzzocrea, S. 2005. Green tea polyphenol extract attenuates ischemia/reperfusion injury of the gut. *Naunyn-Schmiedeberg's Arch Pharmacol*, 371: 364–374.

Nantz, M.P., Rowe, C.A., Bukowski, J.F., and Percival, S.S. 2009. Standardized capsule of *Camellia sinensis* lowers cardiovascular risk factors in a randomized, double-blind, placebo-controlled study. *Nutrition*, 25: 147–154.

Negashi, H., Xu, J.W., Ikeda, K., Njelekela, M., Nara, Y., and Yamori, Y. 2004. Black and green tea polyphenols attenuate blood pressure increases in stroke-prone spontaneously hypertensive rats. *J Nutr*, 134: 38–42.

Raederstorff, D.G., Schlachter, M.F., Elste, V., and Weber P. 2003. Effect of EGCG on lipid absorption and plasma lipid levels in rats. *J Nutr Biochem*, 14: 326–332.

Sabu, M.C., Smitha, K., and Kuttan, R. 2002. Anti-diabetic activity of green tea polyphenols and their role in reducing oxidative stress in experimental diabetes. *J Ethnopharmacol*, 83: 109–116.

Scheline, R.R. 1991. Metabolism of oxygen heterocyclic compounds. In *CRC handbook of mammalian metabolism of plant compounds*, 279–284. Boca Raton, FL: CRC Press.

Skrzydlewska, E., Ostrowska, J., Farbiszewski, R., and Michalak, K. 2002. Protective effect of green tea against lipid peroxidation in the rat liver, blood serum and the brain. *Phytomedicine*, 9: 232–238.

Takano, K., Nakaima, K., Nitta, M., Shibata, F., and Nakagawa, H. 2004. Inhibitory effect of (−)-epigallocatechin 3-gallate, a polyphenol of green tea, on neutrophil chemotaxis *in vitro* and *in vivo*. *J Agric Food Chem*, 52: 4571–4576.

Takizawa, Y., Morota, T., Takeda, S., and Aburada, M. 2003. Pharmacokinetics of (−)-epicatechin-3-*O*-gallate, an active component of Onpi-to, in rats. *Biol Pharm Bull*, 26: 608–612.

Tijburg, L.B., Wiseman, S.A., Meijer, G.W., and Weststrate, J.A. 1997. Effects of green tea, black tea and dietary lipophilic antioxidants on LDL oxidizability and atherosclerosis in hypercholesterolaemic rabbits. *Atherosclerosis*, 135: 37–47.

Vaidyanathan, J.B., and Walle, T. 2002. Glucuronidation and sulfation of the tea flavonoid (−)-epicatechin by the human and rat enzymes. *Drug Metab Dispos*, 30: 897–903.

Van de Waterbeemd, H., Lennernas, H., and Artursson, P. 2003. *Methods and principles in medicinal chemistry*. Vol. 18. Weinheim, Germany: Wiley-VCH Verlag GmbH & Co. KGaA.

Xu, Y., Ho, C.T., Amin, S.G., Han, C., and Chung F.L. 1992. Inhibition of tobacco-specific nitrosamine-induced lung tumorigenesis in A/J mice by green tea and its major polyphenol as antioxidants. *Cancer Res*, 52: 3875–3879.

Yamane, T., Nakatani, H., Kikuoka, N., Matsumoto, H., Iwata, Y., Kitao, Y., Oya, K., and Takahashi, T. 1996. Inhibitory effects and toxicity of green tea polyphenols for gastrointestinal carcinogenesis. *Cancer*, 77: 1662–1667.

Yang, C.S., Lambert, J.D., and Sang, S. 2009. Antioxidant and anti-carcinogenic activities of tea polyphenols. *Arch Toxicol*, 83:11–21.

Yang, M., Wang, C., and Chen, H. 2001. Green, oolong and black tea extracts modulate lipid metabolism in hyperlipidemia rats fed high-sucrose diet. *J Nutr Biochem*, 12: 14–20.

Yokozawa, T., Nakagawa, T., and Kitani, K. 2002. Antioxidative activity of green tea polyphenol in cholesterol-fed rats. *J Agric Food Chem*, 50: 3549–3552.

Zhang, G., Miura, Y., and Yagasaki K. 2002. Effects of dietary powdered green tea and theanine on tumor growth and endogenous hyperlipidemia in hepatoma-bearing rats. *Biosci Biotechnol Biochem*, 66: 711–716.

Zhu, M., Chen, Y., and Li, R.C. 2001. Pharmacokinetics and system linearity of tea catechins in rat. *Xenobiotica*, 31: 51–60.

4

Green Tea Polyphenols for Cancer Risk Reduction

Preclinical and Epidemiological Studies

Naghma Khan, Imtiaz A. Siddiqui, Vaqar M. Adhami, and Hasan Mukhtar

Contents

4.1 Introduction

Green tea accounts for approximately 20% of the world's tea production and for a long time has been the main tea beverage consumed in Japan and parts of China. In recent years, green tea has gained some popularity in other parts of the world. The tea plant, *Camellia sinensis*, has been cultivated in Asia for thousands of years. Currently, more than two-thirds of the world population consumes this popular beverage. Green tea is rich in tea catechins, namely, epigallocatechin gallate (EGCG), epigallocatechin (EGC), epicatechin (EC), and epicatechin gallate (ECG), which have many cancer chemopreventive properties, including antioxidation, anti-inflammatory, antiproliferative, and antiangiogenic. Catechin, gallocatechin, epigallocatechin digallates, epicatechin digallate, 3-*O*-methyl EC and EGC, catechin gallate, and gallocatechin gallate are also present in smaller quantities. Catechins are especially concentrated in green tea, which account for 30 to 40% of the dry weight of the leaves. The polyphenolic constituent EGCG is the major and most effective chemopreventive agent in green tea. EGCG appears to be the most powerful of all the catechins, with an antioxidant activity about 25 to 100 times more potent than that of vitamins C and E (Cao et al., 2002). It has been reported that consumption of green tea polyphenols (GTPs) is related to decreased risk or slower progression of cancer of various organs (Khan and Mukhtar, 2010; Khan et al., 2008; Syed et al., 2007a). During manufacturing, green and black teas are processed differently. Freshly harvested leaves are immediately steamed to prevent fermentation, yielding a dry, stable product to produce green tea. This steaming process destroys the enzymes responsible for breaking down the color pigments in the leaves and allows the tea to maintain its green color during the subsequent rolling and drying processes. These processes preserve natural polyphenols with respect to health-promoting properties. The catechins in green tea are dimerized to form a variety of theaflavins as green tea is fermented to oolong and then to black tea. In this chapter, we discuss the studies of green tea in cell culture, animal models, and humans in the chemoprevention/chemotherapy of cancers of various organs.

4.2 Preclinical Studies

4.2.1 Green Tea and Prostate Cancer

We have earlier reported that oral infusion of GTPs at a human achievable dose equivalent to 6 cups of green tea per day significantly inhibited prostate cancer (PCa) development and increased tumor-free and overall survival of transgenic adenocarcinoma of the mouse prostate (TRAMP) mice. GTP was provided as the sole source of drinking fluid to TRAMP mice from 8 to 32 weeks of age and resulted in significant delay in primary tumor incidence and tumor burden as assessed sequentially by magnetic resonance imaging (MRI), significant decrease in prostate and genitourinary weight, inhibition in serum insulin-like growth factor (IGF)-1 and restoration of IGF binding protein (IGFBP)-3 levels, and reduction in the protein expression of proliferating cell nuclear antigen (PCNA) in the prostate compared with water-fed TRAMP mice (Gupta et al., 2001). A study from our laboratory has recently reported that EGCG is a direct antagonist of androgen action. It was shown by *in silico* modeling and fluorescence resonance energy transfer (FRET)-based competition assay that EGCG physically interacted with the ligand binding domain of an androgen receptor (AR) by replacing a high-affinity labeled ligand. EGCG treatment also repressed the transcriptional activation by a hotspot mutant AR (T877A) expressed ectopically, as well as the endogenous AR mutant. In athymic nude mice, EGCG was found to inhibit nuclear translocation and protein expression of AR, and there was also significant downregulation of androgen-regulated miRNA-21 and upregulation of a tumor suppressor, miRNA-330, in tumors of mice treated with EGCG (Siddiqui et al., 2011). We conducted a study to identify the stage of PCa that is most vulnerable to chemopreventive intervention by GTPs. Treatment with GTP infusion to TRAMP mice was initiated at ages representing different stages of the disease: (1) 6 weeks (group 1, normal prostate), (2) 12 weeks (group 2, prostatic intraepithelial neoplasia), (3) 18 weeks (group 3, well-differentiated adenocarcinoma), and (4) 28 weeks (group 4, moderately differentiated adenocarcinoma). At age 32 weeks, subsets of animals were evaluated by MRI, ultrasound, and prostate weight and for serum IGF-1/IGFBP-3, and IGF signaling. It was found that tumor-free survival was extended to 38 weeks in group 1, 31 weeks in group 2, and 24 weeks in group 3, compared to 19 weeks in water-fed controls. Median life expectancy was 68 weeks in group 1, 63 weeks in group 2, 56 weeks in group 3, and 51 weeks in group 4, compared to 42 weeks in the control mice. When intervention with GTPs was initiated early, when prostatic intraepithelial neoplasia lesions were common, IGF-1 and its downstream targets, including phosphatidylinositol 3-kinase (PI3K), p-Akt, and phosphorylated extracellular signal-regulated kinase (ERK1/2), were significantly inhibited (Adhami et al., 2009). We encapsulated EGCG in polylactic acid–polyethylene glycol nanoparticles and observed that encapsulated EGCG retains its biological effectiveness with >10-fold dose advantage for exerting its proapoptotic and angiogenesis inhibitory effects, critically important determinants of the chemopreventive effects of EGCG in both *in vitro* and *in vivo* systems (Siddiqui et al., 2009). A combination of EGCG and NS-398, a specific cyclooxygenase (COX)-2 inhibitor, resulted in enhanced cell growth inhibition, apoptosis induction, poly(ADP)ribose polymerase (PARP) cleavage, inhibition of peroxisome proliferator-activated receptor (PPAR)-γ, and inhibition of nuclear factor (NF)-κB, compared with the additive effects of the two agents alone, suggesting a possible synergism (Adhami et al., 2007). EGCG sensitized tumor necrosis factor apoptosis-inducing ligand (TRAIL)-resistant PCa LNCaP cells to TRAIL-mediated apoptosis through modulation of intrinsic and extrinsic apoptotic pathways. There was also a synergistic inhibition in the invasion and migration of LNCaP cells through inhibition in the protein expression of vascular endothelial growth factor (VEGF), urokinase plasminogen activator (uPA), and angiopoietins 1 and 2. The activity and protein expression of matrix metalloproteinases (MMPs)-2, -3, and -9 and upregulation of tissue inhibitor of metalloproteinase (TIMP)-1 in cells treated with a combination of EGCG and TRAIL was also observed (Siddiqui et al., 2008a).

It was shown using isogenic cell lines that EGCG activates growth arrest and apoptosis in PCa cells primarily via a p53-dependent pathway that involves the function of both p21 and Bax such that downregulation of either molecule confers a growth advantage to the cells (Hastak et al., 2005). EGCG was also found to inhibit degradation of gelatin, degradation of type IV collagen in reconstituted basement membrane, and activation of MMP-2, but not pro-MMP-9, in a cell-free system (Pezzato et al., 2004). EGCG-induced apoptosis in human PCa LNCaP cells is mediated via modulation of stabilization of p53 by phosphorylation on critical serine residues and p14ARF-mediated downregulation of murine double minute 2 (MDM2) protein, and negative regulation of NF-κB activity, thereby decreasing the expression of Bcl-2. EGCG-induced stabilization of p53 caused an upregulation in its transcriptional activity, resulting in activation of WAF1/p21 and Bax. This altered expression of Bcl-2 family members triggered the activation of caspase-9 and -8, followed by activation of caspase-3 and cleavage of PARP (Hastak et al., 2003). EGCG inhibited the chymotrypsin-like activity of the proteasome *in vitro* and *in vivo* at the concentrations found in the serum of green tea drinkers. The inhibition of the proteasome by EGCG in several tumor and transformed cell lines resulted in the accumulation of two natural proteasome substrates, KIP1/p27 and IκBα, followed by growth arrest in the G1 phase of the cell cycle (Nam et al., 2001).

Continuous GTP infusion for 32 weeks resulted in substantial reduction in expression of NF-κB, IKKα, IKKβ, RANK, NIK, and signal transducers and activators of transcription (STAT) 3 in dorsolateral prostate of TRAMP mice. The level of transcription factor osteopontin was also downregulated, and there was shift in balance between Bax and Bcl2 favoring apoptosis in the dorsolateral prostate of TRAMP mice fed GTPs (Siddiqui et al., 2008b). EGCG treatment inhibited early- but not late-stage PCa in the TRAMP mice. EGCG significantly reduced cell proliferation, induced apoptosis, and decreased AR, IGF-1, IGF-1 receptor (IGF-1R), phospho-ERK1/2, cyclooxygenase (COX)-2, and inducible nitric oxide synthase (iNOS) in the ventral prostate (Harper et al., 2007). Diet containing lysine, proline, arginine, ascorbic acid, and green tea extract caused inhibition of tumor growth, inhibition of MMP-9, VEGF secretion, and mitosis in tissues of athymic nude mice implanted with human PCa PC-3 cells (Roomi et al., 2005). There was reduction in the levels of IGF-1 and increase in the levels of IGFBP-3 in the dorsolateral prostate with an inhibition of protein expression of PI3K, Akt, and ERK1/2 by continuous GTP administration for 24 weeks to TRAMP mice. There was also inhibition of vascular endothelial growth factor (VEGF), uPA, and MMP-2 and -9 (Adhami et al., 2004). Soy phytochemical concentrate (SPC), black tea, and green tea significantly reduced tumorigenicity in a mouse model of orthotopic androgen-sensitive human PCa. The combination of SPC and green tea synergistically inhibited final tumor weight and metastasis and significantly reduced serum concentrations of both testosterone and dihyroxytesterone (DHT) *in vivo*. Inhibition of tumor progression was associated with reduced tumor cell proliferation and tumor angiogenesis (Zhou et al., 2003).

4.2.2 Green Tea and Breast Cancer

EGC inhibited heregulin (HRG)-beta 1-induced migration/invasion of Michigan Cancer Foundation (MCF)-7 human breast carcinoma cells by activation of epidermal growth factor receptor (EGFR)-related protein B2 (ErbB2)/ErbB3, but not Akt. EGCG inhibited this migration/invasion by suppressing the HRG-stimulated activation of ErbB2/ErbB3/Akt (Kushima et al., 2009). Treatment of Polyphenon E inhibited MDA-MB231 breast cancer and human dermal microvascular endothelial (HMVEC) cell migration and the expression of VEGF and MMP-9. Moreover, polyhenon E also inhibited VEGF-induced neovascularization *in vivo*. Polyphenon E also blocked STAT signaling by suppressing interferon-gamma-induced gene transcription and downstream STAT3 activation by inhibiting STAT1 and STAT3 dimerization in MDA-MB231 cells (Leong et al., 2009). EGCG modulated the hepatocyte growth factor (HGF)/Met signaling pathway involved in proliferation, survival, and motility/invasion. EGCG treatment inhibited

HGF-induced Met phosphorylation, and subsequent AKT and ERK activation (Bigelow and Cardelli, 2006). Treatment of the green tea extract has been reported to increase the anticancer effect of *Ganoderma lucidum* extract on cell proliferation, as well as colony formation of breast cancer cells. This was mediated by the downregulation of expression of oncogene c-myc and by the suppression of secretion of uPA in MDA-MB-231 cells (Thyagarajan et al., 2007). The effects of EGCG and tamoxifen, alone or in combination, on mammary carcinogenesis in breast cancer cells and treatment of green tea extract on preneoplastic lesions in C3H/OuJ mice were investigated recently. In the anchorage-independent growth assay, EGCG and tamoxifen exhibited dose-dependent antiproliferative effects on MCF-7 cells. Tumor incidences were decreased in the green tea extract, tamoxifen, and combination groups (Sakata et al., 2011). Administration of Polyphenon E in drinking water delayed tumor onset and suppressed tumor growth in the C3(1)/simian virus 40 (SV40) mouse model as compared to tap water-fed animals, with no adverse side effects. Green tea slowed the progression of ductal lesions to advanced mammary intraepithelial neoplasias, suppressed tumor invasiveness, inhibited the proliferation of ductal epithelial cells and tumors, and overall, disrupted postpubertal ductal growth. It was also shown that green tea inhibited angiogenesis through a decrease in both ductal epithelial and stromal VEGF expression and a decrease in intratumoral microvascular density (Leong et al., 2008).

Green tea and tamoxifen caused a decrease in MCF-7 xenograft tumor size in mice and an increase in apoptosis in tumor tissue, compared with either agent administered alone. Green tea blocked ER-dependent transcription, as well as estradiol-induced phosphorylation and nuclear localization of mitogen-activated protein kinase (MAPK) (Sartippour et al., 2006). It has been shown that there was delayed tumor growth onset, rate of tumor growth, tumor volume, and metastasis after consumption of a GTP mixture in the drinking water of BALB/c mice inoculated with 4T1 mouse mammary carcinoma cells. These effects were found to be related to an increase in the Bax/Bcl$_2$ ratio and caspase-3 activation (Baliga et al., 2005).

4.2.3 Green Tea and Skin Cancer

Since we first proposed that GTPs demonstrate a significant protection against skin tumorigenicity (Wang et al., 1989), the agent has come a long way to demonstrate its efficacy. In one of the subsequent studies, we demonstrated that green tea might have preventive effects against ultraviolet (UV)-induced skin cancers (Wang et al., 1991). Numerous studies indicate that green tea has multiple biological effects that ameliorate the damaging effects of UV radiation. It is effective through topical application as well as oral administration in drinking water. Brewed green tea at concentrations similar to human consumption significantly inhibited UVB- or 12-O-tetradecanoyl phorbol-13-acetate (TPA)-induced tumorigenesis and UVB or 7,12-dimethylbenz(a) anthracene (DMBA)-initiated and UVB- or TPA-promoted carcinogenesis in SKH-1 mice (Wang et al., 1992c). Another study demonstrated that oral administration of green tea to mice not only inhibited skin tumorigenesis, but also reduced fatty tissues in the dermis (Conney et al., 2002). Both oral administration and intraperitoneal injection of GTPs inhibited the growth of UV-induced skin papillomas (Wang et al., 1992b) or TPA-induced COX-2 in rodent models (Kundu et al., 2003). Our laboratory has also demonstrated that in SKH-1 hairless mice, topical application of GTPs results in a significant decrease in UVB-induced skin thickness, skin edema, infiltration of leukocytes, and inhibition of MAPK and NF-κB pathways (Afaq et al., 2003). In a study, Conney et al. (1999) reported that oral administration of various formulations of tea inhibited the growth of well-established skin tumors and, in some cases, also resulted in a tumor regression. Oral administration of GTPs reduced UVB-induced skin tumor incidence, tumor multiplicity, and tumor growth in SKH-1 mice. There was also reduced expression of MMP-2 and MMP-9, CD31, VEGF, and PCNA in the GTP-treated group. Additionally, there were more cytotoxic CD8+T cells and greater activation of caspase-3 in the tumors of the GTP group, indicating the apoptotic death of the tumor cells (Mantena et al., 2005b). In a subsequent study the authors suggested that EGCG inhibits photocarcinogenesis through inhibition of angiogenic factors and activation of CD8+T cells in tumors (Mantena et al., 2005a). Meeran et al. (2006), in a study, determined the effects of EGCG on photocarcinogenesis in interleukin (IL)-12 knockout mice using the formation of cyclobutane pyrimidine dimers as an indicator of extent of UVB-induced DNA damage. The results indicated that EGCG prevented photocarcinogenesis through an EGCG-induced IL-12-dependent DNA repair mechanism.

Schwarz et al. (2008) studied whether GTP induction of IL-12 and DNA repair could also be observed in human cells. Epidermal carcinoma (KB) cells and normal human keratinocytes were exposed to GTPs 5 h before and after UVB. UVB-induced apoptosis was reduced in UVB-exposed cells treated with GTPs, and GTPs induced the secretion of IL-12 in keratinocytes. The reduction in UV-induced cell death by GTPs was almost completely reversed upon addition of an anti-IL-12-antibody, indicating that the reduction of UV-induced cell death by GTPs is mediated via IL-12. The ability of IL-12 to reduce DNA damage and sunburn cells was confirmed in human living skin-equivalent models. Green tea and its individual polyphenols have been shown to protect against many of the other damaging effects of UV radiation. Both systemic and topical application of GTPs and EGCG were demonstrated to protect against the UV-induced sunburn response (Katiyar et al., 1999), photoaging of the skin (Wang et al., 1992a), and immunosuppression (Katiyar et al., 1995, 1999). Direct examination of UV-irradiated skin that had been pretreated *in vivo* with topical EGCG reduced the number of apoptotic keratinocytes (Chung et al., 2003; Elmets et al., 2001). It was also demonstrated that oral administration of green tea to SKH-1 hairless mice enhanced a UV-induced increase in the number of p53- and p21-positive cells in the epidermis following UV exposure (Lu et al., 2000) suggesting that the photoprotective effect of green tea on UV-induced carcinogenesis may be mediated through stimulation of UV-induced increases in the levels of p53 and p21.

The *in vivo* observations are well supported by *in vitro* studies in which normal human epidermal keratinocytes (NHEKs) were exposed to UVB radiation (Xia et al., 2005). In contrast to its effect on NHEKs, EGCG is known to stimulate apoptosis in UV-induced premalignant papillomas and invasive squamous cell carcinomas (Chen et al., 1998; Chung et al., 2003). EGCG was also shown to protect against UV-induced oxidative stress in humans. When it was applied to the skin just before exposure to a 4× minimal erythema dose of UVB radiation, EGCG significantly decreased the production of hydrogen peroxide and nitric oxide, as well as lipid peroxidation, in human dermis and epidermis (Katiyar et al., 2001). In a study, EGCG treatment was observed to induce a dose-dependent decrease in the viability and growth of malignant melanoma and metastatic melanoma cell lines (Nihal et al., 2005). In an *in vitro* study using cultured human cells (lung fibroblasts, skin fibroblasts, and epidermal keratinocytes), EGCG resulted in a dose-dependent reduction in UV-induced DNA damage (Morley et al., 2005). GTPs also significantly inhibited the UVB-induced DNA damage when applied topically to the mouse epidermis, using a ^{32}P postlabeling technique (Chatterjee et al., 1996). EGCG was also shown to selectively decrease both proliferation and survival of primary cultures of ornithine decarboxylase (ODC) overexpressing transgenic keratinocytes but not keratinocytes from normal littermates or Ras-infected keratinocytes (Paul et al., 2005).

Hsu et al. (2007) determined whether MAPK pathways are required for GTP-induced caspase-14 expression in NHEKs, and if GTP can modulate the expression of pathological markers in the psoriasiform lesions that develop in the flaky skin mouse. The results indicated that the p38 and JNK MAPK pathways are required for EGCG-induced expression of caspase-14 in NHEKs. Importantly, topical application of 0.5% GTPs significantly reduced the symptoms of epidermal pathology in the flaky skin mice, associated with efficient caspase-14 processing and reduction in proliferating cell nuclear antigen levels. A recent study determined whether GTPs in drinking water could prevent UV-induced immunosuppression and the potential mechanisms of this effect in mice. Compared with untreated mice, GTP-treated mice (0.2%, w/v) had a reduced number of cyclobutane pyrimidine dimer-positive (CPD(+)) cells in the skin, showing faster repair of UV-induced DNA damage, and had a reduced migration of CPD(+) cells from the skin to draining lymph nodes, which was associated with elevated levels of nucleotide excision repair (NER) genes. GTPs did not prevent UV-induced immunosuppression in NER-deficient mice but significantly prevented it in NER-proficient mice; immunohistochemical analysis of CPD(+) cells indicated that GTPs reduced the numbers of UV-induced CPD(+) cells in NER-proficient mice but not in NER-deficient mice. Overall, this study showed a novel NER mechanism by which drinking GTPs prevents UV-induced immunosuppression, and that inhibiting UV-induced immunosuppression may underlie the chemopreventive activity of GTPs against photocarcinogenesis (Katiyar et al., 2010).

A study (Nandakumar et al., 2011) investigated whether tea catechins, particularly EGCG, would modify epigenetic events to regulate DNA methylation-silenced tumor suppressor genes in skin cancer cells. DNA methylation, histone modifications, and tumor suppressor gene expressions were studied in detail using human epidermoid carcinoma A431 cells as an *in vitro* model after EGCG treatment. The study demonstrated that EGCG treatment decreased global DNA methylation levels in a dose-dependent manner. EGCG also decreased the levels of 5-methylcytosine, DNA methyltransferase (DNMT) activity, messenger RNA (mRNA), and protein levels of DNMT1, DNMT3a, and DNMT3b. EGCG subsequently decreased histone deacetylase activity and increased levels of acetylated lysine 9 and 14 on histone H3 (H3-Lys 9 and 14) and acetylated lysine 5, 12, and 16 on histone H4, but decreased levels of methylated H3-Lys 9. Additionally, EGCG treatment resulted in reexpression of the mRNA and proteins of silenced tumor suppressor genes, p16INK4a and Cip1/p21.

4.2.4 Green Tea and Lung Cancer

The growth of tumors in a xenograft study was dose-dependently inhibited by EGCG given in the diet. EGCG treatment increased tumor cell apoptosis and oxidative DNA damage as assessed by the formation of 8-hydroxy-2'-deoxyguanosine (8-OHdG) and phosphorylated histone 2A variant X. The growth of viable H1299 cells was also dose-dependently reduced *in vitro* by

EGCG. Treatment with EGCG also caused the generation of intracellular reactive oxygen species (ROS) and mitochondrial ROS (Li et al., 2010). In a study reported by Shimizu et al. (2010), water containing green tea catechins was freely given to SAMP10 mice, and the chemopreventive effect of green tea catechins intake on tumor metastasis was examined. Natural killer cell activity was maintained by green tea catechins intake. The accumulation of lung-metastatic K1735M2 melanoma cells in lungs after intravenous injection of the cells at 6 and 24 h was significantly suppressed, and the number of lung-metastatic colonies was significantly reduced, in comparison with those in control mice (Shimizu et al., 2010). EGCG has been reported to inhibit cell proliferation in erlotinib-sensitive and -resistant cell lines, including those with c-Met overexpression and acquired resistance to erlotinib. The combination of erlotinib and EGCG caused more inhibition of cell proliferation and colony formation than either agent alone, and EGCG treatment also inhibited ligand-induced c-Met and EGFR phosphorylation. The triple combination of EGCG/erlotinib/c-Met inhibitor SU11274 resulted in greater inhibition of proliferation than EGCG and erlotinib. The combination of EGCG and erlotinib treatment caused inhibition of the growth of H460 xenografts (Milligan et al., 2009).

Polyphenon E treatment did not significantly inhibit average tumor multiplicity but reduced per animal tumor load in A/J mice injected with benzo(a)pyrene (B(a)P). Analysis of tumor pathology revealed a specific inhibition of carcinomas, with the largest carcinomas significantly decreased in Polyphenon E-treated animals. Aerosolized difluoromethylornithine (DFMO) did not have a significant effect on lung tumor progression. MRI of B(a)P-induced lung tumors confirmed the presence of a subset of large, rapidly growing tumors in untreated mice (Anderson et al., 2008). Study from our laboratory has reported that EGCG caused suppression of NF-κB/PI3K/Akt/mammalian target of rapamycin (mTOR) and MAPKs in normal human bronchial epithelial cells, which may contribute to its ability to suppress inflammation, proliferation, and angiogenesis induced by cigarette smoke (Syed et al., 2007b). Administration of Polyphenon E reduced the NNK-induced lung tumor incidence and multiplicity in female A/J mice and inhibited cell proliferation, enhanced apoptosis in adenocarcinomas and adenomas, and lowered levels of c-Jun and ERK1/2 phosphorylation (Lu et al., 2006). Treatment with green tea caused reduction in tumor incidence and multiplicity in N-methyl-N9-nitro-N-nitrosoguanidine (MNNG)-induced lung cancers and precancerous lesions in LACA mice (Luo and Li, 1992). Administration of GTPs prior to challenge with carcinogen afforded significant protection against both diethylnitrosamine (DEN)- and B(a)P-induced forestomach and lung tumorigenesis in A/J mice. There was a decrease in numbers of tumors/mouse in GTP-fed groups compared to non-GTP-fed controls. No adenocarcinomas were observed in GTP-fed groups compared to mice treated with DEN and B(a)P alone-treated controls (Katiyar et al., 1993).

4.2.5 Green Tea and Liver Cancer

Decaffeinated green and black teas were both observed to significantly reduce the number of hepatic tumors after 40 weeks of treatment in diethyl-nitrosamine-treated C3H mice (Cao et al., 1996). In another study polyphe-nols from green tea were shown to significantly decrease the number and total volume of glutathione-S-transferase (GST)-P-positive prenioplastic foci of hepatocacinogenesis (Matsumoto et al., 1996). In a study utilizing AFB1 and CCl4 as carcinogens in rats, green tea inhibited the number and total volume of GST-P- and gamma-glutamyl transpeptidase (GGT)-positive foci, as well as inhibiting cell proliferation (Qin et al., 2000). In another study of a rat model of AFB1-induced hepatocarcinogenesis, green tea reduced the size and number of GGT- and GST-P-positive preneoplastic foci (Chen et al., 1987; Qin et al., 1997). Nishida et al. (1994) demonstrated that EGCG inhibits the growth and secretion of alpha-fetoprotein by human hepatoma-derived PLC/PRF/5 cells without decreasing their viability. This study further showed that EGCG reduced the incidence of hepatoma-bearing mice and also reduced the average number of hepatomas per mouse. Green tea catechins have also been shown to reduce the number and total volume of preneoplastic foci in both 2-amino-6-methyldipyrido imidazole-induced and heterocyclic amine-induced models of hepatocarcinogenesis (Hirose et al., 1999a, 1999b).

Singh et al. (2009) reported that green and black tea extracts inhibit HMG-CoA reductase and activate AMP kinase to decrease cholesterol synthesis in hepatoma cells. The study measured cholesterol synthesis in cultured rat hepatoma cells in the presence of green and black tea extracts. Both formulations decreased cholesterol synthesis as measured by a 3 h incorporation of radiolabeled acetate. Inhibition was much less evident when radiolabeled mevalonate was used, suggesting that the inhibition was medi-ated largely at or above the level of HMG-CoA reductase. Both extracts also increased AMP-kinase phosphorylation and HMG-CoA reductase phosphor-ylation by 2.5- to 4-fold, but with different time courses: maximal phosphor-ylation with green tea was evident within 30 min of treatment, whereas with black tea phosphorylation was slower to develop, with maximal phosphory-lation occurring ≥3 h after treatment. Another study investigated the effect of EGCG on thrombin-PAR1/PAR4-mediated hepatocellular carcinoma cell invasion and p42/p44 MAPK activation. This study found that stimulation of hepatocellular carcinoma (HCC) cells with thrombin, the PAR1-selective activating peptide, TFLLRN-NH2, and the PAR4-selective activating peptide, AYPGKF-NH2, increased cell invasion and stimulated activation of p42/p44 MAPK phosphorylation. The effects on both p42/p44 MAPKs and cell invasiveness induced by thrombin and the PAR1/4 subtype-selective agonist peptides were effectively blocked by EGCG (Kaufmann et al., 2009). The suppression of cytochrome P450 1A1 (CYP1A1) expression was examined in mouse hepatoma cells treated with serum prepared from EGCG and green tea extract-administered rats. Catechins were found in the rat plasma after the administration. In Hepa-1c1c7 cells, 2,3,7,8-tetrachlorodibenzo-*p*-dioxin-induced CYP1A1 expression was suppressed by treatment with the rat serum. This study concluded that catechins could possibly modulate CYP1A1 expression (Fukuda et al., 2009). Shirakami et al. (2009) examined the effects of EGCG on the activity of the VEGF–vascular endothelial growth factor receptor (VEGFR) axis in human HCC cells. The study concluded that EGCG suppresses the growth of human hepatocellular carcinoma cells by inhibiting activation of VEGF-VEGFR axis. Carbonyl reductase 1 was recently found to be a novel target of EGCG against hepatocellular carcinoma. The study reported that EGCG is a promising inhibitor of CBR1. EGCG was observed to directly interact with CBR1 and acts as a noncompetitive inhibitor with respect to the cofactor-reduced nicotinamide adenine dinucleotide phosphate and the substrate isatin (Huang et al., 2010). Green tea catechins were observed to augment the antitumor activity of doxorubicin in an *in vivo* mouse model for chemoresistant liver cancer (Liang et al., 2010). The study showed that ECG or EGCG at higher doses had a slight inhibitory effect on cell proliferation in the resistant human HCC cell line *in vitro* and *in vivo*, whereas the administration of DOX with these compounds at lower doses significantly inhibited HCC cell proliferation *in vitro* and hepatoma growth in a xenograft mouse model, compared with treatment with either agent alone at the same dose. The intracellular retention of rhodamine 123, a P-gp substrate, was increased, and the level of P-gp was decreased in cells concurrently treated with DOX and ECG or EGCG. Treatment with EGCG also increased topo II expression, but did not alter GST protein levels in tumor xenografts. Shimizu et al. (2011) examined the effects of EGCG on the development of diethylnitrosamine (DEN)-induced liver tumorigenesis in C57BL/KsJ-db/db (db/db) obese mice. EGCG significantly inhibited the development of liver cell adenomas in comparison with the control group. EGCG also inhibited the phosphorylation of the IGF-1R,

ERK, Akt, GSK-3β, STAT3, and JNK proteins in the livers of experimental mice. The serum levels of insulin, IGF-1, IGF-2, free fatty acid, and tumor necrosis factor (TNF)-α were all decreased by drinking EGCG, which also decreased the expression of TNF-α, IL-6, IL-1β, and IL-18 mRNAs in the livers. In addition, EGCG improved liver steatosis and activated the AMP-activated kinase protein in the liver. In summary, there is considerable evidence to suggest that green tea and its individual polyphenols have benefits in both prevention and treatment of hepatocarcinogenesis.

4.3 Epidemiological and Intervention Studies

4.3.1 Green Tea and Prostate Cancer

PCa is one of the most common cancers in men in the United States and is the second leading cause of male cancer death worldwide after lung cancer. The effects of green tea consumption against PCa have been reported in humans. Recently, an open label trial was conducted to determine the effects of short-term Polyphenon E supplementation on serum biomarkers in patients with PCa. Daily doses of Polyphenon E were given to 26 men with positive prostate biopsies scheduled for radical prostatectomy until the time of prostatectomy. There were positive changes in serum levels of cytokines and growth factors, including prostate-specific antigen (PSA), HGF, VEGF, and IGF axis by Polyphenon E intervention (McLarty et al., 2009). The green tea consumption habit of men aged 40 to 69 years was recorded in the Japan Public Health Center-Based Prospective Study. Among these men, 404 were newly diagnosed with PCa, of which 114 had advanced cases, 271 were localized, and 19 were of an undetermined stage during that time. It was found that green tea was not associated with localized PCa; while consumption was associated with a dose-dependent decrease in the risk of advanced PCa. The multivariate relative risk was 0.52 for men drinking 5 or more cups per day compared with less than 1 cup per day (Kurahashi et al., 2008). In high-grade prostate intraepithelial neoplasia (HG-PIN) patients, prostate mapping was done in a subset of these patients after a 2-year follow-up. The mean follow-up from the end of green tea catechins (GTCs) dosing was 23.3 months for placebo arm and 19.1 months for GTC arm. The third prostate mapping was done in only 9 from the placebo arm and 13 from the GTC arm. There was an appearance of three further cancer diagnoses during follow-up, two in the placebo arm and one in the GTC arm. Long-lasting inhibition of PCa progression was achieved in these subjects after 1 year of GTC administration. The treatment effect on early lesions suggested the early emergence of benefit observed at 6 months. There was an almost 80% reduction in PCa diagnosis, from 53% to 11% on treatment with GTC (Brausi et al., 2008).

In HG-PIN volunteers, a proof-of-principle clinical trial was conducted to assess the safety and efficacy of GTC for the chemoprevention of PCa. They were treated every day with three GTC capsules, 200 mg each. One tumor was diagnosed among the 30 GTC-treated men with an incidence of 3%, while nine cancers were found among the 30 placebo-treated men with an incidence of 30% after 1 year. GTC-treated men showed values constantly lower with respect to placebo-treated ones; however, there was no significant change in total PSA between the two arms. The International Prostate Symptom Score and quality of life scores of GTC-treated men with coexistent benign prostatic hyperplasia (BPH) improved, reaching statistical significance in the case of International Prostate Symptom Scores. No side effects were reported, and administration of GTC was found to reduce lower urinary tract symptoms (Bettuzzi et al., 2006).

In patients with histologically confirmed adenocarcinoma of the prostate, a case-control study was conducted in Hangzhou, southeast China. Hospital inpatients without PCa or any other malignant diseases were considered as controls and matched to the age of cases. Information on duration, quantity, and frequency of usual tea consumption, as well as the number of new batches brewed per day, was collected by personal interview conducted using a structured questionnaire. The risk of PCa declined with increasing frequency, duration, and quantity of green tea consumption, and the dose-response relationships were also significant, suggesting that green tea is protective against PCa (Jian et al., 2004). In a prospective clinical trial, patients with hormone refractory prostate cancer (HRPCa) were given green tea capsules, 250 mg twice a day, and the effect of green tea and its toxicity was evaluated during monthly visits. The primary endpoint was PSA or measurable disease progression after a minimum of 2 months of therapy. The treatment was generally well tolerated. In the study, 12 patients reported at least one side effect, but only two of these were of moderate or severe grade among 19 patients enrolled in the study. Four patients did not complete the minimum 2 months of therapy, and 15 patients completed at least 2 months of therapy. Nine of these patients had progressive disease within 2 months of starting therapy, and six patients developed progressive disease after an additional 1 to 4 months of therapy (Choan et al., 2005). In a phase II trial in patients with androgen-independent prostate carcinoma, 42 patients who were asymptomatic and had manifested progressive PSA elevation with hormone therapy were evaluated. Six grams of green tea per day was given orally in six divided doses, and patients were monitored monthly for response and toxicity. Decline in the baseline PSA value occurred in a single patient and was not sustained beyond 2 months. The median change in the PSA value from baseline for the cohort increased by 43% at the end of the first month. In 69% of patients, green tea toxicity, usually grade 1 or 2, and six episodes of grade 3 toxicity and one episode of grade 4 toxicity occurred (Jatoi et al., 2003).

4.3.2 Green Tea and Breast Cancer

Breast cancer is the most common malignancy in women worldwide. Although many *in vitro* and animal studies have suggested a protective effect of green tea against breast cancer, findings from epidemiological studies have been inconsistent. In a population-based prospective cohort study in Japan, 581 cases of breast cancer were newly diagnosed in 53,793 women during 13.6 years of follow-up from the baseline survey in 1990 to 1994. In 43,639 women during 9.5 years of follow-up, 350 cases were newly diagnosed after the 5-year follow-up survey in 1995 to 1998. The frequency of total green tea drinking was assessed in the baseline questionnaire, while the 5-year follow-up questionnaire assessed two other types of green tea, sencha and bancha/genmaicha, separately. The adjusted hazard ratio (HR) for women who drank 5 or more cups per day was 1.12, compared with women who drank less than 1 cup of green tea per week in the baseline data. Adjusted HRs for women who drank 10 or more cups per day were 1.02 for sencha and 0.86 for bancha or genmaicha, compared with women who drank <1 cup of sencha, bancha, or genmaicha per week (Iwasaki et al., 2010). Recently, Ogunleye et al. (2010) conducted a systematic search of five databases and performed a meta-analysis of studies of breast cancer risk and recurrence published between 1998 and 2009, encompassing 5617 cases of breast cancer. They identified two studies of breast cancer recurrence and seven studies of breast cancer incidence. Increased green tea consumption, that is, ≥3 cups a day, was inversely associated with breast cancer recurrence. Combining all studies of breast cancer incidence resulted in significant heterogeneity. They concluded that available epidemiologic evidence supports the hypothesis that increased green tea consumption may be inversely associated with risk of breast cancer recurrence. However, the association between green tea consumption and breast cancer incidence remains unclear (Ogunleye et al., 2010). In a population-based case-control study conducted in Shanghai, China, during 1996 to 2005, it was investigated whether regular green tea consumption was associated with breast cancer risk among 3454 incident cases and 3474 controls aged 20 to 74 years. All participants were interviewed in person about green tea consumption habits, including age of initiation, duration of use, brew strength, and quantity of tea. Regular drinking of green tea was associated with a slightly decreased risk for breast cancer compared with nondrinkers. Reduced risk was observed for years of green tea drinking, and a dose-response relationship with the amount of tea consumed per month was also observed among premenopausal women (Shrubsole et al., 2009).

A case-control study was conducted in southeast China between 2004 and 2005 in patients with histologically confirmed breast cancer. Green tea consumption was associated with a reduced risk of breast cancer after adjusting established and potential confounders. Similar dose-response relationships were observed for the duration of drinking green tea, number of cups consumed, and new batches prepared per day (Zhang et al., 2007). In the Hospital-Based Epidemiologic Research Program at Aichi Cancer Center, a total of 1160 new surgical cases of female invasive breast cancers diagnosed between June 1990 and August 1998 were followed up through December 1999, and the risk of recurrence was assessed with reference to daily green tea consumption. During 5264 person-years of follow-up, 133 subjects were documented to suffer recurrence of breast cancer. A decreased HR for recurrence adjusted for stage was observed with consumption of ≥3 daily cups of green tea. These results suggest the possibility that regular green tea consumption may be preventive against recurrence of breast cancer in early-stage cases (Inoue et al., 2001).

4.3.3 Green Tea and Lung Cancer

Li et al. (2008) examined the risk of lung cancer in relation to green tea consumption in a population-based cohort study in Japan. During the follow-up period of 7 years, 302 cases of lung cancer were identified. The multivariable-adjusted HRs of lung cancer incidence for green tea consumption of 1 or 2, 3 or 4, and ≥5 cups per day, as compared to <1 cup per day, were 1.14, 1.18, and 1.17, respectively. In a population-based case-control study, the association between past consumption of green tea and the risk of lung cancer was identified using the population-based Shanghai Cancer Registry. Consumption of green tea was associated with a reduced risk of lung cancer, and the risks decreased with increasing consumption among nonsmoking women; however, there was little association among smoking women (Zhong et al., 2001). A phase II randomized controlled tea intervention trial was designed to study the effect of high consumption, that is, 4 cups per day of decaffeinated green or black tea on oxidative DNA damage as measured by urinary 8-hydroxydeoxyguanosine (8-OHdG) among smokers over a 4-month period. There was an increase in plasma and urinary levels of catechins in the green tea group compared with the other two groups. Assessment of urinary 8-OHdG after adjustment for baseline measurements and other potential confounders revealed a highly significant decrease in urinary 8-OHdG after 4 months of drinking decaffeinated green tea, while there was no change in urinary 8-OHdG among smokers given black tea. These data suggest that drinking regular green tea might protect smokers from oxidative damages and could reduce cancer risk or other diseases caused by free radicals associated with smoking (Hakim et al., 2003).

4.3.4 Green Tea and Liver Cancer

Association of coffee and green tea consumption with a reduced risk of liver cancer by hepatitis virus infection status was investigated in the Japan Public Health Center-Based Prospective Study Cohort II. Increased coffee consumption was associated with a reduced risk of liver cancer in all subjects, compared with almost never drinkers. However, no association was observed between green tea consumption and the risk of liver cancer in all subjects (Inoue et al., 2009). A population-based case-control study was conducted in Taixing, Jiangsu province, to explore the role of green tea in decreasing the risks of gastric cancer, liver cancer, and esophageal cancer among alcohol drinkers or cigarette smokers. Green tea drinking decreased 81, 78, and 39% risk for the development of gastric cancer, liver cancer, and esophageal cancer, respectively, among alcohol drinkers. It also decreased 16, 43, and 31% risks of developing the three kinds of cancers among cigarette smokers. Interaction assessment showed that drinking green tea could significantly decrease the risk of gastric cancer and liver cancer among alcohol drinkers. A habit of drinking green tea also had protective effects on the development of both gastric and liver cancer among alcohol drinkers, while green tea also had some protective effect on esophageal cancer among alcohol drinkers and on three kinds of cancers among cigarette smokers (Mu et al., 2003). A cross-sectional study was conducted to investigate the association between consumption of green tea and various serum markers in a Japanese population, with special reference to preventive effects of green tea against cardiovascular diseases and disorders of the liver. Increased consumption of green tea was associated with decreased serum concentrations of total cholesterol, triglycerides, and an increased proportion of high-density lipoprotein, together with a decreased proportion of low-density lipoprotein, which resulted in a decreased atherogenic index. Moreover, increased consumption of green tea, especially ≥10 cups per day, was related to decreased concentrations of hepatological markers like aspartate aminotransferase, alanine transferase, and ferritin in serum. Therefore, this inverse association between consumption of green tea and various serum markers shows that green tea may protect against cardiovascular diseases and disorders of the liver (Imai and Nakachi, 1995).

4.3.5 Green Tea and Digestive Tract Cancer

A total of 343 patients with squamous cell carcinoma (SCC) of the esophagus and 755 cancer-free control subjects were recruited for a study from 1996 to 2005. It was found that there was a significant inverse relation between the frequency of tea consumption and esophageal SCC risk, with a 0.5-fold lower risk associated with the intake of unfermented tea like green tea, oolong tea, or jasmine tea (Chen et al., 2009). In a large population-based case-control study of biliary tract disease in Shanghai, China, the effects of tea consumption on the risk of biliary tract cancers and biliary stones were evaluated. The study included 627 incident cases with biliary tract cancer, 1037 cases with biliary stones, and 959 randomly selected controls. Forty-one percent of the controls were ever tea drinkers, consuming at least 1 cup of tea per day for at least 6 months. Ever tea drinkers among women had significantly reduced risks of biliary stones and gallbladder cancer. The inverse relationship between tea consumption and gallbladder cancer risk was independent of gallstone disease. In men, tea drinkers were more likely to be cigarette smokers, and the risk estimates were generally <1.0, but were not statistically significant (Zhang et al., 2006). There was an inverse association between green tea drinking and chronic gastritis and stomach cancer risks in a population-based case-control study in Yangzhong, China, with 133 stomach cancer cases, 166 chronic gastritis cases, and 433 healthy controls (Setiawan et al., 2001). A comparative case-referent study in Nagoya, Japan, comprised 1706 histologically diagnosed cases of digestive tract and a total of 21,128 noncancer outpatients aged ≥40 years. The odds ratio (OR) of stomach cancer decreased to 0.69 with high intake of green tea, that is, ≥7 cups or more per day. A decreased risk was also observed for rectal cancer with ≥3 cups daily intake of coffee (Inoue et al., 1998).

4.4 Conclusion and Future Perspectives

Many of the beneficial effects of green tea are related to its major polyphenol, EGCG. Epidemiological and animal studies have provided evidence
that green tea and tea polyphenols reduce the risk of certain types of cancer.
Careful mechanistic studies in animal models representing different stages
of carcinogenesis, and an integrated approach to analyze the data from these
studies, will be essential in understanding the effects of green tea and its
constituents. A major challenge of cancer prevention is to integrate new
molecular findings into clinical practice. Identification of molecular targets
or biomarkers, whose changes are associated with inhibition of malignantly
transformed cell properties, is paramount to cancer prevention and treatment
by green tea and will greatly assist in a better understanding of anticancer
mechanisms by green tea. Future research should define the mechanisms of
action for the cancer preventive activities of green tea and establishing the
safe range of tea consumption associated with these benefits.

Acknowledgment

The original work from an author's (HM) laboratory was supported by the
Department of Defense Idea Development Award W81XWH-10–1-0245
(USA).

References

Adhami, V.M., Malik, A., Zaman, N., Sarfaraz, S., Siddiqui, I.A., Syed, D.N.,
    Afaq, F., Pasha, F.S., Saleem, M., and Mukhtar, H. 2007. Combined inhibitory
    effects of green tea polyphenols and selective cyclooxygenase-2 inhibitors on
    the growth of human prostate cancer cells both *in vitro* and in vivo. *Clin Cancer
    Res*, 13: 1611–1619.
Adhami, V.M., Siddiqui, I.A., Ahmad, N., Gupta, S., and Mukhtar, H. 2004. Oral
    consumption of green tea polyphenols inhibits insulin-like growth factor-I-
    induced signaling in an autochthonous mouse model of prostate cancer. *Cancer
    Res*, 64: 8715–8722.
Adhami, V.M., Siddiqui, I.A., Sarfaraz, S., Khwaja, S.I., Hafeez, B.B., Ahmad, N.,
    and Mukhtar, H. 2009. Effective prostate cancer chemopreventive intervention
    with green tea polyphenols in the TRAMP model depends on the stage of the
    disease. *Clin Cancer Res*, 15: 1947–1953.

Afaq, F., Ahmad, N., and Mukhtar, H. 2003. Suppression of UVB-induced phosphorylation of mitogen-activated protein kinases and nuclear factor kappa B by green tea polyphenol in SKH-1 hairless mice. *Oncogene*, 22: 9254–9264.

Anderson, M.W., Goodin, C., Zhang, Y., Kim, S., Estensen, R.D., Wiedmann, T.S., Sekar, P., Buncher, C.R., Khoury, J.C., Garbow, J.R., You, M., and Tichelaar, J.W. 2008. Effect of dietary green tea extract and aerosolized difluoromethylornithine during lung tumor progression in A/J strain mice. *Carcinogenesis*, 29: 1594–1600.

Baliga, M.S., Meleth, S., and Katiyar, S.K. 2005. Growth inhibitory and antimetastatic effect of green tea polyphenols on metastasis-specific mouse mammary carcinoma 4T1 cells *in vitro* and *in vivo* systems. *Clin Cancer Res*, 11: 1918–1927.

Bettuzzi, S., Brausi, M., Rizzi, F., Castagnetti, G., Peracchia, G., and Corti, A. 2006. Chemoprevention of human prostate cancer by oral administration of green tea catechins in volunteers with high-grade prostate intraepithelial neoplasia: A preliminary report from a one-year proof-of-principle study. *Cancer Res*, 66: 1234–1240.

Bigelow, R.L., and Cardelli, J.A. 2006. The green tea catechins, (–)-epigallocatechin-3-gallate (EGCG) and (–)-epicatechin-3-gallate (ECG), inhibit HGF/Met signaling in immortalized and tumorigenic breast epithelial cells. *Oncogene*, 25: 1922–1930.

Brausi, M., Rizzi, F., and Bettuzzi, S. 2008. Chemoprevention of human prostate cancer by green tea catechins: Two years later. A follow-up update. *Eur Urol*, 54: 472–473.

Cao, J., Xu, Y., Chen, J., and Klaunig, J.E. 1996. Chemopreventive effects of green and black tea on pulmonary and hepatic carcinogenesis. *Fundam Appl Toxicol*, 29: 244–250.

Cao, Y., Cao, R., and Brakenhielm, E. 2002. Antiangiogenic mechanisms of diet-derived polyphenols. *J Nutr Biochem*, 13: 380–390.

Chatterjee, M.L., Agarwal, R., and Mukhtar, H. 1996. Ultraviolet B radiation-induced DNA lesions in mouse epidermis: An assessment using a novel 32P-postlabelling technique. *Biochem Biophys Res Commun*, 229: 590–595.

Chen, Y.K., et al. 2009. Food intake and the occurrence of squamous cell carcinoma in different sections of the esophagus in Taiwanese men. *Nutrition*, 25: 753–761.

Chen, Z.P., Schell, J.B., Ho, C.T., and Chen, K.Y. 1998. Green tea epigallocatechin gallate shows a pronounced growth inhibitory effect on cancerous cells but not on their normal counterparts. *Cancer Lett*, 129: 173–179.

Chen, Z.Y., Yan, R.Q., Qin, G.Z., and Qin, L.L. 1987. Effect of six edible plants on the development of AFB1-induced gamma-glutamyltranspeptidase-positive hepatocyte foci in rats. *Zhonghua Zhong Liu Za Zhi*, 9: 109–111.

Choan, E., Segal, R., Jonker, D., et al. 2005. A prospective clinical trial of green tea for hormone refractory prostate cancer: An evaluation of the complementary/alternative therapy approach. *Urol Oncol*, 23: 108–113.

Chung, J.H., Han, J.H., Hwang, E.J., et al. 2003. Dual mechanisms of green tea extract (EGCG)-induced cell survival in human epidermal keratinocytes. *FASEB J*, 17: 1913–1915.

Conney, A.H., Lu, Y.P., Lou, Y.R., and Huang, M.T. 2002. Inhibitory effects of tea and caffeine on UV-induced carcinogenesis: Relationship to enhanced apoptosis and decreased tissue fat. *Eur J Cancer Prev*, 11(Suppl. 2): S28–S36.

Conney, A.H., Lu, Y., Lou, Y., Xie, J., and Huang, M. 1999. Inhibitory effect of green and black tea on tumor growth. *Proc Soc Exp Biol Med*, 220: 229–233.

Elmets, C.A., Singh, D., Tubesing, K., Matsui, M., Katiyar, S., and Mukhtar, H. 2001. Cutaneous photoprotection from ultraviolet injury by green tea polyphenols. *J Am Acad Dermatol*, 44: 425–432.

Fukuda, I., Tsutsui, M., Sakane, I., and Ashida, H. 2009. Suppression of cytochrome P450 1A1 expression induced by 2,3,7,8-tetrachlorodibenzo-*p*-dioxin in mouse hepatoma hepa-1c1c7 cells treated with serum of (–)-epigallocatechin-3-gallate- and green tea extract-administered rats. *Biosci Biotechnol Biochem*, 73: 1206–1208.

Gupta, S., Hastak, K., Ahmad, N., Lewin, J.S., and Mukhtar, H. 2001. Inhibition of prostate carcinogenesis in TRAMP mice by oral infusion of green tea polyphenols. *Proc Natl Acad Sci USA*, 98: 10350–10355.

Hakim, I.A., Harris, R.B., Brown, S., et al. 2003. Effect of increased tea consumption on oxidative DNA damage among smokers: A randomized controlled study. *J Nutr*, 133: 3303S–3309S.

Harper, C.E., Patel, B.B., Wang, J., Eltoum, I.A., Lamartiniere, C.A. 2007. Epigallocatechin-3-gallate suppresses early stage, but not late stage prostate cancer in TRAMP mice: Mechanisms of action. *Prostate*, 67: 1576–1589.

Hastak, K., Agarwal, M.K., Mukhtar, H., and Agarwal, M.L. 2005. Ablation of either p21 or Bax prevents p53-dependent apoptosis induced by green tea polyphenol epigallocatechin-3-gallate. *FASEB J*, 19: 789–791.

Hastak, K., Gupta, S., Ahmad, N., Agarwal, M.K., Agarwal, M.L., and Mukhtar, H. 2003. Role of p53 and NF-kappaB in epigallocatechin-3-gallate induced apoptosis of LNCaP cells. *Oncogene*, 22: 4851–4859.

Hirose, M., Takahashi, S., Ogawa, K., Futakuchi, M., and Shirai, T. 1999a. Phenolics: Blocking agents for heterocyclic amine-induced carcinogenesis. *Food Chem Toxicol*, 37: 985–992.

Hirose, M., Takahashi, S., Ogawa, K., Futakuchi, M., Shirai, T., Shibutani, M., Uneyama, C., Toyoda, K., Iwata, H. 1999b. Chemoprevention of heterocyclic amine-induced carcinogenesis by phenolic compounds in rats. *Cancer Lett*, 143: 173–178.

Hsu, S., Dickinson, D., Borke, J., et al. 2007. Green tea polyphenol induces caspase 14 in epidermal keratinocytes via MAPK pathways and reduces psoriasiform lesions in the flaky skin mouse model. *Exp Dermatol*, 16: 678–684.

Huang, W., et al. 2010. Carbonyl reductase 1 as a novel target of (–)-epigallocatechin gallate against hepatocellular carcinoma. *Hepatology*, 52: 703–714.

Imai, K., and Nakachi, K. 1995. Cross sectional study of effects of drinking green tea on cardiovascular and liver diseases. *BMJ*, 310: 693–696.

Inoue, M., Kurahashi, N., Iwasaki, M., et al. 2009. Effect of coffee and green tea consumption on the risk of liver cancer: Cohort analysis by hepatitis virus infection status. *Cancer Epidemiol Biomarkers Prev*, 18: 1746–1753.

Inoue, M., Tajima, K., Hirose, K., et al. 1998. Tea and coffee consumption and the risk of digestive tract cancers: Data from a comparative case-referent study in Japan. *Cancer Causes Control*, 9: 209–216.

Inoue, M., Tajima, K., Mizutani, M., et al. 2001. Regular consumption of green tea and the risk of breast cancer recurrence: Follow-up study from the Hospital-Based Epidemiologic Research Program at Aichi Cancer Center (HERPACC). *Jpn Cancer Lett*, 167: 175–182.

Iwasaki, M., Inoue, M., Sasazuki, S., et al. 2010. Green tea drinking and subsequent risk of breast cancer in a population to based cohort of Japanese women. *Breast Cancer Res*, 12: R88.

Jatoi, A., et al. 2003. A phase II trial of green tea in the treatment of patients with androgen independent metastatic prostate carcinoma. *Cancer*, 97: 1442–1446.

Jian, L., Xie, L.P., Lee, A.H., and Binns, C.W. 2004. Protective effect of green tea against prostate cancer: A case-control study in southeast China. *Int J Cancer*, 108: 130–135.

Katiyar, S.K., Afaq, F., Perez, A., and Mukhtar, H. 2001. Green tea polyphenol (−)-epigallocatechin-3-gallate treatment of human skin inhibits ultraviolet radiation-induced oxidative stress. *Carcinogenesis*, 22: 287–294.

Katiyar, S.K., Agarwal, R., and Mukhtar, H. 1993. Protective effects of green tea polyphenols administered by oral intubation against chemical carcinogen-induced forestomach and pulmonary neoplasia in A/J mice. *Cancer Lett*, 73: 167–172.

Katiyar, S.K., Challa, A., McCormick, T.S., Cooper, K.D., and Mukhtar, H. 1999. Prevention of UVB-induced immunosuppression in mice by the green tea polyphenol (−)-epigallocatechin-3-gallate may be associated with alterations in IL-10 and IL-12 production. *Carcinogenesis*, 20: 2117–2124.

Katiyar, S.K., Elmets, C.A., Agarwal, R., and Mukhtar, H. 1995. Protection against ultraviolet-B radiation-induced local and systemic suppression of contact hypersensitivity and edema responses in C3H/HeN mice by green tea polyphenols. *Photochem Photobiol*, 62: 855–861.

Katiyar, S.K., Vaid, M., van Steeg, H., and Meeran, S.M. 2010. Green tea polyphenols prevent UV-induced immunosuppression by rapid repair of DNA damage and enhancement of nucleotide excision repair genes. *Cancer Prev Res (Phila)*, 3: 179–189.

Kaufmann, R., Henklein, P., and Settmacher, U. 2009. Green tea polyphenol epigallocatechin-3-gallate inhibits thrombin-induced hepatocellular carcinoma cell invasion and p42/p44-MAPKinase activation. *Oncol Rep*, 21: 1261–1267.

Khan, N., Afaq, F., and Mukhtar, H. 2008. Cancer chemoprevention through dietary antioxidants: Progress and promise. *Antioxidants Redox Signaling*, 10: 475–510.

Khan, N., and Mukhtar, H. 2010. Cancer and metastasis: Prevention and treatment by green tea. *Cancer Metastasis Rev*, 29: 435–445.

Kundu, J.K., Na, H.K., Chun, K.S., et al. 2003. Inhibition of phorbol ester-induced COX-2 expression by epigallocatechin gallate in mouse skin and cultured human mammary epithelial cells. *J Nutr*, 133: 3805S–3810S.

Kurahashi, N., Sasazuki, S., Iwasaki, M., Inoue, M., and Tsugane, S. 2008. Green tea consumption and prostate cancer risk in Japanese men: A prospective study. *Am J Epidemiol*, 167: 71–77.

Kushima, Y., Iida, K., Nagaoka, Y., Kawaratani, Y., Shirahama, T., Sakaguchi, M., Baba, K., Hara, Y., and Uesato, S. 2009. Inhibitory effect of (−)-epigallocatechin and (−)-epigallocatechin gallate against heregulin beta1-induced migration/invasion of the MCF-7 breast carcinoma cell line. *Biol Pharm Bull*, 32: 899–904.

Leong, H., Mathur, P.S., and Greene, G.L. 2008. Inhibition of mammary tumorigenesis in the C3(1)/SV40 mouse model by green tea. *Breast Cancer Res Treat*, 107: 359–369.

Leong, H., Mathur, P.S., and Greene, G.L. 2009. Green tea catechins inhibit angiogenesis through suppression of STAT3 activation. *Breast Cancer Res Treat*, 117: 505–515.

Li, G.X., et al. 2010. Pro-oxidative activities and dose-response relationship of (−)-epigallocatechin-3-gallate in the inhibition of lung cancer cell growth: A comparative study *in vivo* and *in vitro*. *Carcinogenesis*, 31: 902–910.

Li, Q., Kakizaki, M., Kuriyama, S., Sone, T., Yan, H., Nakaya, N., Mastuda-Ohmori, K., and Tsuji, I. 2008. Green tea consumption and lung cancer risk: The Ohsaki study. *Br J Cancer*, 99: 1179–1184.

Liang, G., Tang, A., Lin, X., Li, L., Zhang, S., Huang, Z., Tang, H., and Li, Q.Q. 2010. Green tea catechins augment the antitumor activity of doxorubicin in an *in vivo* mouse model for chemoresistant liver cancer. *Int J Oncol*, 37: 111–123.

Lu, G., Liao, J., Yang, G., Reuhl, K.R., Hao, X., and Yang, C.S. 2006. Inhibition of adenoma progression to adenocarcinoma in a 4-(methylnitrosamino)-1-(3-pyridyl)-1-butanone-induced lung tumorigenesis model in A/J mice by tea polyphenols and caffeine. *Cancer Res*, 66: 11494–11501.

Lu, Y.P., Lou, Y.R., Li, X.H., Xie, J.G., Brash, D., Huang, M.T., and Conney, A.H. 2000. Stimulatory effect of oral administration of green tea or caffeine on ultraviolet light-induced increases in epidermal wild-type p53, p21(WAF1/CIP1), and apoptotic sunburn cells in SKH-1 mice. *Cancer Res*, 60: 4785–4791.

Luo, D., and Li, Y. 1992. Preventive effect of green tea on MNNG-induced lung cancers and precancerous lesions in LACA mice. *Hua Xi Yi Ke Da Xue Xue Bao*, 23: 433–437.

Mantena, S.K., Meeran, S.M., Elmets, C.A., and Katiyar, S.K. 2005b. Orally administered green tea polyphenols prevent ultraviolet radiation-induced skin cancer in mice through activation of cytotoxic T cells and inhibition of angiogenesis in tumors. *J Nutr*, 135: 2871–2877.

Mantena, S.K., Roy, A.M., and Katiyar, S.K. 2005a. Epigallocatechin-3-gallate inhibits photocarcinogenesis through inhibition of angiogenic factors and activation of CD8+ T cells in tumors. *Photochem Photobiol*, 81: 1174–1179.

Matsumoto, N., Kohri, T., Okushio, K., and Hara, Y. 1996. Inhibitory effects of tea catechins, black tea extract and oolong tea extract on hepatocarcinogenesis in rat. *Jpn J Cancer Res*, 87: 1034–1038.

McLarty, J., Bigelow, R.L., Smith, M., Elmajian, D., Ankem, M., and Cardelli, J.A. 2009. Tea polyphenols decrease serum levels of prostate-specific antigen, hepatocyte growth factor, and vascular endothelial growth factor in prostate cancer patients and inhibit production of hepatocyte growth factor and vascular endothelial growth factor *in vitro*. *Cancer Prev Res (Phila Pa)*, 2: 673–682.

Meeran, S.M., Mantena, S.K., Elmets, C.A., and Katiyar, S.K. 2006. (-)-Epigallocatechin-3-gallate prevents photocarcinogenesis in mice through interleukin-12-dependent DNA repair. *Cancer Res*, 66: 5512–5520.

Milligan, S.A., Burke, P., Coleman, D.T., Bigelow, R.L., Steffan, J.J., Carroll, J.L., Williams, B.J., and Cardelli, J.A. 2009. The green tea polyphenol EGCG potentiates the antiproliferative activity of c-Met and epidermal growth factor receptor inhibitors in non-small cell lung cancer cells. *Clin Cancer Res*, 15: 4885–4894.

Morley, N., Clifford, T., Salter, L., Campbell, S., Gould, D., and Curnow, A. 2005. The green tea polyphenol (−)-epigallocatechin gallate and green tea can protect human cellular DNA from ultraviolet and visible radiation-induced damage. *Photodermatol Photoimmunol Photomed*, 21: 15–22.

Mu, L.N., Zhou, X.F., Ding, B.G., Wang, R.H., Zhang, Z.F., Chen, C.W., Wei, G.R., Zhou, X.M., Jiang, Q.W., and Yu, S.Z. 2003. A case-control study on drinking green tea and decreasing risk of cancers in the alimentary canal among cigarette smokers and alcohol drinkers. *Zhonghua Liu Xing Bing Xue Za Zhi*, 24: 192–195.

Nam, S., Smith, D.M., and Dou, Q.P. 2001. Ester bond-containing tea polyphenols potently inhibit proteasome activity *in vitro* and *in vivo*. *J Biol Chem*, 276: 13322–13330.

Nandakumar, V., Vaid, M., and Katiyar, S.K. 2011. (–)-Epigallocatechin-3-gallate reactivates silenced tumor suppressor genes, Cip1/p21 and p16INK4a, by reducing DNA methylation and increasing histones acetylation in human skin cancer cells. *Carcinogenesis*, 32: 537–544.

Nihal, M., Ahmad, N., Mukhtar, H., and Wood, G.S. 2005. Anti-proliferative and proapoptotic effects of (–)-epigallocatechin-3-gallate on human melanoma: Possible implications for the chemoprevention of melanoma. *Int J Cancer*, 114: 513–521.

Nishida, H., Omori, M., Fukutomi, Y., Ninomiya, M., Nishiwaki, S., Suganuma, M., Moriwaki, H., and Muto, Y. 1994. Inhibitory effects of (–)-epigallocatechin gallate on spontaneous hepatoma in C3H/HeNCrj mice and human hepatoma-derived PLC/PRF/5 cells. *Jpn J Cancer Res*, 85: 221–225.

Ogunleye, A.A., Xue, F., and Michels, K.B. 2010. Green tea consumption and breast cancer risk or recurrence: A meta-analysis. *Breast Cancer Res Treat*, 119: 477–484.

Paul, B., Hayes, C.S., Kim, A., Athar, M., and Gilmour, S.K. 2005. Elevated polyamines lead to selective induction of apoptosis and inhibition of tumorigenesis by (–)-epigallocatechin-3-gallate (EGCG) in ODC/Ras transgenic mice. *Carcinogenesis*, 26: 119–124.

Pezzato, E., Sartor, L., Dell'Aica, I., Dittadi, R., Gion, M., Belluco, C., Lise, M., and Garbisa, S. 2004. Prostate carcinoma and green tea: PSA-triggered basement membrane degradation and MMP-2 activation are inhibited by (–)epigallocatechin-3-gallate. *Int J Cancer*, 112: 787–792.

Qin, G., Gopalan-Kriczky, P., Su, J., Ning, Y., and Lotlikar, P.D. 1997. Inhibition of aflatoxin B1-induced initiation of hepatocarcinogenesis in the rat by green tea. *Cancer Lett*, 112: 149–154.

Qin, G., Ning, Y., and Lotlikar, P.D. 2000. Chemoprevention of aflatoxin B1-initiated and carbon tetrachloride-promoted hepatocarcinogenesis in the rat by green tea. *Nutr Cancer*, 38: 215–222.

Roomi, M.W., Ivanov, V., Kalinovsky, T., Niedzwiecki, A., and Rath, M. 2005. *In vivo* antitumor effect of ascorbic acid, lysine, proline and green tea extract on human colon cancer cell HCT 116 xenografts in nude mice: Evaluation of tumor growth and immunohistochemistry. *Oncol Rep*, 13: 421–425.

Sakata, M., Ikeda, T., Imoto, S., Jinno, H., and Kitagawa, Y. 2011. Prevention of mammary carcinogenesis in C3H/OuJ mice by green tea and tamoxifen. *Asian Pac J Cancer Prev*, 12: 567–571.

Sartippour, M.R., et al. 2006. The combination of green tea and tamoxifen is effective against breast cancer. *Carcinogenesis*, 27: 2424–2433.

Schwarz, A., Maeda, A., Gan, D., Mammone, T., Matsui, M.S., and Schwarz, T. 2008. Green tea phenol extracts reduce UVB-induced DNA damage in human cells via interleukin-12. *Photochem Photobiol*, 84: 350–355.

Setiawan, V.W., et al. 2001. Protective effect of green tea on the risks of chronic gastritis and stomach cancer. *Int J Cancer*, 92: 600–604.

Shimizu, K., Shimizu, N., Hakamata, W., Unno, K., Asai, T., and Oku, N. 2010. Preventive effect of green tea catechins on experimental tumor metastasis in senescence-accelerated mice. *Biol Pharm Bull*, 33: 117–121.

Shimizu, M., et al. 2011. Preventive effects of (–)-epigallocatechin gallate on diethyl-nitrosamine-induced liver tumorigenesis in obese and diabetic C57BL/KsJ-db/db mice. *Cancer Prev Res (Phila)*, 4: 396–403.

Shirakami, Y., Shimizu, M., Adachi, S., Sakai, H., Nakagawa, T., Yasuda, Y., Tsurumi, H., Hara, Y., and Moriwaki, H. 2009. (–)-Epigallocatechin gallate suppresses the growth of human hepatocellular carcinoma cells by inhibiting activation of the vascular endothelial growth factor–vascular endothelial growth factor receptor axis. *Cancer Sci*, 100: 1957–1962.

Shrubsole, M.J., et al. 2009. Drinking green tea modestly reduces breast cancer risk. *J Nutr*, 139: 310–316.

Siddiqui, I.A., Adhami, V.M., Bharali, D.J., Hafeez, B.B., Asim, M., Khwaja, S.I., Ahmad, N., Cui, H., Mousa, S.A., and Mukhtar, H. 2009. Introducing nano-chemoprevention as a novel approach for cancer control: Proof of principle with green tea polyphenol epigallocatechin-3-gallate. *Cancer Res*, 69: 1712–1716.

Siddiqui, I.A., Asim, M., Hafeez, B.B., Adhami, V.M., Tarapore, R.S., and Mukhtar, H. 2011. Green tea polyphenol EGCG blunts androgen receptor function in prostate cancer. *FASEB J*, 25: 1198–1207.

Siddiqui, I.A., Malik, A., Adhami, V.M., Asim, M., Hafeez, B.B., Sarfaraz, S., and Mukhtar, H. 2008a. Green tea polyphenol EGCG sensitizes human prostate carcinoma LNCaP cells to TRAIL-mediated apoptosis and synergistically inhibits biomarkers associated with angiogenesis and metastasis. *Oncogene*, 27: 2055–2063.

Siddiqui, I.A., Shukla, Y., Adhami, V.M., Sarfaraz, S., Asim, M., Hafeez, B.B., and Mukhtar, H. 2008b. Suppression of NFkappaB and its regulated gene products by oral administration of green tea polyphenols in an autochthonous mouse prostate cancer model. *Pharm Res*, 25: 2135–2142.

Singh, D.K., Banerjee, S., and Porter, T.D. 2009. Green and black tea extracts inhibit HMG-CoA reductase and activate AMP kinase to decrease cholesterol synthesis in hepatoma cells. *J Nutr Biochem*, 20: 816–822.

Syed, D.N., Afaq, F., Kweon, M.H., Hadi, N., Bhatia, N., Spiegelman, V.S., and Mukhtar, H. 2007b. Green tea polyphenol EGCG suppresses cigarette smoke condensate-induced NF-kappaB activation in normal human bronchial epithe-lial cells. *Oncogene*, 26: 673–682.

Syed, D.N., Khan, N., Afaq, F., and Mukhtar, H. 2007a. Chemoprevention of prostate cancer through dietary agents: Progress and promise. *Cancer Epidemiol Biomarkers Prev*, 16: 2193–2203.

Thyagarajan, A., Zhu, J., and Sliva, D. 2007. Combined effect of green tea and *Ganoderma lucidum* on invasive behavior of breast cancer cells. *Int J Oncol*, 30: 963–969.

Wang, Z.Y., Agarwal, R., Bickers, D.R., and Mukhtar, H. 1991. Protection against ultraviolet B radiation-induced photocarcinogenesis in hairless mice by green tea polyphenols. *Carcinogenesis*, 12: 1527–1530.

Wang, Z.Y., Hong, J.Y., Huang, M.T., Reuhl, K.R., Conney, A.H., and Yang, C.S. 1992a. Inhibition of N-nitrosodiethylamine- and 4-(methylnitrosamino)-1-(3-pyridyl)-1-butanone-induced tumorigenesis in A/J mice by green tea and black tea. *Cancer Res*, 52: 1943–1947.

Wang, Z.Y., Huang, M.T., Ferraro, T., Wong, C.Q., Lou, Y.R., Reuhl, K., Iatropoulos, M., Yang, C.S., and Conney, A.H. 1992c. Inhibitory effect of green tea in the drinking water on tumorigenesis by ultraviolet light and 12-*O*-tetradecanoylphorbol-13-acetate in the skin of SKH-1 mice. *Cancer Res*, 52: 1162–1170.

Wang, Z.Y., Huang, M.T., Ho, C.T., Chang, R., Ma, W., Ferraro, T., Reuhl, K.R., Yang, C.S., and Conney, A.H. 1992b. Inhibitory effect of green tea on the growth of established skin papillomas in mice. *Cancer Res*, 52: 6657–6665.

Wang, Z.Y., Khan, W.A., Bickers, D.R., and Mukhtar, H. 1989. Protection against polycyclic aromatic hydrocarbon-induced skin tumor initiation in mice by green tea polyphenols. *Carcinogenesis*, 10: 411–415.

Xia, J., Song, X., Bi, Z., Chu, W., and Wan, Y. 2005. UV-induced NF-kappaB activation and expression of IL-6 is attenuated by (–)-epigallocatechin-3-gallate in cultured human keratinocytes *in vitro*. *Int J Mol Med*, 16: 943–950.

Zhang, M., Holman, C.D., Huang, J.P., and Xie, X. 2007. Green tea and the prevention of breast cancer: A case-control study in southeast China. *Carcinogenesis*, 28: 1074–1078.

Zhang, X.H., et al. 2006. Tea drinking and the risk of biliary tract cancers and biliary stones: A population-based case-control study in Shanghai, China. *Int J Cancer*, 118: 3089–3094.

Zhong, L., Goldberg, M.S., Gao, Y.T., Hanley, J.A., Parent, M.E., and Jin, F. 2001. A population-based case-control study of lung cancer and green tea consumption among women living in Shanghai, China. *Epidemiology*, 12: 695–700.

Zhou, J.R., Yu, L., Zhong, Y., and Blackburn, G.L. 2003. Soy phytochemicals and tea bioactive components synergistically inhibit androgen-sensitive human prostate tumors in mice. *J Nutr*, 133: 516–521.

5

Chemopreventive Action of Green Tea Polyphenols (Molecular-Biological Mechanisms)

Vijay S. Thakur and Sanjay Gupta

Contents

5.1 Introduction

Tea drinking has been a tradition for centuries and remains the most widely consumed beverage today, next only to water (Mukhtar et al., 1992). The history of tea drinking dates back more than 4000 years and has been documented in China as early as 1911 in *Kissa Yojoki* (Book of Tea) by Aisai, a Zen priest. Tea is derived from the leaves of the *Camellia sinensis* plant, which is grown in the eastern part of the world, predominantly in China, India, Japan, Sri Lanka, and Thailand. Based on the technology of manufacturing, tea is classified as green, black, or oolong tea. The different methods used for its production alter the chemical composition of the dried tea leaves. Green tea, which accounts for 20% of the world tea consumption, is prepared by steaming the tea leaves. This process preserves the characteristic of tea catechins, which constitute 10 to 15% of the weight of the dried leaves by inactivating polyphenol oxidase enzyme. The major catechins present in green tea are (–)-epigallocatechin-3-gallate (EGCG), (–)-epigallocatechin (EGC), (–)-epicatechin-3-gallate (ECG), and (–)-epicatechin (EC). EGCG contributes more than two-fifths of the total phenols present in green tea and has been widely reported for its health benefits (Zaveri, 2006). A typical cup of green tea, brewed with 2.5 g of tea leaves in 250 ml hot water, contains 620 to 880 mg of water-extractable solids, of which about one-third are catechins. The presumed beneficial effects of tea in the prevention of cancer, as well as cardiovascular, neurodegenerative, obesity, and other diseases, have been extensively studied.

5.2 Cancer: Risk Factors and Field Effects

Cancer can simply be defined as abnormal growth of cells, which proliferate and spread in an uncontrolled fashion. Normal cells are strictly regulated for their growth through the cell cycle. Abnormalities at any stage of the cell cycle can lead to uncontrolled growth or growth arrest. Capabilities acquired during unfavorable conditions for survival, such as insensitivity to growth inhibitory signals, evasion of apoptosis, unlimited replication potential, and increased blood supply to ensure uninterrupted availability of nutrients and growth factors, offer unrestricted growth advantages to cells to transform into neoplastic cells, and with the loss of tumor suppressor function, converts to cancer cells. These malignant cells invade other organs in the body through achievement of migration and metastasis developed during carcinogenesis (Herceg and Hainaut, 2007).

A number of risk factors have been identified in cancer development. These are tobacco, sunlight, ionizing radiation, certain chemicals and substances, viruses and bacteria, hormones, family history, alcohol ingestion, poor diet, inactivity, and being overweight. Some of these are common risk factors for several human cancers, whereas others are potential risk factors for organ-specific cancers. Many of these risk factors can be avoided, whereas others, such as family history, are unavoidable (Brockmöller et al., 2000; Stein and Colditz, 2004). Alterations in risk factors may cause genotoxicity by oxidative stress or by other mechanisms leading to gene mutations.

These mutations may activate oncogenes with dominant gain of function or tumor suppressor genes with recessive loss of function, leading to altered gene expression. In addition to the direct effect on gene structure leading to mutations, these risk factors may also lead to modification of the genome, ultimately resulting in epigenetic modifications. Alterations in gene expression provide selection advantage to the cell and ability to acquire neoplastic transformation. During the selection process limited cells adapt to survive and might attain the phenotype of cancer cells (Dreher and Junod, 1996).

Numerous studies have shown a critical role of paracrine signaling between tumor and host cells in the local microenvironment, which is mediated by cytokines, chemokines, and growth factors such as IL (interleukin) 6, IL8, platelet-derived growth factor (PDGF), vascular endothelial growth factor (VEGF), transforming growth factor (TGF)-β, hepatocyte growth factor (HGF), and so on. These factors are secreted by cancer cells as well as tumor stroma-associated cells, such as fibroblasts, adipocytes, myofibroblasts, pericytes, and immune and endothelial cells. This creates an effective field around cancer cells, which is conducive to their survival, growth and metastasis (McAllister and Weinberg, 2010). Also, cells from noninvolved tissue from cancer-containing organs are reported to have lower apoptotic indices and altered expression of growth factors, growth factor receptors, enzymes of xenobiotic system, and apoptotic molecules, which are modulators of cell survival and inducers of mitogenic activity (Ananthanarayanan et al., 2006; Hassan and Walker, 1998). Due to the alteration in the microenvironment around cancer cells, normal cells tend to change their behavior in the organ-developing cancer, a phenomenon known as field effect. The concept of field effect was further expanded to include the molecular abnormalities in histologically normal-appearing tissues surrounding cancer. Fearon and Vogelstein (1990) proposed a multistep carcinogenesis model. According to this model, genetic alterations take place in an incremental fashion such that a clone with proliferative advantage proliferates and acquires more genetic alterations and undergoes another selection for survival and growth. This eventually leads to cancer development. According to this model, precancerous cells in the proximity of cancer cells should have at least some, but not all, the genetic alterations needed for those cells to become cancerous (Fearon and Vogelstein, 1990). Such field effect biomarkers have now been reported in the carcinoma of various organs. Although the exact underlying mechanisms of the field effect in cancer are not yet fully understood, two popular hypotheses have been proposed. One hypothesis implicates genetic alterations that occur in a stepwise fashion (initiation, promotion, and progression): a clone gains growth advantage and acquires more genetic alterations, eventually resulting in cancer. Another hypothesis emphasizes epigenetic alterations, including hypermethylation of the DNA promoter of certain tumor suppressor genes, posttranslational changes in histones, nonhistone proteins, and microRNAs leading to abnormal regulation of genes, which may contribute to the field effects (Chai and Brown, 2009).

There are a number of cell loci within a tumor consisting of cells with different degrees and types of alteration at the molecular level, which may eventually manifest in a cancer phenotype. This heterogeneity makes it difficult to treat cancers. Almost all available treatments kill both normal cells and cancer cells and might lead to unwarranted side effects, forcing clinicians to sometimes terminate the treatment prematurely. An ideal chemotherapeutic and chemopreventive agent, therefore, would be one that not only differentiates between normal cells and cancer cells, but also affects multiple targets, which have relevance in cancer. A vast volume of literature available on green tea and cancer suggests that green tea can act on a variety of molecular targets and prevent the genesis of cancer. Green tea polyphenols have the ability to eliminate cells, which have already transformed into cancer cells. They also affect the pathways, which could prevent growth and metastasis of the cancer cells and prevent angiogenesis needed for invasion and migration. An added advantage of tea polyphenols is the differential response in normal cells versus cancer cells, which makes them a more preferred drug for chemoprevention as well as chemotherapy. This chapter focuses on the effects of green tea in various molecular biological mechanisms in the prevention and treatment of human cancers.

5.3 Modification of Oxidative Stress by Green Tea Polyphenols

Various risk factors for cancer can lead to genetic or epigenetic aberrations in the cell by generation of oxidative stress. Di- or trihydroxyl groups on the B-ring and the meta-5,7-dihydroxyl groups on the A-ring provide tea catechins with a very strong antioxidant activity, and the presence of the trihydroxyle structure in their gallate ring (D-ring), as in EGCG, further enhances this activity (Rice-Evans, 1999; Wiseman et al., 1997). The B-ring appears to be the principal site of antioxidant reactions (Sang et al., 2002; Valcic et al., 2000). Tea catechins have been shown to react with most of the reactive oxygen species (ROS) generated in a cell. The polyphenolic structure allows electron delocalization, conferring high reactivity to quench free radicals. Several oxidation products are formed during the reaction of tea polyphenols with free radicals (Sang et al., 2007).

Lipid oxidation is a major event in the oxidation of biological systems that occurs as a chain reaction involving lipids and molecular oxygen as substrates, and metals as catalysts. Inhibition of lipid oxidation chain reaction may be considered one of the most important mechanisms explaining the antioxidant actions of green tea polyphenols. Conceptually, chain-breaking antioxidants inhibit or retard lipid oxidation by interfering with chain-propagating reactions. Generation of hydroxyl, peroxyl, or alcoxyl radicals by decomposition of hydrogen peroxide and lipid hydroperoxides is very slow, but this conversion may increase due to catalytical activity of metals like iron and copper present in cells, thus making this conversion biologically relevant. Therefore, removal of iron or copper by sequestration to prevent metal-catalyzed free radical formation is considered an antioxidant strategy (Brown et al., 1998; Guo et al., 1996; Morel et al., 1998). Studies demonstrate that green tea polyphenols can act as antioxidants by metal sequestration, as catechol moieties in polyphenol molecules possess a high affinity for metal ions. The catechin-metal stability constant, a basic parameter to analyze the capacity of polyphenols for metal chelation, is similar to or even higher than the values of their widely used chelators for iron and copper (Galleano et al., 2010). Many polyphenol groups sharing the chemical characteristics of catechin could be used as chelating agents. Regardless of the large differences observed in the metal-chelating capacity of different polyphenols, some molecular aspects are common, including the (1) presence of 3'-4' hydroxyl groups in the B-ring, (2) presence of hydroxyl groups at positions 3 and 5, and (3) presence of the 4-oxo function. Thus flavonoid-metal complexes could favor the removal of the metal from the reaction milieu, thereby depriving the reaction from a catalyst (Perron and Brumaghim, 2009; van Acker et al., 1996). A recent report also indicated that metals bound to polyphenols can have "superoxide dismutase-like" activity. This might explain why certain flavonoids show higher antioxidant activity than expected by virtue of their sequestration capacity (Kostyuk et al., 2004).

Despite numerous evidence reported in the scientific literature about the antioxidant properties of tea polyphenols, few studies have presented evidence of their pro-oxidant activity caused by the metabolic conversion of polyphenols to pro-oxidant derivatives (Suh et al., 2010). Metals like copper and iron can also cause phenolic oxidation of tea catechins to yield ROS. Tea polyphenols are unstable in cell culture conditions and can generate ROS by undergoing auto-oxidation reactions (Sang et al., 2005). Some studies have shown that production of ROS during auto-oxidation of tea polyphenols plays an important role in inducing proapoptotic effects of tea polyphenols (Elbling et al., 2005; Nakagawa et al., 2004; Vittal et al., 2004; Yang et al.,

2000). Exogenous addition of catalase or superoxide dismutase significantly reduced ROS and apoptosis in selected cancer cell lines (Hou et al., 2006). Some studies, however, confirmed that tea polyphenols can cause apoptosis of cancer cells independent of ROS generation (Thakur et al., 2010). One study showed that EGCG produces more hydrogen peroxide (H_2O_2) in culture media in the absence of cancer cells than in the presence of cancer cells. The levels of H_2O_2 decreased to basal levels much faster in the presence of HT-29 human colon cancer cells, possibly because of cell-derived catalase eliminating H_2O_2 (Hong et al., 2002). Reduction in EGCG and increase in the oxidative dimer of EGCG, theasinesin A, are also accompanied by H_2O_2 production (Hong et al., 2002). The stability of EGCG is dependent on its concentration, pH, oxygen pressure, availability of metals, and ambient temperature (Sang et al., 2005). Most of these factors are well controlled in *in vivo* systems. The concentration of tea polyphenols is very low in *in vivo* systems compared with *in vitro*; therefore, it is important to determine if these factors make any significant contribution to tea polyphenol catabolism and production of ROS in *vivo*. To date there is no convincing data that suggest that pro-oxidant effects of tea polyphenols can occur *in vivo*.

Tea polyphenols have been shown to modulate the activity of various xenobiotic metabolic enzymes. Decreased activity of cytochrome P450 (CYP) 3A4, 2A6, 2C19, and 2E1 and increased activities of the isoforms 1A2 and 2B in humans and rodents have been described after green tea supplementation (Dudka et al., 2005; Jang et al., 2005; Muto et al., 2001). In a recent study of 42 nonsmoking healthy individuals, supplementation of green tea polyphenols significantly increased both the activity and the levels of glutathione S-transferase Pi (GSTP1), a phase 2 enzyme, in individuals with a low-baseline enzyme activity/level, suggesting tea polyphenol intervention may enhance the detoxification of carcinogens in individuals with low-baseline detoxification capacity (Chow et al., 2007). The results are encouraging and indicative of the antioxidant capacity imparted by regular consumption of tea to decrease the risk of certain diseases, including cancer.

5.4 Epigenetic Regulation by Green Tea Polyphenols

Until recently, it was hypothesized that mutations in genes consequent to internal or external insult to the cell lead to the generation of structurally and functionally abnormal proteins progressing to cancer. Because of an improved understanding of epigenetic processes, it is now evident that gene expression can be switched "on" or "off" without abrupt changes in the nucleotide sequences (Bird, 2002). This can be achieved by modifying the functional moieties either on the nucleotides or on histone proteins or activity of enzymes, especially DNA methyltransferases (DNMTs), histone acetyltransferases (HATs), and histone deacetylases (HDACs). These changes control higher organizational genetic information and determine gene expression changes, which explain how a cell can differentiate into many different cell types and organs by epigenetic switching. These processes are highly regulated in normal cells, but loss of this regulation leads to the genesis of disease processes, including carcinogenesis.

DNA methylation is the only genetically programmed DNA modification in mammals, and perhaps the most studied epigenetic mechanism. This postreplication modification is almost exclusively found on the 5′ position of the pyrimidine ring of cytosine in the context of the dinucleotide sequence CpG. 5-Methylcytosine accounts for 1% of all bases, varying slightly in different tissue types, and the majority (75%) of CpG dinucleotides are methylated in mammalian genomes. The composition of the genome is reflected in and dictates the epigenetic machinery to establish particular local and global epigenetic patterns using CpG spacing, sequence motifs, and DNA structure (Bock et al., 2006; Jia et al., 2007). A global hypomethylation of the genome is also observed in cancers (Feinberg and Vogelstein, 1983). This decreased methylation may initiate oncogenesis by causing chromosome instabilities and transcriptional activation of oncogenes and genes involved in the metastasis process (Ehrlich, 2002). A region- and gene-specific increase in the methylation of multiple CpG islands is observed along with global genomic hypomethylation in malignant cells (Jones and Baylin, 2007; Laird, 2005). In contrast, hypermethylation of CpG islands in the promoter region of a tumor suppressor or otherwise cancer-related gene is often associated with transcriptional gene silencing. A number of genes involved in DNA repair, cell cycle regulation, apoptosis, and other physiologic processes are modulated by hypermethylation of respective CpG islands present on promoter regions. A recent study suggested that methylation of multiple genes plays an important role in the prognosis of patients with breast cancer. This study not only described the association of methylation-mediated silencing of multiple genes with the severity of disease, but also speculated that the molecular cross talk between genes or genetic pathways is regulated by them individually (Sharma et al., 2009).

Several studies indicate that DNA methylation can be reversed by intake of multiple food components, including tea polyphenols (Ross, 2003).

For example, methyl-deficient diets have led to changes in the methylation patterns consistent with alterations observed during transformation of
normal cells to neoplasms (Pogribny et al., 2006). Thus, changes in these
methylation patterns by tea polyphenols may be responsible for their chemopreventive action. Together, silencing and unsilencing of genes can occur
through modification of histones, as well as by changes in the DNA methylation. In addition to factors that govern the overall recruitment and release
of histones (histone occupancy), there is a complex interplay of reversible
histone modifications that govern gene expression, including histone acetylation, methylation, phosphorylation, ubiquitination, and biotinylation.
Modification of histone deacetylase has also surfaced as a strategy for changing tumor behavior (Glozak and Seto, 2007; Holloway and Oakford, 2007;
Myzak and Dashwood, 2006).

Earlier studies demonstrated that tea polyphenols bind to DNA and RNA,
accumulate through multiple administrations of green tea beverage, and
play a significant role in human cancer prevention (Kuzuhara et al., 2006,
2007). It was reported that catechol-containing dietary polyphenol inhibited
enzymatic DNA methylation *in vitro* largely by increasing the formation of
S-adenosyl-L-homocysteine (a potent noncompetitive inhibitor of DNMT)
during the catechol-O-methyltransferase-mediated O-methylation of the
dietary catechol (Lee et al., 2005a). EGCG and EGC repressed telomerase
mRNA in lung, oral cavity, thyroid, and liver cancer cells might be linked to
inhibition of cell growth (Lin et al., 2006). EGCG also demonstrated anti-
neoplastic activity by suppressing the telomerase activity of digestive cancer
cells (Ran et al., 2005). EGCG can inhibit DNMT activity and reactivate
methylation-silenced genes p16/INK4a, retinoic acid receptor β (RAR β),
O-6-methylguanine methyltransferase (MGMT), and human mutL homolog 1 (hMLH1) genes in human colon, esophageal, and prostate cancer cells
(Fang et al., 2003). In another study, methylation of caudal-type homeobox
transcription factor 2 (CDX2) and other genes involved in gastric carcinogenesis was investigated in relation to the clinicopathologic and selected lifestyle factors of patients with gastric cancer. An inverse association of CDX2
methylation was observed with the intake of green tea (Yuasa et al., 2005).
Decreased annexin-I expression is a common event in early-stage bladder
cancer development. In part, green tea induced the expression of mRNA
and protein levels of annexin-I through demethylation of its promoter
and actin remodeling (Xiao et al., 2007). EGCG, an efficient inhibitor of
human dihydrofolate reductase, altered the p16 methylation pattern from
methylated to unmethylated after folic acid deprivation, resulting in growth

inhibition of human colon carcinoma cells. This same study demonstrated that through disruption of purine metabolism, EGCG caused adenosine release from the cells modulating different signaling pathways by binding to adenosine-specific receptors (Navarro-Perán et al., 2007). Treatment of oral cancer cells with EGCG partially reversed the hypermethylation status of the reversion-inducing cysteine-rich protein with kazal motifs (RECK) gene and significantly enhanced RECK mRNA expression (Kato et al., 2008). Another study illustrates that tissue factor pathway inhibitor-2 (TFPI-2), a member of the Kunitz-type serine proteinase inhibitor family, is inversely related to an increasing degree of malignancy. EGCG inhibited growth and induced apoptosis in renal cell carcinoma through TFPI-2 mRNA and protein overexpression (Gu et al., 2009a). Epigenetic silencing of GSTP1 by hypermethylation is recognized as being a molecular hallmark of human prostate cancer. Recently our laboratory reported that exposure of human prostate cancer LNCaP cells to green tea polyphenols (GTPs) at physiologically attainable concentrations caused demethylation in the proximal GSTP1 promoter and regions distal to the transcription factor binding sites. GTP exposure caused a concentration and time-dependent reexpression of GSTP1 and inhibition of DNMT1. GTP exposure also caused decreased mRNA and protein levels of methyl binding domain proteins (MBD1, MBD4, MeCP20) and HDAC 1–3, and increased levels of acetylated histone H3 (LysH9/18) and H4. In addition, green tea polyphenols reduced MBD2 association with accessible Sp1 binding sites causing increased binding and transcriptional activation of the GSTP1 gene. More importantly, GTP treatment did not result in global hypomethylation and promoted maintenance of genomic integrity. Unlike 5-aza-2′-deoxycitidine treatment, GTP exposure did not activate prometastatic gene S100 calcium-binding protein P (S100P). This study demonstrates the dual potential of tea polyphenols at physiologically attainable nontoxic doses to alter DNA methylation and chromatin modeling, the two global epigenetic mechanisms of gene regulation (Pandey et al., 2009). Another report demonstrated a significant reduction in the number of newly formed tumors in Apc (Min/+) mice treated with azoxymethane treatment, and supplementation with a solution of green tea caused downregulation of retinoid X receptor (RXR) α. These results correlated with decreased CpG methylation in the promoter region of the RXR α gene (Volate et al., 2009).

Recent reports demonstrate that treatment of breast cancer and promyelocytic leukemia cells with EGCG resulted in a decrease in E2F-1 binding sites, human telomerase reverse transcriptase (hTERT) promoter methylation, and ablation of histone H3Lys9 acetylation, causing an increase in binding of E2F-1 repressor at the hTERT promoter, resulting in cell death (Berletch et al., 2008). The Polycomb group (PcG) proteins are epigenetic repressors of gene expression, and their repression is achieved via action of two multi-protein PcG complexes: Polycomb group repressive multiprotein complex (PRC) 2 (eed) and PRC1 (Bmi-1). These complexes increase histone methylation and reduce acetylation, leading to a closed chromatin conformation. BMI1 Polycomb ring finger oncogene (Bmi-1) is overexpressed in breast, prostate, colon, pancreatic, and non-small-cell lung cancers. EGCG treatment caused suppression of two key PcG proteins, Bmi-1 and Ezh2, and led to global reduction in histone H3K27 trimethylation. This caused reduced expression of key cell cycle-regulated proteins, that is, cdk1, cdk2, cdk4, cyclin D1, cyclin E, cyclin A, and cyclin B1, and increased expression of cell cycle kinase inhibitors, that is, p21/waf1 and p27/kip1. EGCG treatment also resulted in induction of apoptosis through increased expression of BCL2-associated X protein (Bax), caspases-9, -8 and -3, and poly-ADP-ribose polymerase (PARP) cleavage, along with a decrease in B-cell lymphoma–extra large (Bcl-xL) expression (Balasubramanian et al., 2009).

Another important epigenetic regulation occurs via modifications of microRNA (miRNA) expression. Limited studies are available in the literature that explored the effect of tea polyphenols on the expression of miRNAs in human cancers. A recent study demonstrated that EGCG treatment altered the expression of miRNAs in human hepatocellular carcinoma HepG2 cells. Thirteen miRNAs were upregulated and 48 were downregulated. Among the miRNAs upregulated by EGCG, some target genes included rat sarcoma oncogene (RAS), Bcl2, E2F, transforming growth factor β receptor II (TGFBR2), and c-Kit. Among those miRNAs downregulated by EGCG are the target genes comprised of HOX family proteins, including the phosphatase and tensin homolog (PTEN), Sma- and Mad-related proteins (SMAD), myeloid leukemia cell differentiation protein (MCL) 1, solute carrier family 16 member 1 (SLC16A1), TTK, phosphoribosyl pyrophosphate synthetase 1 (PRPS1), zinc finger protein 513 (ZNF513), and sorting nexin 19 (SNX19) with diversified functions. Treatment with EGCG downregulated Bcl-2, an antiapoptotic protein, and transfection with anti-miR-16 inhibitor suppressed miR-16 expression and counteracted the EGCG effects on Bcl-2 downregulation and induced apoptosis in these cells (Tsang and Kwok, 2010). These results suggest that polyphenols from green tea, especially EGCG, may exert their biologic functions through epigenetic modifications.

5.5 Alteration of Growth Signals by Green Tea Polyphenols

Growth factors and growth factor receptors play an important physiological role in the normal process of growth and differentiation. It is evident that growth factors prevent cell death and their deficiency leads to apoptosis. Growth factors exert their effect by binding to the receptors mostly present on the cell membrane, leading to receptor tyrosine kinase dimerization, cross-phosphorylation, and receptor activation. Activated receptors phosphorylate a series of cytoplasmic proteins, which in turn sets off a cascade of events leading to the activation of transcription factors in the nucleus (Fantl et al., 1993). Peptide and polypeptide growth factors are produced in a variety of tissues throughout the body, and their action is not necessarily restricted to a single tissue type; these factors can act by an intracrine, autocrine, juxtacrine, paracrine, or endocrine process. Aberrations in the growth factor signaling pathways can lead to abnormal growth and development. Unregulated expression of growth factors or components of their signaling pathways can lead to neoplastic transformation or cancer. Cancer cells often produce growth factors at higher concentrations than their normal counterparts. These growth factors act by binding to the receptor tyrosine kinase family of receptors, and many of them execute their survival functions using common intracellular signaling molecules and pathways. Tea polyphenols modulate the activity of membrane-associated receptor tyrosine kinases, which play an important role in the control of many fundamental cellular processes. Overexpression of the human epidermal growth factor receptor 2 (HER-2/neu) is associated with poor prognosis in patients with breast cancer and head and neck squamous cell carcinoma (Masuda et al., 2001, 2002, 2003). EGCG and Polyphenon E® have been shown to cause a decrease in phosphorylation of epidermal growth factor receptor (EGFR) and HER-2 proteins, resulting in activation of Akt and extracellular signal-regulated kinase (ERK) proteins, and transcriptional activity of activator protein 1 (AP-1) and nuclear factor κB (NF-κB) promoters in HT29 human colon cancer cells (Shimizu et al., 2005a). EGCG inhibited HER-3 signaling, cyclooxygenase-2 (COX-2) transcription, and prostaglandin E2 (PGE-2) production in human colon cancer cell lines (Shimizu et al., 2005b). EGCG causes downregulation of EGFR by inducing mitogen-activated protein kinase (MAPK), which phosphorylates EGFR at sites critical for receptor internalization (Adachi et al., 2009). EGCG also inhibits the activation of platelet-derived growth factor receptors (PDGFRs) and fibroblast growth factor receptors (FGFRs) in human epidermoid carcinoma cells (Liang et al., 1997), and PDGFRs in human glioblastoma and hepatic stellate cells (Sachinidis et al., 2000). EGCG has shown responsiveness to cancer cells by affecting the 67 KDa laminin receptor present in the lipids raft on the membrane, thereby modulating the activity of tyrosine kinases (Sakata et al., 2004).

The phosphatidylinositol 3-kinase (PI3K)/Akt survival signaling pathway plays an important role in many human cancers (Jiang and Liu, 2008). The regulatory subunit of PI3Kinase (p85) binds to autophosphorylated tyrosines on the receptor by its SH2 domain. The catalytic unit of PI3K phosphorylates phosphatidylinositol (3,4)-bisphosphate (PIP2) to phosphatidylinositol (3,4,5)-trisphosphate (PIP3), which then recruits 3-phosphoinositide-dependent protein kinase (PDK) 1 and Akt, resulting in phosphorylation at Thr308 of Akt by PDK1 and subsequent phosphorylation at Ser493 in the regulatory domain of Akt for full-length activation. Once activated, Akt can phosphorylate a large number of downstream target molecules, which results in the modulation of a number of pathways regulating cell survival (Sen et al., 2003; Woodgett, 2005). An important downstream target of Akt is Bad, which is phosphorylated at Ser136, facilitating phosphorylation of Bad at Ser112 and S155 by other kinases. Docking of 14-3-3 proteins on phosphorylated Bad transports it from the mitochondria to the cytoplasm. This allows the antiapoptotic proteins Bcl2/Bcl-XL to be free to bind and inactivate Bax/Bad and inhibit mitochondrial-mediated apoptotic pathway (Datta et al., 1997; Masters et al., 2001). Akt also phosphorylates both α and β isoforms of glycogen synthase kinase 3 (GSK-3) at Ser21 and Ser9 on the N-terminal. Phosphorylated GSK-3 blocks apoptotic signaling by regulating both pro- and antiapoptotic Bcl2 family members (Beurel and Jope, 2006). GSK-3β can induce mitochondrial-mediated apoptosis by activating Bax by its phosphorylation at Ser163 (Linseman et al., 2004). GSK-3β also induces apoptosis by enhancing the expression of Bim, which degrades antiapoptotic protein Mcl2 (Maurer et al., 2006). More importantly, Akt phosphorylates the mammalian target of rapamycin (mTOR) on Ser2448, which inhibits its ability to phosphorylate Bcl2 on Thr69, Ser70, and Ser80, thereby losing its ability to block apoptosis, and increases proteosomal degradation (Asnaghi et al., 2004). Akt is known to phosphorylate, the FOXO family of proteins at Thr32, Ser253, and Ser315. Phosphorylation at these sites results in binding with 14-3-3 proteins and its exclusion from nucleus to cytoplasm, causing a decrease in expression of the proapoptotic signal and antioxidant enzymes (Van der Heide et al., 2004). Other than its role in inhibiting apoptotic pathways, Akt also enhances the NF-κB survival pathway by phosphorylating IκB kinases (IKKs), which in turn free NF-κB from inhibitors of κB (IκB). Free NF-κB translocates to the nucleus and upregulates a number of survival genes (Bai et al., 2009). EGCG inhibits activation of Akt by inhibiting HER-2/neu receptors and exerts antiproliferative and antiangiogenic activities by inhibiting signal transducer and activator of transcription (STAT) 3 and NF-κB activation (Masuda et al., 2001, 2002, 2003). Green tea polyphenols inhibit the PI3K/Akt pathway, independent of their antioxidant activity. Activated PI3K/Akt in various cancers is reported to be suppressed by tea polyphenols, leading to apoptosis caused by the activation of downstream proapoptotic signals. EGCG inhibits cell growth and proliferation by inhibiting the PI3K/Akt pathway in breast (Thangapazham et al., 2007), bladder (Qin et al., 2007), prostate (Siddiqui et al., 2004), and cervical (Sah et al., 2004) cancers. EGCG inhibited PDGF-β-induced proliferation and migration of rat pancreatic stellate cells by inhibiting PDGF-mediated activation of ERK and PI3K/Akt pathways (Masamune et al., 2005). In human cervical carcinoma (HeLA) and hepatoma (HepG2) cells, green tea extract and EGCG significantly inhibited serum-induced hypoxia-inducible transcription factor 1 α (HIF-1α) protein and VEGF expression by interfering with the PI3-K/Akt/mTOR signaling pathways in human cervical carcinoma and hepatoma cells. Chemoattractant- and hypoxia-stimulated HeLa cell migration was abolished by green tea extract and EGCG (Zhang et al., 2006). Treatment with EGCG inhibited the constitutive activation of EGFR, STAT3, and Akt in YCU-H891 head and neck squamous cell carcinoma (HNSCC) and MDA-MB-231 breast carcinoma cells (Masuda et al., 2002). Furthermore, EGCG inhibited PI3K/Akt activation that, in turn, resulted in modulation of Bcl-2 family proteins, leading to enhanced apoptosis of bladder cancer T24 cells (Qin et al., 2007). These findings suggest that the PI3K/Akt pathway could be a major target for intervention with green tea polyphenols.

Studies demonstrate that phosphorylation at the serine and threonine sites is the major phosphorylation event in the signal transduction of the PI3K/Akt signaling pathway. Phosphatases such as protein phosphatase 1 α (PP1α), PP2A, PP2B, and PP2C are the major dual-specificity phosphatases linked to the Akt pathway (Klumpp and Krieglstein, 2002; Tonks, 2006). Both PP1α and PP2A can dephosphorylate Akt and Bax, whereas PP2B can dephosphorylate GSK-3β. PP2A is a key Akt target and dephosphorylates Akt at Thr308 and Ser493, blocking this pathway, whereas PP2C can dephosphorylate Akt only at Ser493. These dual-specificity phosphatases are sensitive to redox modifications. Hydrogen peroxide treatment has been shown to reversibly block dual-phosphate activity of PP2A and PP1α (Klumpp et al., 2004; Lee et al., 2005b; Wu et al., 2006). The major regulator of the PI3K/Akt pathway is PTEN (phosphatase and tensin homolog deleted on chromosome 10); a member of phosphotyrosine phosphatase catalyzes the conversion of PIP3 to PIP2, preventing the binding and recruitment of –PH containing proteins to the plasma membrane, resulting in decreased translocation of survival signals. This important function of PTEN led to its classification as a tumor suppressor. PTEN has been reported to be deleted, mutated, repressed, or epigenetically silenced in several forms of human cancers (Priulla et al., 2007). More recent studies have demonstrated that oxidation results in modifications of cysteine residues, leading to inhibition or blockage of phosphatase activity. In the oxidized environment the two cysteine residues in PTEN (Cys124 and Cys71) form a sulfur cross-link resulting in a decrease in the phosphatase activity (Leslie et al., 2003). Green tea polyphenols, by virtue of their antioxidant activity, may be able to protect the oxidation of cysteine residues sustaining the phosphatase activity and preventing activation of the PI3K/Akt survival signaling pathway.

NF-κB is a family of closely related protein dimers that bind a common sequence motif in DNA called the κB site, originally discovered in B cells. It belongs to the category of rapid-acting primary transcription factors that allows NF-κB to act as a first responder to free radicals, inflammatory stimuli, cytokines, carcinogens, tumor promoters, endotoxin radiation, ultraviolet light, and x-rays. NF-κB dimers reside in the cytoplasm bound to inhibitors of κB (IκBs). Phosphorylation of IκBs by IκB kinase (IKK) causes ubiquitination and degradation of IκB, thus releasing NF-κB, which then translocates to the nucleus. Phosphorylation and activation of IκB kinase are controlled by NF-κB-inducing kinases, and there is cross talk between activation of the MAPK, ERK pathway, and NF-κB-inducing kinase/IκB kinase/NF-κB pathway. Once in the nucleus, NF-κB induces the expression of several regulated genes (Nishikori, 2005). Many of the activated target genes are critical to the establishment of the early and late stages of aggressive cancers, including expression of cyclin D1, Bcl-2, Bcl-XL, matrix metalloproteinases (MMPs), and VEGF. EGCG has been shown to inhibit NF-κB activity in the human colon and prostate cancer cells. Treatment of normal human epidermal keratinocytes with EGCG was found to inhibit UVB-mediated activation of NF-κB. The cleavage of the RelA/p65 subunit of NF-κB was blocked by a pan caspase inhibitor, N-benzyloxycarbonyl-Val-Ala-Asp(OMe)-fluoromethylketone (Z-VAD-FMK), during EGCG-mediated apoptosis. EGCG can suppress NF-κB activation as well as other pro-survival pathways, such as PI3K/Akt/mTOR and MAPKs in human bronchial epithelial cells, which may contribute to its ability to suppress inflammation, proliferation, and angiogenesis induced by cigarette smoke. Thus, modulation of the NF-κB pathway by EGCG may contribute to its chemopreventive potential.

The insulin-like growth factor–receptor (IGF/IGF-1R) system also plays an important role in the development and growth of various types of cancer. EGCG inhibited IGF-1R levels and activity, increased expression of TGF-β2 and IGFBP-3, and decreased levels of MMP-7 and -9 mRNAs in colon cancer cells (Shimizu et al., 2005c). EGCG caused apoptosis of malignant brain tumor and hepatocellular carcinoma cells by inhibiting IGF-1 (Shimizu et al., 2008; Yokoyama et al., 2001). Oral infusion of a green tea polyphenol mixture inhibited the development and progression in a transgenic adenocarcinoma of the mouse prostate (TRAMP) model. Green tea polyphenol intake reduced IGF-1 levels and decreased the activation of Akt, ERK, VEGF, MMP-2, and MMP-9 levels in the dorsoventral prostate of these mice. The effectiveness of green tea polyphenols was most prominent during the early stage of intervention in significantly inhibiting IGF-1 signaling (Adhami et al., 2004, 2009; Harper et al., 2007). In a recent study, 26 men with positive prostate biopsies scheduled for radical prostatectomy were supplemented with 800 mg of Polyphenon E capsules until time of radical prostatectomy showed a significant reduction in serum levels of prostate-specific antigen (PSA), HGF, and VEGF (McLarty et al., 2009).

In most somatic human cells telomeres shorten in length with every cell division and eventually reach a critical length. This shortening causes a cell to lose its proliferative ability, and the cell undergoes permanent cell cycle arrest or senescence. Unlike somatic cells, hematopoietic stem cells, keratinocytes in the basal layers of the epidermis, uterine endometrial cells, germ cells, and various tumors possess unlimited replicative potential because of their ability to maintain constant telomere length and presence of active telomerase activity. Increased telomerase activity is correlated with upregulation of the human telomerase reverse transcriptase (hTERT) gene and increased levels of hTERT mRNA (Ćukušić et al., 2008). Studies demonstrate that green tea polyphenols block telomerase activity as a major mechanism for limiting the growth of human cancer cells (Naasani et al., 2003). Tea polyphenols significantly inhibited telomerase activity of HepG2 cells compared with the control group (Jia et al., 2004). EGCG treatment caused inhibition of proliferation and telomerase activity and induction of apoptosis in cervical adenocarcinoma cells (Noguchi et al., 2006). Tea polyphenols cause repression of hTERT mRNA expression in the lung, oral cavity, thyroid, and liver carcinoma cells (Lin et al., 2006). Another study using tongue cancer cell lines indicated that tea polyphenols reduced hTERT activity in a time- and dose- dependent manner, disabling telomerase activity and terminating unlimited cell proliferation (Hua et al., 2006). EGCG induced telomere fragmentation in human cervical HeLa and transformed human embryonic kidney (HEK) 293 cells, but not normal human lung fibroblasts MRC-5 cells. This could be relevant to the apoptosis-inducing effect of EGCG on cancerous cells, but not on normal cells (Li et al., 2005).

5.6 Regulation of Cell Cycle and Apoptosis by Green Tea Polyphenols

Cell division is accomplished through a sequence of reproducible events called the cell cycle (Vermeulen et al., 2003). The cell cycle consists of an S phase (DNA replication) and an M phase (mitosis) separated temporally by G1 and G2 phases. Cyclin-dependent kinases (CDKs) are the key regulatory enzymes of the cell cycle. CDKs regulate the progression of cell division through different phases of the cell cycle by modulating the activity of key substrates. CDKs' downstream targets are E2F and its regulator, retinoblastoma. Activation and inactivation of CDKs at specific points in the cell cycle are required for orderly division of the cell. The cell cycle is negatively regulated by cyclin-CDK inhibitors (CKIs), such as p16, p15, p27, and p21. CDKs along with cyclins are major control switches for the cell cycle to move from G1 to the S phase or G2 to the M phase of the cell cycle. In eukaryotes, cells respond to DNA damage by activating those signaling pathways that promote cell cycle arrest and DNA repair. In response to DNA damage at the checkpoint, ataxia telangiectasia mutated protein kinase (ATM) phosphorylates and activates serine/threonine-protein kinase (Chk) 2, which in turn directly phosphorylates and activates tumor suppressor p53, resulting in activation of transcriptional targets such as p21, which functions to block the cell cycle if the DNA is damaged. This provides time to repair DNA by blocking the cell cycle. If the damage is severe, p53 can cause apoptosis (cell death) by activating its proapoptotic transcriptional targets, such as Bax, Bak, p53 upregulated modulator of apoptosis (PUMA), and so on. p53 and its transcriptional targets play an important role in both G1 and G2 checkpoints. Mutations in the p53 gene are frequently observed in most forms of human cancers. ATR-Chk1-mediated protein degradation of cell division cycle 25 homolog A (Cdc25A) protein phosphatase is also a mechanism conferring intra-S-phase checkpoint activation (Falck et al., 2002).

During carcinogenesis, cell cycle checkpoint controls are lost, and these cells also attain the ability to escape apoptosis. The uncontrolled and unchecked cell division along with the ability to escape apoptosis makes cancer cells immortal and difficult to kill.

Studies indicate that tea polyphenols can induce cell cycle arrest and apoptosis in various cancer cell lines. EGCG treatment of human prostate cancer cells (LNCaP and DU145) resulted in significant dose- and time-dependent upregulation of p21, p27, p16, and p18 proteins along with downmodulation of cyclins D1 and E, and cdk 2, 4, and 6 expressions. Increase in the binding of cyclin D1 toward p21 and p27 and decreased binding of cyclin E toward cdk2 was observed after exposure of cells to green tea polyphenols (Gupta et al., 2003). In pancreatic cancer cells, EGCG caused G1 growth arrest through regulation of cyclin D1, cdk4, cdk6, p21/WAF1, and p27/KIP1 and induced apoptosis through activation of caspases-3 and -9. EGCG inhibited the expressions of antiapoptotic Bcl-2 and Bcl-XL and induced the expressions of proapoptotic BCl2 family members Bax, Bak, Bcl-XS, and PUMA. Mouse embryonic fibroblasts (MEFs) derived from Bax and Bak double-knockout mice exhibited greater protection against EGCG-induced apoptosis than wild-type or single-knockout MEFs. Activation of Bax in p53$^{-/-}$ MEFs by EGCG suggests that EGCG can induce apoptosis in the absence of p53 (Shankar et al., 2007). In a recent study, tea polyphenols significantly inhibited the proliferation signal of colorectal cancer cells by inhibiting the expression of hairy and enhancer of split 1 (HES1), jagged 1 protein (JAG1), metallothionen 2A (MT2A), and v-maf musculoaponeurotic fibrosarcoma oncogene homolog A (MAFA) genes, and upregulating the expression of Bax and inhibiting p38 expression (Jin et al., 2010). Furthermore, activities of Ras, Raf-1, and ERK1/2 were inhibited by EGCG, whereas activities of MEKK1, c-Jun N-terminal kinase (JNK) 1/2, and p38 MAPKs were induced by EGCG treatment. Inhibition of cRaf-1 or ERK enhanced EGCG-induced apoptosis, whereas inhibition of JNK or p38 MAPK inhibited EGCG-induced apoptosis. EGCG inhibited activation of p90 ribosomal protein S6 kinase, and induced activation of cJUN, suggesting that EGCG induces growth arrest and apoptosis through multiple mechanisms (Shankar et al., 2007). A study with prostate cancer cell lines (LNCaP and PC-3) showed that EGCG activates growth arrest and apoptosis primarily via the p53-dependent pathway that involves the function of both p21 and Bax such that downregulation of either molecule confers a growth advantage to the cells (Hastak et al., 2005). Treatment of EGCG resulted in the inhibition of cellular proliferation and cell viability in human breast cancer MDA-MB-468 cells. Increased expression and phosphorylation at Ser15 residue of tumor suppressor protein p53 and its target Bax and decreased expression Bcl-2 were observed after EGCG treatment, resulting in activation of downstream targets and subsequent apoptosis (Roy et al., 2005). EGCG also inhibited protein synthesis, lipogenesis, and cell cycle progression through activation of 5' AMP-activated protein kinase (AMPK) and inhibition of mTOR in p53-positive and -negative human hepatoma cells (Huang et al., 2009).

EGCG has also been shown to induce apoptosis by activating p53 and influencing the ratio of pro/antiapoptotic factors in favor of apoptosis in various human carcinoma cells (Beltz et al., 2006). Apoptosis induced by green tea polyphenols in most human cancer cells is caspase-3 dependent (Hsu et al., 2003; Qanungo et al., 2005), which aligns with their ability to inhibit proteasome activity and accumulate Bax (Chen et al., 2005). Green tea polyphenol-induced apoptosis of lung cancer cells and H-ras-transformed bronchial epithelial cells is prevented by addition of catalase (Yang et al., 1998), indicating that polyphenols have the ability to stimulate H_2O_2 generation in cancer cells. EGCG also induced apoptosis via inhibiting nonsteroidal anti-inflammatory drug-activated gene (NAG)-1 in a p53-independent manner in colon cancer cells (Baek et al., 2004). Polyphenols including EGCG have also been shown to induce cell cycle arrest and apoptosis in cells lacking functional p53 by activating p73, a closely related p53 family member that expresses a subset of p53 target genes, including p21 and MDM2 (Amin et al., 2007; Shammas et al., 2006). Recent studies demonstrated that EGCG can sensitize tumor necrosis factor apoptosis-inducing ligand (TRAIL)-mediated apoptosis in prostate, malignant glioma, and pancreatic hepatocellular carcinoma cells (Basu and Haldar, 2009; Nishikawa et al., 2006; Siddiqui et al., 2008; Siegelin et al., 2008). Upregulation of proapoptotic BH3-only protein PUMA by tea polyphenols caused both p53-dependent and -independent apoptosis of human colorectal cancer cells (Thakur et al., 2010; Wang et al., 2008). TFPI-2 overexpression in renal cell carcinoma or downregulation of inhibitor of DNA binding 2, a dominant negative helix-loop-helix protein in prostate cancer cells, inhibits cell growth and induces apoptosis after EGCG treatment (Gu et al., 2009a; Luo et al., 2009). Another study, however, demonstrated that green tea polyphenols imparts a protective effect on PC12 cells by inhibiting 6-hydroxydopamine-induced cell apoptosis (Nie et al., 2002). Drinking green tea has been shown to significantly reduce the incidence of chemical-induced lung carcinoma or mouse skin carcinogenesis and also upregulate p53 and Bax expression and deregulate Bcl2 and survivin (Gu et al., 2009b; Manna et al., 2009; Ran et al., 2007;

Roy et al., 2009; Wu et al., 2009). On the contrary, a recent study showed that EGCG protects non-small-cell lung carcinoma (NSCLC) cells from apoptosis induced by serum deprivation via Akt activation, and this effect may limit clinical use of EGCG in the treatment and prevention of NSCLC (Kim et al., 2009).

5.7 Inhibition of Invasion, Metastasis, and Angiogenesis by Green Tea Polyphenols

Invasion and metastasis are the most life-threatening aspects of cancer. Metastasis involves multiple processes, including invasion of tumor cells into the adjacent tissue, entry of tumor cells in the systemic circulation (intravasation), survival in circulation, extravasation to distant organs, and ultimate growth to produce tumor at the secondary site (Stracke and Liotta, 1992). The progression of human tumors involves the matrix metalloproteinase (MMP) family. Two members in particular, gelatinases A and B (MMP-2 and MMP-9), seem to play an important role in tumor invasion and metastasis released by most epithelial and endothelial cells. The MMPs are involved in the turnover of basement membrane collagen under basal conditions and of other matrix proteins during angiogenesis, tissue remodeling, and repair. Since these enzymes play an essential role in the homeostasis of the extracellular matrix, an imbalance in their expression or activity may have important consequences in the development of human cancers. MMPs are zinc-dependent endopeptidases produced as proMMPs and require activation to perform their activities. In addition, MMPs can form complexes with tissue inhibitor of metalloproteinases (TIMP), TIMP-2 for MMP-2 and TIMP-1 for MMP-9. One key element for the activation of these MMPs is the sequential proteolysis of the propeptide, which blocks the active site. In this respect, membrane-associated (MT) MMPs were identified as an important modulator of MMP-2 activity. In particular, MT1-MMP was shown to form a complex with TIMP2, which is essential for MMP-2 activation. Besides, MMPs, urokinase plasminogen activator (uPA), and its receptor (uPAR)-mediated signaling increase the activity of urokinase enzyme, which facilitates the mobility of cancer cells to their neighboring tissues. uPA has been implicated in tumor cell invasion, survival, and metastasis in a variety of human cancers (Li and Cozzi, 2007). Green tea polyphenols affect cell invasion by affecting the expression of these molecules. EGCG-mediated inhibition of uPA reduced tumor size or caused complete remission of tumors in mice (Jankun et al., 1997). Green tea polyphenols inhibited invasion of breast cancer cells by inhibiting AP-1 and NF-κB and suppressed uPA secretion. Invasion of human oral cancer cells was blocked by EGCG by inhibiting uPA and MMP-2 and MMP-9 (Ho et al., 2007). EGCG and theaflavin strongly suppressed the invasion of human fibrosarcoma cells by modulating MMP-2 and MMP-9 (Maeda-Yamamoto et al., 1999). EGCG inhibited MMP-2 and MMP-9 while inducing the activity of their inhibitors, TIMP-1 and TIMP-2, in neuroblastoma, fibrosarcoma (Garbisa et al., 2001), glioblastoma (Annabi et al., 2002), prostate (Vayalil and Katiyar, 2004), endothelial (Fassina et al., 2004), and human gastric (Kim et al., 2004) cancer cells. In the TRAMP model of prostate carcinogenesis, green tea polyphenol infusion resulted in marked inhibition of effectors of angiogenesis and metastasis, notably VEGF, uPA, MMP-2, and MMP-9 (Adhami et al., 2004). Among them, the inhibition of pro-MMP-2 secretion was crucial for EGCG-induced tumor cell inactivation during angiogenesis (Sartor et al., 2004). Moreover, EGCG directly inhibited metallothionein-1 (MT1)-MMP activity in HT-1080 human fibrosarcoma cells and human umbilical vein endothelial cells (HUVECs), leading to accumulation of nonactivated MMP-2 at the cell surface (Dell'Aica et al., 2002, Yamakawa et al., 2004). EGCG treatment of human breast cancer cells reduced activity, protein expression, and mRNA expression levels of MMP-2 (Sen et al., 2009). On the contrary, EGCG enhanced the production of pro-MMP-7 via generation of reactive oxygen species and activation of JNK1/2 and c-JUN/c-FOS induction, as well as AP-1 transactivation, in human colon cancer cells (Kim et al., 2005).

In addition to basement degradation, tumor cells need a continuous supply of nutrients to maintain their high metabolic activities, which could be achieved either by chemotaxis toward a preexisting vascular network or by infiltration of vascular endothelial cells in cancer structures leading to neovascularization. In this respect, urokinases, VEGFs, FGFs, transforming growth factors-β (TGF-β), PGDFs, endothelin-1, extracellular matrix (ECM) proteolytic enzymes, MMPs, and tissue inhibitor of metalloproteinases (TIMPs) play important roles. EGCG treatment has been shown to reduce the expression of focal adhesion kinase (FAK), membrane type 1–matrix metalloproteinase (MT1-MMP), NF-κB, and VEGF, and reduces the adhesion of cells to ECM, fibronectin, and vitronectin (Sen et al., 2009). EGCG significantly enhanced the expression of reversion-inducing cysteine-rich protein with Kazal motifs (RECK) mRNA and inhibition of MMP-2 and MMP-9 levels in oral squamous cell carcinoma cells (Kato et al., 2008; Murugan et al., 2009).

5.8 Modulation of Other Signaling Pathways by Green Tea Polyphenols

The cellular signaling pathways that are known to be important in cancer development and progression include Wnt, Hedgehog, Notch, and Met/HGF. Both Wnt and Hedgehog signaling pathways and their cross talk are critical to the processes of epithelial-to-mesenchymal transition and considered to be accountable for tumor recurrence, invasion, and metastasis (Dodge and Lum, 2011; Wu and Zhou, 2008). The Wnt signaling pathway involves a large number of proteins that are required for basic developmental processes, such as cell fate specification, progenitor cell proliferation, and the control of asymmetric cell division, in many different species and organs. The canonical Wnt signaling pathway is induced when Wnt proteins bind to cell surface receptors of the Frizzled family, causing the receptors to activate disheveled family proteins, and ultimately resulting in a change in the amount of β-catenin that reaches the nucleus. The major effect of Wnt ligand binding to its receptor is the stabilization of cytoplasmic β-catenin through inhibition of the β-catenin degradation complex. The free β-catenin enters the nucleus and activates Wnt-regulated genes through its interaction with the T-cell factor (TCF) family of transcription factors and concomitant recruitment of coactivators. The noncanonical Wnt signaling pathway includes the planar cell polarity (PCP) pathway and the Wnt/Ca2+ pathway. The inhibitory effects of EGCG on Wnt signaling have been reported in various human cancers. β-Catenin expression and β-catenin/TCF-4 reporter activity were inhibited at physiologically relevant concentrations of EGCG in HEK293 cells (Dashwood et al., 2002). EGCG treatment caused a decrease in cytosolic β-catenin protein levels and inhibited the T cell specific transcription factor/lymphoid enhancer-binding factor-1 (TCF/LEF) reporter in human lung cancer H460 and A549 cells. Further studies demonstrate that EGCG downregulates the canonical Wnt pathway by reactivating Wnt antagonist gene Wnt inhibitory factor 1 (WIF-1) through the reversal of its promoter methylation (Gao et al., 2009). In another study, treatment of the HT29 colon cancer cell line with EGCG caused a significant inhibition of GSK3-α and GSK-3β activity. A concentration-dependent decrease in phosphorylated β-catenin levels and overall reduction in β-catenin and its mRNA levels were also observed after EGCG treatment (Pahlke et al., 2006). In another study with HEK293 cells and physiologically relevant concentrations of green tea polyphenol, EGCG facilitated the trafficking of β-catenin into lysosomes, presumably as a mechanism for sequestering β-catenin and circumventing further nuclear transport and activation of β-catenin/TCF/LEF signaling (Dashwood et al., 2005). EGCG has also been reported to inhibit canonical Wnt signaling in antler progenitor cells at the TCF/LEF level (Mount et al., 2006). Microarray gene expression profiling analysis revealed that EGCG treatment inhibits Wnt and Id signaling involved in cell proliferation (Liu et al., 2008). Wnt signaling has also been found to be inhibited in breast cancer cells in a dose-dependent manner by EGCG. This effect may be mediated by induction of the HBP1 protein, a suppressor of Wnt signaling. EGCG treatment induced HBP1 mRNA stability and increased HBP1 transcriptional repressor levels in breast cancer cells and in an HBP1-dependent manner. Also, knockdown of HBP1 reduced sensitivity to EGCG in the suppression of Wnt and reduction in cell proliferation and invasiveness (Kim et al., 2006).

The Hedgehog (Hh) signaling pathway plays an important role in human embryogenesis, but is largely inactive in adult tissues under normal conditions. The Hh signal is relayed by a number of key proteins, including Patched (PTCH) and Smoothened (SMO) at the cell surface. In the absence of Hh ligand, PTCH acts as a suppressor of SMO. Upon ligand binding to PTCH, SMO is translocated to the primary cilium, where the transcription factor glioma-associated oncogene (GLI) is activated, translocates to the nucleus, and leads to the expression of Hh target genes. GLI mediates the expression of genes involved in cell growth and differentiation. Activation of the Hh pathway has been implicated in the development of cancers in various organs, including brain, lung, mammary gland, prostate, and skin (Evangelista et al., 2006; Theunissen and de Sauvage, 2009). Abnormal activation of the pathway probably leads to development of disease through transformation of adult stem cells into cancer stem cells that give rise to the tumor. The inhibitory effects of EGCG on the Hedgehog signaling have been reported. It has been demonstrated that EGCG could significantly inhibit the growth of prostate cancer cells *in vitro* and *in vivo* in TRAMP mice by inhibiting Gli-1 reporter activity and mRNA expression (Slusarz et al., 2010). EGCG also inhibited cellular proliferation and induced apoptosis of human chondrosarcoma SW1353 and CRL-7891 cells by inhibition of Indian Hedgehog pathway, and downregulation of PTCH and Gli-1 expression. The caspase-3 protein expression remained unchanged, but there was a significant decrease in Bcl-2 and increase in Bax levels after EGCG treatment in these cells. This indicates that EGCG can be an effective growth inhibitor in chondrosarcoma by modifying the Hedgehog (Hh) signaling pathway (Tang et al., 2010).

Hepatocyte growth factor (HGF) and its receptor (HGFR), also known as c-Met, play an important role in angiogenesis and tumor growth.

N-Methyl-N'-nitro-N-nitroso-guanidine (MNNG) HOS transforming gene product (MET) is a tyrosine kinase receptor that, upon binding to its natural ligand, HGF, is phosphorylated and subsequently activates different signaling pathways involved in tumor proliferation, motility, migration, and invasion. MET activation by its ligand HGF induces MET kinase catalytic activity, which triggers phosphorylation of the tyrosines 1234 and 1235. These tyrosines engage various signal transducers, thus initiating a whole spectrum of biological activities driven by MET, collectively known as the invasive growth program. The transducers interact with the intracellular multisubstrate docking site of MET either directly, such as growth receptor-bound protein 2 (GRB2), Src homology-2-containing (SHC), v-src sarcoma (Schmidt-Ruppin A-2) viral oncogene homolog (SRC), and the p85 regulatory subunit of phosphatidylinositol-3 kinase (PI3K), or indirectly through the GRB2-associated binding protein 1 (Gab1). MET engagement activates multiple signal transduction pathways. It induces RAS activation, and prolonged MAPK activity activates PI3K either downstream of RAS or directly by recruiting through the multiple docking site, STAT pathway, β-catenin pathway, and NOTCH pathway (Cecchi et al., 2010). Studies have demonstrated that EGCG can prevent the metastatic spread of breast and hypopharyngeal carcinoma cell by inhibiting HGF/Met signaling (Bigelow and Cardelli, 2006; Lim et al., 2008). Green tea polyphenols prevent activation of the c-Met receptor by preventing its phosphorylation on tyrosines 1234/1235 in the kinase domain by altering the structure or function of lipid rafts in prostate cancer cells (Duhon et al., 2010). Pretreatment of the nontumorigenic breast cell line, MCF10A, and the invasive breast carcinoma cell line, MDA-MB-231, with EGCG inhibited HGF-induced Met phosphorylation and downstream activation of Akt and ERK and blocked the ability of HGF to induce cell motility and invasion. This study also suggested that the R1 galloyl and R2 hydroxyl groups are important in mediating the green tea catechins' inhibitory effect toward HGF/Met signaling. EGCG also potentiated antiproliferative activity of c-Met (Bigelow and Cardelli, 2006). Furthermore, EGCG treatment markedly suppressed Met activation in the presence of HGF in human colon cancer cell lines HCT116 and HT29 (Larsen and Dashwood, 2009). At physiologically relevant concentration (1 μM), EGCG suppressed HGF-induced tumor motility and MMP-9 and uPA activities in hypopharyngeal carcinoma FaDu cells. EGCG further suppressed the Akt and Erk pathway in these cells, suggesting that these could be a downstream effect or pathways of HGF/Met signaling inhibited by EGCG (Lim et al., 2008).

5.9 Potential Limitation of the Work

The concentrations of green tea polyphenols used for *in vitro* studies range from 20 to 100 µM or even higher. These levels cannot be achieved *in vivo*, especially inside or surrounding cancer cells. Therefore, it may be unwise to extrapolate the results of *in vitro* studies to *in vivo* situations. The levels that can be achieved in blood after 2 to 3 cups of tea range from 0.1 to 0.6 µM, and even after 7 to 9 cups, it is less than 1 µM. Furthermore, the concentrations of catechins in the tissue depend on the duration of tea intake; this might be the reason that the results from clinical trials have not been encouraging. Green tea polyphenols are partially absorbed as demonstrated through a study that 0.012% of EGCG was absorbed after 30 min of an oral dose of 56 mg in rats and 0.32% after 60 min of an oral dose of 97 mg in humans (Nakagawa and Miyazawa, 1997). This low absorption of polyphenols is probably because of increased breakdown in the intestines at high pH, as tea catechins have been shown to be more stable at low pH (Lam et al., 2004). Tea polyphenols have a half-life of less than 2 h, which can be extended multiple folds by addition of superoxide dismutase or catalase. Green tea polyphenols are considered natural antioxidants but have been shown to oxidize to form pro-oxidant species, depending on the physiologic conditions. The extent of this conversion depends on the partial pressure of oxygen in *in vivo* and *in vitro* conditions (Tachibana, 2009). Hence, elucidation of *in vitro* results to the *in vivo* situations needs further investigation, especially in the light of reports describing that under physiologic conditions, biotransformation reactions, such as methylation, can modify green tea polyphenols and therefore limit their *in vivo* cancer preventive activity (Landis-Piwowar et al., 2008).

Tachibana (2009) determined an effective concentration of EGCG to inhibit the biologic activity of many proteins by competitive or noncompetitive mechanisms. Interestingly, these effective concentrations ranged from 0.003 µM to 10 mM. To achieve higher concentrations of polyphenols in the tissues, either higher intake of tea polyphenols or derivatives of tea polyphenols with increased stability and bioavailability may be required. However, the increased bioavailability and effectiveness of tea polyphenols derivatives and increased concentrations of natural tea polyphenols are difficult to achieve and may be toxic to normal cells. Therefore, development of synthetic derivatives is needed, or new technologies to achieve higher concentrations of tea polyphenols in the systemic circulation and target

FIGURE 5.1
Pathways altered by green tea polyphenols. Numerous pathways are deregulated in cancer cells that include oxidative stress regulators, epigenetic modifiers, molecules regulating survival and death, oncogene activation, and overexpression of molecules regulating growth signals, angiogenesis, invasion, and metastasis. Green tea polyphenols have been shown to affect epigenetic and various signaling pathways. –, demonstrates regulation; ⊥, demonstrates inhibition.

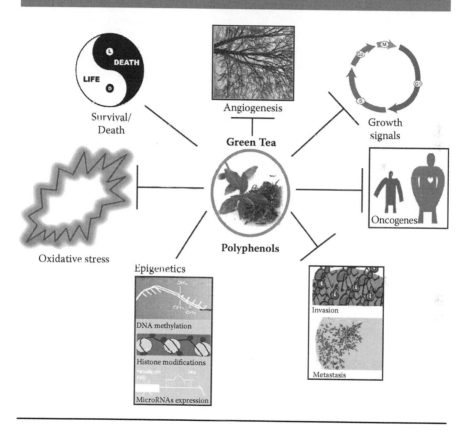

tissue are required. One such effort using encapsulated green tea polyphenol epigallocatechin-3-gallate (EGCG) in polylactic acid–polyethylene glycol nanoparticles showed that encapsulated EGCG retains its biologic effectiveness with a more than 10-fold dose advantage for exerting its proapoptotic and angiogenesis inhibitory effects (Siddiqui et al., 2009).

5.10 Conclusion and Future Directions

Green tea polyphenols have shown potential to decrease the risk of various human diseases, including cancer, by targeting various signaling pathways. Therefore, tea polyphenols remain an excellent candidate for chemoprevention, and further research is warranted to understand the involvement of particular mechanisms and pathways. A schematic representation of targets and pathways modulated by green tea polyphenols is shown in Figure 5.1. A better understanding of the mechanisms of epigenetic changes with tea polyphenols is required for the precise use in chemoprevention regimens. Recent reports on the additive or synergistic effects of tea polyphenols to increase the efficacy of chemo- or radiation therapy is a promising area of research, and emphasis on more well-designed and well-planned studies is needed. This approach has the potential to enhance the efficacy of conventional therapies by adjustment of dose and time periods needed to achieve optimum results. In addition, development of new derivatives of tea polyphenols with improved bioavailability and efficacy is required. Studies to develop techniques like nanotechnology should be encouraged, which could lead to sustained and precise delivery of polyphenols to the target site(s), thereby achieving improved efficacy and effective treatments to improve patient outcomes as an ultimate goal.

References

Adachi, S., Shimizu, M., Shirakami, Y., Yamauchi, J., Natsume, H., Matsushima-Nishiwaki, R., To, S., Weinstein, I.B., Moriwaki, H., and Kozawa, O. 2009. (−)-Epigallocatechin gallate downregulates EGF receptor via phosphorylation at Ser1046/1047 by p38 MAPK in colon cancer cells. *Carcinogenesis*, 30: 1544–1552.

Adhami, V.M., Siddiqui, I.A., Ahmad, N., Gupta, S., and Mukhtar, H. 2004. Oral consumption of green tea polyphenols inhibits insulin-like growth factor-I-induced signaling in an autochthonous mouse model of prostate cancer. *Cancer Res*, 64: 8715–8722.

Adhami, V.M., Siddiqui, I.A., Sarfaraz, S., et al. 2009. Effective prostate cancer chemopreventive intervention with green tea polyphenols in the RAMP model depends on the stage of the disease. *Clin Cancer Res*, 15: 1947–1953.

Amin, A.R., Thakur, V.S., Paul, R.K., et al. 2007. SHP-2 tyrosine phosphatase inhibits p73-dependent apoptosis and expression of a subset of p53 target genes induced by EGCG. *Proc Natl Acad Sci USA*, 104: 5419–5424.

Ananthanarayanan, V., Deaton, R.J., Yang, X.J., et al. 2006. Alteration of proliferation and apoptotic markers in normal and premalignant tissue associated with prostate cancer. *BMC Cancer*, 6: 73.

Annabi, B., Lachambre, M.P., Bousquet-Gagnon, N., et al. 2002. Green tea polyphenol (–)-epigallocatechin 3-gallate inhibits MMP-2 secretion and MT1-MMP-driven migration in glioblastoma cells. *Biochim Biophys Acta*, 1542: 209–220.

Asnaghi, L., Bruno, P., and Priulla, M. 2004. mTOR: A protein kinase switching between life and death. *Pharmacol Res*, 50: 545–549.

Baek, S.J., Kim, J.S., Jackson, F.R., et al. 2004. Epicatechin gallate-induced expression of NAG-1 is associated with growth inhibition and apoptosis in colon cancer cells. *Carcinogenesis*, 25: 2425–2432.

Bai, D., Ueno, L., and Vogt, P.K. 2009. Akt-mediated regulation of NFkappaB and the essentialness of NFkappaB for the oncogenicity of PI3K and Akt. *Int J Cancer*, 125: 2863–2870.

Balasubramanian, S., Adhikary, G., and Eckert, R.L. 2009. The Bmi-1 Polycomb protein antagonizes the (–)-epigallocatechin-3-gallate dependent suppression of skin cancer cell survival. *Carcinogenesis*, 31: 496–503.

Basu, A., and Haldar, S. 2009. Combinatorial effect of epigallocatechin-3-gallate and TRAIL on pancreatic cancer cell death. *Int J Oncol*, 34: 281–286.

Beltz, L.A., Bayer, D.K., Moss, A.L., et al. 2006. Mechanisms of cancer prevention by green and black tea polyphenols. *Anticancer Agents Med Chem*, 6: 389–406.

Berletch, J.B., Liu, C., Love. W.K., et al. 2008. Epigenetic and genetic mechanisms contribute to telomerase inhibition by EGCG. *J Cell Biochem*, 103: 509–519.

Beurel, E., and Jope, R.S. 2006. The paradoxical pro- and antiapoptotic actions of GSK3 in the intrinsic and extrinsic apoptosis signaling pathways, *Prog Neurobiol*, 79:173–189.

Bigelow, R.L.H., and Cardelli, J.A. 2006. The green tea catechins, (–)-epigallocatechin-3-gallate (EGCG) and (–)-epicatechin-3-gallate (ECG), inhibit HGF/Met signaling in immortalized and tumorigenic breast epithelial cells. *Oncogene*, 25: 1922–1930.

Bird, A. 2002. DNA methylation patterns and epigenetic memory. *Genes Dev*, 16: 6–21.

Bock, C., Paulsen, M., Tierling, S., et al. 2006. CpG island methylation in human lymphocytes is highly correlated with DNA sequence, repeats, and predicted DNA structure. *PLoS Genet*, 2: e26

Brockmöller, J., Cascorbi, I., Henning, S., et al. 2000. Molecular genetics of cancer susceptibility. *Pharmacology*, 61: 212–227.

Brown, K.E., Kinter, M.T., Oberley, T.D., et al. 1998. Enhanced gamma-glutamyl transpeptidase expression and selective loss of CuZn superoxide dismutase in hepatic iron overload. *Free Radic Biol Med*, 24: 545–555.

Cecchi, F., Rabe, D.C., and Bottaro, D.P. 2010. Targeting the HGF/Met signalling pathway in cancer. *Eur J Cancer*, 461260–461270.

Chai, H., and Brown, R.E. 2009. Field effect in cancer—An update. *Ann Clin Lab Sci*, 39: 331–337.

Chen, D., Daniel, K.G., Chen, M.S., et al. 2005. Dietary flavonoids as proteasome inhibitors and apoptosis inducers in human leukemia cells. *Biochem Pharmacol*, 69: 1421–1432.

Chow, H.H.S., Hakim, I.A., Vining, D.R., et al. 2007. Modulation of human glutathione s-transferases by Polyphenon E intervention. *Cancer Epidemiol Biomarkers Prev*, 16: 1662–1666.

Ćukušić, A., Škrobot Vidaček, N., Sopta, M., et al. 2008. Telomerase regulation at the crossroads of cell fate. *Cytogenet Genome Res*, 122: 263–272.

Dashwood, W.M., Carter, O., Al-Fageeh, M., et al. 2005. Lysosomal trafficking of β-catenin induced by the tea polyphenol epigallocatechin-3-gallate. *Mutat Res*, 591: 161–212.

Dashwood, W.M., Orner, G.A., and Dashwood, R.H. 2002. Inhibition of beta-catenin/Tcf activity by whitetea, green tea, and epigallocatechin-3-gallate (EGCG): Minor contribution of H(2)O(2) at physiologically relevant EGCG concentrations. *Biochem Biophys Res Commun*, 296: 584–588.

Datta, S.R., Dudek, H., Tao, X., et al. 1997. Akt phosphorylation of BAD couples survival signals to the cell-intrinsic death machinery. *Cell*, 91: 231–241.

Dell'Aica, I., Dona, M., Sartor, L., et al. 2002. (–)Epigallocatechin-3-gallate directly inhibits MT1-MMP activity, leading to accumulation of nonactivated MMP-2 at the cell surface. *Lab Invest*, 82: 1685–1693.

Dodge, M.E., and Lum, L. 2011. Drugging the cancer stem cell compartment: Lessons learned from the Hedgehog and Wnt signal transduction pathways. *Annu Rev Pharmacol Toxicol*, 51: 289–310.

Dreher, D., and Junod, A.F. 1996. Role of oxygen free radicals in cancer development. *Eur J Cancer*, 32: 30–38.

Dudka, J., Jodynis-Liebert, J., Korobowicz, E., et al. 2005. Activity of NADPH-cytochrome P-450 reductase of the human heart, liver and lungs in the presence of (–)-epigallocatechin gallate, quercetin and resveratrol: An *in vitro* study. *Basic Clin Pharmacol Toxicol*, 97: 74–79.

Duhon, D., Bigelow, R.L., Coleman, D.T., et al. 2010. The polyphenol epigallocatechin-3-gallate affects lipid rafts to block activation of the c-Met receptor in prostate cancer cells. *Mol Carcinog*, 49: 739–749.

Ehrlich, M. 2002. DNA methylation in cancer: Too much, but also too little. *Oncogene*, 21: 5400–5413.

Elbling, L., Weiss, R.M., Teufelhofer, O., et al. 2005. Green tea extract and (–)-epigallocatechin-3-gallate, the major tea catechin, exert oxidant but lack antioxidant activities. *FASEB J*, 19: 807–809.

Evangelista, M., Tian, H., et al. 2006. The Hedgehog signaling pathway in cancer. *Clin Cancer Res*, 12: 5924–5928.

Falck, J., Petrini, J.H.J., Williams, B.R., et al. 2002. The DNA damage-dependent intra–S phase checkpoint is regulated by parallel pathways. *Nat Genet*, 30: 290–294.

Fang, M., Wang, Y., Ai, N., et al. 2003. Tea polyphenol (–)-epigallocatechin-3-gallate inhibits DNA methyltransferase and reactivates methylation-silenced genes in cancer cell lines. *Cancer Res*, 63: 7563–7570.

Fantl, W.J., Johnson, D.E., Williams, L.T. 1993. Signalling by receptor tyrosine kinases. *Ann Rev Biochem*, 62: 453–481.

Fassina, G., Vene, R., Morini, M., et al. 2004. Mechanisms of inhibition of tumor angiogenesis and vascular tumor growth by epigallocatechin-3-gallate. *Clin Cancer Res*, 10: 4865–4873.

Fearon, E.R., and Vogelstein, B. 1990. A genetic model for colorectal tumorigenesis. *Cell*, 61: 759–767.

Feinberg, A.P., and Vogelstein, B. 1983. Hypomethylation distinguishes genes of some human cancers from their normal counterparts. *Nature*, 301: 89–92.

Galleano, M., Verstraeten, S.V., Oteiza, P.I., et al. 2010. Antioxidant actions of flavonoids: Thermodynamic and kinetic analysis. *Arch Biochem Biophys*, 501: 23–30.

Gao, Z., Xu, Z., Hung, M.S., et al. 2009. Promoter demethylation of WIF-1 by epigallocatechin-3-gallate in lung cancer cells. *Anticancer Res*, 29: 2025–2030.

Garbisa, S., Sartor, L., Biggin, S., et al. 2001. Tumor gelatinases and invasion inhibited by the green tea flavanol epigallocatechin-3-gallate. *Cancer*, 91: 822–832.

Glozak, M.A., and Seto, E. 2007. Histone deacetylases and cancer. *Oncogene* 26: 5420–5432.

Gu, B., Ding, Q., Xia, G., et al. 2009a. EGCG inhibits growth and induces apoptosis in renal cell carcinoma through TFPI-2 over expression. *Oncol Rep*, 21: 635–640.

Gu, Q., Hu, C., Chen, Q., et al. 2009b. Development of a rat model by 3,4-benzopyrene intra-pulmonary injection and evaluation of the effect of green tea drinking on p53 and bcl-2 expression in lung carcinoma. *Cancer Detect Prev*, 32: 444–451.

Guo, Q., Zhao, B., Li, M., et al. 1996. Studies on protective mechanisms of four components of green tea polyphenols against lipid peroxidation in synaptosomes. *Biochim Biophys Acta*, 1304: 210–222.

Gupta, S., Hussain, T., and Mukhtar, H. 2003. Molecular pathway for (–)-epigallocatechin-3-gallate-induced cell cycle arrest and apoptosis of human prostate carcinoma cells. *Arch Biochem Biophys*, 410: 177–185.

Harper, C.E., Patel, B.B., Wang, J., et al. 2007. Epigallocatechin-3-gallate suppresses early stage, but not late stage prostate cancer in TRAMP mice: Mechanisms of action. *Prostate*, 67: 1576–1589.

Hassan, H.I., and Walker, R.A. 1998. Decreased apoptosis in non-involved tissue from cancer-containing breasts. *J Pathol*, 184: 258–264.

Hastak, K., Agarwal, M.K., Mukhtar, H., et al. 2005. Ablation of either p21 or Bax prevents p53-dependent apoptosis induced by green tea polyphenol epigallocatechin-3-gallate. *FASEB J*, 19: 789–791.

Herceg, Z., and Hainaut, P. 2007. Genetic and epigenetic alterations as biomarkers for cancer detection, diagnosis and prognosis. *Mol Oncol*, 1: 26–41.

Ho, Y.C., Yang, S.F., Peng, C.Y., et al. 2007. Epigallocatechin-3-gallate inhibits the invasion of human oral cancer cells and decreases the productions of matrix metalloproteinases and urokinase-plasminogen activator. *J Oral Pathol Med*, 36: 588–593.

Holloway, A.F., and Oakford, P.C. 2007. Targeting epigenetic modifiers in cancer. *Curr Med Chem*, 14: 2540–2547.

Hong, J., Lu, H., Meng, X., et al. 2002. Stability, cellular uptake, biotransformation, and efflux of tea polyphenol (–)-epigallocatechin-3-gallate in HT-29 human colon adenocarcinoma cells. *Cancer Res*, 62: 7241–7246.

Hou, Z., Xiao, H., Lambert, J., et al. 2006. Green tea polyphenol, (–)-epigallocatechin-3-gallate, induces oxidative stress and DNA damage in cancer cell lines, xenograft tumors, and mouse liver (Abstract 4896). *AACR Meeting Abstracts*, 47: 1150–1151.

Hsu, S., Lewis, J., Singh, B., et al. 2003. Green tea polyphenol targets the mitochondria in tumor cells inducing caspase 3-dependent apoptosis. *Anticancer Res*, 23: 1533–1539.

Hua, Y., Jianhua, L., Qiuliang, W., et al. 2006. Effects of tea polyphenols on telomerase activity of a tongue cancer cell line: A preliminary study. *Int J Oral Maxillofac Surg*, 35: 352–355.

Huang, C.H., Tsai, S.J., Wang, Y.J., et al. 2009. EGCG inhibits protein synthesis, lipogenesis, and cell cycle progression through activation of AMPK in p53 positive and negative human hepatoma cells. *Mol Nutr Food Res*, 53: 1156–1165.

Jang, E.H., Choi, J.Y., Park, C.S., et al. 2005. Effects of green tea extract administration on the pharmacokinetics of clozapine in rats. *J Pharm Pharmacol*, 57: 311–316.

Jankun, J., Keck, R.W., Skrzypczak-Jankun, E., et al. 1997. Inhibitors of urokinase reduce size of prostate cancer xenografts in severe combined immunodeficient mice. *Cancer Res*, 57: 559–563.

Jia, D., Jurkowska, R.Z., Zhang, X., et al. 2007. Structure of Dnmt3a bound to Dnmt3L suggests a model for *de novo* DNA methylation. *Nature*, 449: 248–251.

Jia, X.D., Han, C., and Chen, J.S. 2004. Effects of tea polyphenols and tea pigments on telomerase activity of HepG2 cells. *Zhonghua Yu Fang Yi Xue Za Zhi*, 38: 159–161.

Jiang, B.H., and Liu, L.Z. 2008. PI3K/PTEN signaling in tumorigenesis and angiogenesis. *Biochim Biophys Acta*, 1784: 150–158.

Jin, H., Tan, X., Liu, X., et al. 2010. The study of effect of tea polyphenols on microsatellite instability colorectal cancer and its molecular mechanism. *Int J Colorectal Dis*, 25:1407–1415.

Jones, P.A., and Baylin, S.B. 2007. The epigenomics of cancer. *Cell*, 128: 683–692.

Kato, K., Long, N.K., Makita, H., et al. 2008. Effects of green tea polyphenol on methylation status of RECK gene and cancer cell invasion in oral squamous cell carcinoma cells. *Br J Cancer*, 99: 647–654.

Kim, H.S., Kim, M.H., Jeong, M., et al. 2004. EGCG blocks tumor promoter-induced MMP-9 expression via suppression of MAPK and AP-1 activation in human gastric AGS cells. *Anticancer Res*, 24: 747–753.

Kim, J., Zhang, X., Rieger-Christ, K.M., et al. 2006. Suppression of Wnt signaling by the green tea compound (–)-epigallocatechin 3-gallate (EGCG) in invasive breast cancer cells. Requirement of the transcriptional repressor HBP1. *JBC*, 281: 10865–10875.

Kim, M., Murakami, A., Kawabata, K., et al. 2005. (–)-Epigallocatechin-3-gallate promotes pro-matrix metalloproteinase-7 production via activation of the JNK1/2 pathway in HT-29 human colorectal cancer cells. *Carcinogenesis*, 26:1553–1562.

Kim, M.J., Kim, H.I., Chung, J., et al. 2009. (–)-Epigallocatechin-3-gallate (EGCG) increases the viability of serum-starved A549 cells through its effect on Akt. *Am J Chin Med*, 37:723–734.

Klumpp, S., and Krieglstein, J. 2002. Serine/threonine protein phosphatases in apoptosis. *Curr Opin Pharmacol*, 2: 458–462.

Klumpp, S., Maurer, A., and Zhu, Y. 2004. Protein kinase CK2 phosphorylates BAD at threonine-117. *Neurochem Int*, 45: 747–752.

Kostyuk, V.A., Potapovich, A.I., Strigunova, E.N., et al. 2004. Experimental evidence that flavonoid metal complexes may act as mimics of superoxide dismutase. *Arch Biochem Biophys*, 428: 204–208.

Kuzuhara, T., Sei, Y., Yamaguchi, K., et al. 2006. DNA and RNA as new binding targets of green tea catechins. *J Biol Chem*, 81: 17446–17456.

Kuzuhara, T., Tanabe, A., Sei, Y., et al. 2007. Synergistic effects of multiple treatments, and both DNA and RNA direct bindings on, green tea catechins. *Mol Carcinog*, 46: 640–645.

Laird, P.W. 2005. Cancer epigenetics. *Hum Mol Genet*, 14: R65–R76.

Lam, W.H., Kazi, A., Kuhn, D.J., et al. 2004. A potential prodrug for a green tea polyphenol proteasome inhibitor: Evaluation of the peracetate ester of (–)-epi-gallocatechin gallate [(–)-EGCG]. *Bioorg Med Chem*, 12: 5587–5593.

Landis-Piwowar, K.R., Milacic, V., and Dou, Q.P. 2008. Relationship between the methylation status of dietary flavonoids and their growth-inhibitory and apoptosis-inducing activities in human cancer cells. *Cell Biochem*, 105: 514–523.

Larsen, C.A., and Dashwood, R.H. 2009. Suppression of Met activation in human colon cancer cells treated with (–)-epigallocatechin-3-gallate: Minor role of hydrogen peroxide. *Biochem Biophys Res Commun*, 389: 527–530.

Lee, W.J., Shim, J.Y., and Zhu, B.T. 2005a. Mechanisms for the inhibition of DNA methyltransferases by tea catechins and bioflavonoids. *Mol Pharmacol*, 68: 1018–1030.

Lee, Y.I., Seo, M., and Kim, Y. 2005b. Membrane depolarization induces the undulating phosphorylation/dephosphorylation of glycogen synthase kinase 3beta, and this dephosphorylation involves protein phosphatases 2A and 2B in SH-SY5Y human neuroblastoma cells. *J Biol Chem*, 280: 22044–22052.

Leslie, N.R., Bennett, D., and Lindsay, Y.E. 2003. Redox regulation of PI 3-kinase signaling via inactivation of PTEN. *EMBO J*, 22: 5501–5510.

Li, W.G., Li, Q.H., and Tan, Z. 2005. Epigallocatechin gallate induces telomere fragmentation in HeLa and 293 but not in MRC-5 cells. *Life Sci*, 76: 1735–1746.

Li, Y., and Cozzi, P.J. 2007. Targeting uPA/uPAR in prostate cancer. *Cancer Treat Rev*, 33: 521–527.

Liang, Y.C., Lin-shiau, S.Y., Chen, C.F., et al. 1997. Suppression of extracellular signals and cell proliferation through EGF receptor binding by (–)-epigallocatechin gallate in human A431 epidermoid carcinoma cells. *J Cell Biochem*, 67: 55–65.

Lim, Y.C., Park, H.Y., Hwang, H.S., et al. 2008. (–)-Epigallocatechin-3-gallate (EGCG) inhibits HGF-induced invasion and metastasis in hypopharyngeal carcinoma cells. *Cancer Lett*, 271: 140–152.

Lin, S.C., Li, W.C., Shih, J.W., et al. 2006. The tea polyphenols EGCG and EGC repress mRNA expression of human telomerase reverse transcriptase (hTERT) in carcinoma cells. *Cancer Lett*, 236: 80–88.

Linseman, D.A., Butts, B.D., Precht, T.A., et al. 2004. Glycogen synthase kinase-3beta phosphorylates Bax and promotes its mitochondrial localization during neuronal apoptosis. *J Neurosci*, 24: 9993–10002.

Liu, L., Lai, C.Q., Nie, L., et al. 2008. The modulation of endothelial cell gene expression by green tea polyphenol-EGCG. *Mol Nutr Food Res*, 52:1182–1192.

Luo, K.L., Luo, J.H., and Yu, Y.P. 2009. (–)-Epigallocatechin-3-gallate induces Du145 prostate cancer cell death via downregulation of inhibitor of DNA binding 2, a dominant negative helix-loop-helix protein. *Cancer Sci*, 101: 707–712.

Maeda-Yamamoto, M., Kawahara, H., Tahara, N., et al. 1999. Effects of tea polyphenols on the invasion and matrix metalloproteinases activities of human fibrosarcoma HT1080 cells. *J Agric Food Chem*, 47: 2350–2354.

Manna, S., Mukherjee, S., Roy, A., et al. 2009. Tea polyphenols can restrict benzo[a] pyrene-induced lung carcinogenesis by altered expression of p53-associated genes and H-ras, c-myc and cyclin D1. *J Nutr Biochem*, 20: 337–349.

Masamune, A.K., Kikuta, M., Satoh, N., et al. 2005. Green tea polyphenol epigallocatechin-3-gallate blocks PDGF-induced proliferation and migration of rat pancreatic stellate cells. *World J Gastroenterol*, 11: 3368–3374.

Masters, S.C., Yang, H., Datta, S.R., Greenberg M.E., and Fu, H. 2001. 14-3-3 inhibits Bad-induced cell death through interaction with serine-136. *Mol Pharmacol*, 60: 1325–1331.

Masuda, M., Suzui, M., and Lim, J.T. 2003. Epigallocatechin-3-gallate inhibits activation of HER-2/neu and downstream signaling pathways in human head and neck and breast carcinoma cells. *Clin Cancer Res*, 9: 3486–3491.

Masuda, M., Suzui, M., Lim, J.T., et al. 2002. Epigallocatechin-3-gallate decreases VEGF production in head and neck and breast carcinoma cells by inhibiting EGFR-related pathways of signal transduction. *J Exp Ther Oncol*, 2: 350–359.

Masuda, M., Suzui, M., and Weinstein, I.B. 2001. Effects of epigallocatechin-3-gallate on growth, epidermal growth factor receptor signaling pathways, gene expression, and chemosensitivity in human head and neck squamous cell carcinoma cell lines. *Clin Cancer Res*, 7: 4220–4229.

Maurer, U., Charvet, C., Wagman, A.S., et al. 2006. Glycogen synthase kinase-3 regulates mitochondrial outer membrane permeabilization and apoptosis by destabilization of MCL-1. *Mol Cell*, 21:749–760.

McAllister, S.S., and Weinberg, R.A. 2010. Tumor-host interactions: A far-reaching relationship. *J Clin Oncol*, 28: 4022–4028.

McLarty, J., Bigelow, R.L., and Smith, M. 2009. Tea polyphenols decrease serum levels of prostate-specific antigen, hepatocyte growth factor, and vascular endothelial growth factor in prostate cancer patients and inhibit production of hepatocyte growth factor and vascular endothelial growth factor *in vitro*. *Cancer Prev Res (Phila)*, 2: 673–682.

Morel, I., Abalea, V., Sergent, O., et al. 1998. Involvement of phenoxyl radical intermediates in lipid antioxidant action of myricetin in iron-treated rat hepatocyte culture. *J Cillard Biochem Pharmacol*, 55:1399–1404.

Mount, J.G., Muzylak, M., Allen, S., et al. 2006. Evidence that the canonical Wnt signaling pathway regulates deer antler regeneration. *Dev Dyn*, 235: 1390–1399.

Mukhtar, H., Wang, Z.Y., Katiyar, S.K., et al. 1992. Tea components: Antimutagenic and anticarcinogenic effects. *Prev Med*, 21: 351–360.

Murugan, R.S., Vinothini, G., Hara, Y., et al. 2009. Black tea polyphenols target matrix metalloproteinases, RECK, proangiogenic molecules and histone deacetylase in a rat hepatocarcinogenesis model. *Anticancer Res*, 29: 2301–2305.

Muto, S., Fujita, K., Yamazaki, Y., et al. 2001. Inhibition by green tea catechins of metabolic activation of procarcinogens by human cytochrome P450. *Mut Res*, 479: 197–206.

Myzak, M., and Dashwood, R.H. 2006. Histone deacetylases as targets for dietary cancer preventive agents: Lessons learned with butyrate, diallyl disulfide, and sulforaphane. *Curr Drug Targets*, 7: 443–452.

Naasani, I., Oh-hashi, F., Oh-hara, T., et al. 2003. Blocking telomerase by dietary polyphenols is a major mechanism for limiting the growth of human cancer cells *in vitro* and *in vivo*. *Cancer Res*, 63: 824–830.

Nakagawa, H., Hasumi, K., Woo, J.T., et al. 2004. Generation of hydrogen peroxide primarily contributes to the induction of Fe(II)-dependent apoptosis in Jurkat cells by (–)-epigallocatechin gallate. *Carcinogenesis*, 25:1567–1574.

Nakagawa, K., and Miyazawa, T. 1997. Chemiluminescence–high-performance liquid chromatographic determination of tea catechin, (–)-epigallocatechin 3-gallate, at picomole levels in rat and human plasma. *Anal Biochem*, 248: 41–49.

Navarro-Perán, E., Cabezas-Herrera, J., Campo, L.S., et al. 2007. Effects of folate cycle disruption by the green tea polyphenol epigallocatechin-3-gallate. *Int J Biochem Cell Biol*, 39: 2215–2225.

Nie, G., Jin, C., Cao, Y., et al. 2002. Distinct effects of tea catechins on 6-hydroxy-dopamine-induced apoptosis in PC12 cells. *Arch Biochem Biophys*, 397: 84–90.

Nishikawa, T., Nakajima, T., Moriguchi, M., et al. 2006. A green tea polyphenol, epigalocatechin-3-gallate, induces apoptosis of human hepatocellular carcinoma, possibly through inhibition of Bcl-2 family proteins. *J Hepatol*, 44: 1074–1082.

Nishikori, M. 2005. Classical and alternative NF-κB activation pathways and their roles in lymphoid malignancies. *J Clin Exp Hematopathol*, 45: 15–24.

Noguchi, M., Yokoyama, M., Watanabe, S., et al. 2006. Inhibitory effect of the tea polyphenol, (–)-epigallocatechin gallate, on growth of cervical adenocarcinoma cell lines. *Cancer Lett*, 234: 135–142.

Pahlke, G., Ngiewih, Y., Kern, M., et al. 2006. Impact of quercetin and EGCG on key elements of the Wnt pathway in human colon carcinoma cells. *J Agric Food Chem*, 54: 7075–7082.

Pandey, M., Shukla, S., and Gupta, S. 2009. Promoter demethylation and chromatin remodeling by green tea polyphenols leads to re-expression of GSTP1 in human prostate cancer cells. *Int J Cancer*, 126: 2520–2533.

Perron, N.R., and Brumaghim, J.L. 2009. A review of the antioxidant mechanisms of polyphenol compounds related to iron binding. *Cell Biochem Biophys*, 53: 75–100.

Pogribny, I.P., Ross, S.A., Wise, C., et al. 2006. Irreversible global DNA hypomethylation as a key step in hepatocarcinogenesis induced by dietary methyl deficiency. *Mut Res*, 593: 80–87.

Priulla, M., Calastretti, A., and Bruno, P. 2007. Preferential chemosensitization of PTEN-mutated prostate cells by silencing the Akt kinase. *Prostate*, 67: 782–789.

Qanungo, S., Das, M., Haldar, S., et al. 2005. Epigallocatechin-3-gallate induces mitochondrial membrane depolarization and caspase-dependent apoptosis in pancreatic cancer cells. *Carcinogenesis*, 26: 958–967.

Qin, J., Xie, L.P., Zheng, X.Y., et al. 2007. A component of green tea, (–)-epigallocatechin-3-gallate, promotes apoptosis in T24 human bladder cancer cells via modulation of the PI3K/Akt pathway and Bcl-2 family proteins. *Biochem Biophys Res Commun*, 354: 852–857.

Ran, Z.H., Xu, Q., Tong, J.L., et al. 2007. Apoptotic effect of epigallocatechin-3-gallate on the human gastric cancer cell line MKN45 via activation of the mitochondrial pathway. *World J Gastroenterol*, 13: 4255–4259.

Ran, Z.H., Zou, J., and Xiao, S.D. 2005. Experimental study on anti-neoplastic activity of epigallocatechin-3-gallate to digestive tract carcinomas. *Chin Med J (Engl)*, 118: 1330–1337.

Rice-Evans, C. 1999. Implications of the mechanisms of action of tea polyphenols as antioxidants *in vitro* for chemoprevention in humans. *Proc Soc Exp Biol Med*, 220: 262–266.

Ross, S.A. 2003. Diet and DNA methylation interactions in cancer prevention. *Ann NY Acad Sci*, 983: 197–207.

Roy, A.M., Baliga, M.S., and Katiyar, S.K. 2005. Epigallocatechin-3-gallate induces apoptosis in estrogen receptor negative human breast carcinoma cells via modulation in protein expression of p53 and Bax and caspase-3 activation. *Mol Cancer Ther*, 4: 81–90.

Roy, P., Nigam, N., George, J., Srivastava S., Shukla Y., et al. 2009. Induction of apoptosis by tea polyphenols mediated through mitochondrial cell death pathway in mouse skin tumors. *Cancer Biol Ther*, 8:1281–1287.

Sachinidis, A., Seul, C., Seewald, S., et al. 2000. Green tea compounds inhibit tyrosine phosphorylation of PDGF beta-receptor and transformation of A172 human glioblastoma. *FEBS*, 471: 51–55.

Sah, J.F., Balasubramanian, S., Eckert, R.L., et al. 2004. Epigallocatechin-3-gallate inhibits epidermal growth factor receptor signaling pathway. Evidence for direct inhibition of ERK1/2 and AKT kinases. *J Biol Chem*, 279: 12755–12762.

Sakata, R., Ueno, T., Nakamura, T., et al. 2004. Green tea polyphenol epigallo-catechin-3-gallate inhibits platelet-derived growth factor-induced proliferation of human hepatic stellate cell line LI90. *J Hepatol*, 40: 52–59.

Sang, S., Yang, I., Buckley, B., et al. 2007. Autoxidative quinine formation *in vitro* and metabolite formation *in vivo* from tea polyphenol (–)-epigallocatechin-3-gallate: Studied by real-time mass spectrometry combined with tandem mass ion mapping. *Free Radic Biol Med*, 43:362–371.

Sang, S.M., Lee, M.J., Hou, Z., et al. 2005. Stability of tea polyphenol (–)-epi-gallocatechin-3-gallate and formation of dimers and epimers under common experimental conditions. *J Agric Food Chem*, 53: 9478–9484.

Sang, S.M., Tian, S., Meng, X., et al. 2002. Theadibenzotropolone A, a new type pigment from enzymatic oxidation of (–)-epicatechin and (–)-epigallocatechin gallate and characterized from black tea using LC/MS/MS. *Tetrahedron Lett*, 43: 7129–7133.

Sartor, L., Pezzato, E., Dona, M., et al. 2004. Prostate carcinoma and green tea: (–) Epigallocatechin-3-gallate inhibits inflammation-triggered MMP-2 activation and invasion in murine TRAMP model. *Int J Cancer*, 112: 823–829.

Sen, P., Mukherjee, S., Ray, D., et al. 2003. Involvement of the Akt/PKB signaling pathway with disease processes. *Mol Cell Biochem*, 253: 241–246.

Sen, T., Moulik, S., Dutta, A., et al. 2009. Multifunctional effect of epigallocate-chin-3-gallate (EGCG) in downregulation of gelatinase-A (MMP-2) in human breast cancer cell line MCF-7. *Life Sci*, 84: 194–204.

Shammas, M.A., Neri, P., Koley, H., et al. 2006. Specific killing of multiple myeloma cells by (–)-epigallocatechin-3-gallate extracted from green tea: Biologic activity and therapeutic implications. *Blood*, 108: 2804–2810.

Shankar, S., Suthakar, G., and Srivastava, R.K. 2007. Epigallocatechin-3-gallate inhibits cell cycle and induces apoptosis in pancreatic cancer. *Front Biosci*, 12: 5039–5051.

Sharma, G., Mirza, S., Yang, Y.H., et al. 2009. Prognostic relevance of promoter hypermethylation of multiple genes in breast cancer patients. *Cell Oncol*, 31: 487–500.

Shimizu, M., Deguchi, A., Hara, Y., et al. 2005a. EGCG inhibits activation of the insulin-like growth factor-1 receptor in human colon cancer cells. *Biochem Biophys Res Commun*, 334: 947–953.

Shimizu, M., Deguchi, A., Joe, A.K., et al. 2005b. EGCG inhibits activation of HER3 and expression of cyclooxygenase-2 in human colon cancer cells. *J Exp Ther Oncol*, 5: 69–78.

Shimizu, M., Deguchi, A., Lim, J.T., et al. 2005c. (–)-Epigallocatechin gallate and Polyphenon E inhibit growth and activation of the epidermal growth factor receptor and human epidermal growth factor receptor-2 signaling pathways in human colon cancer cells. *Clin Cancer Res*, 11: 2735–2746.

Shimizu, M., Shirakami, Y., Sakai, H., et al. 2008. EGCG inhibits activation of the insulin-like growth factor (IGF)/IGF-1 receptor axis in human hepatocellular carcinoma cells. *Cancer Lett*, 262: 10–18.

Siddiqui, I.A., Adhami, V.M., Afaq, F., et al. 2004. Modulation of phosphatidylino-sitol-3-kinase/protein kinase B- and mitogen-activated protein kinase-pathways by tea polyphenols in human prostate cancer cells. *J Cell Biochem*, 91: 232–242.

Siddiqui, I.A., Adhami, V.M., Bharali, D.J., et al, 2009. Introducing nanochemo-prevention as a novel approach for cancer control: Proof of principle with green tea polyphenol epigallocatechin-3-gallate. *Cancer Res*, 69: 1712–1716.

Siddiqui, I.A., Malik, A., Adhami, V.M., et al. 2008. Green tea polyphenol EGCG sensitizes human prostate carcinoma LNCaP cells to TRAIL-mediated apoptosis and synergistically inhibits biomarkers associated with angiogenesis and metastasis. *Oncogene*, 27: 2055–2063.

Siegelin, M.D., Habel, A., and Gaiser, T. 2008. Epigalocatechin-3-gallate (EGCG) down regulates PEA15 and thereby augments TRAIL-mediated apoptosis in malignant glioma. *Neurosci Lett*, 448: 161–165.

Slusarz, A., Shenouda, N.S., Sakla, M.S., et al. 2010. Common botanical compounds inhibit the Hedgehog signaling pathway in prostate cancer. *Cancer Res*, 70: 3382–3390.

Stein, C.J., and Colditz, G.A. 2004. Modifiable risk factors for cancer. *Br J Cancer*, 90: 299–303.

Stracke, M.L., and Liotta, L.A. 1992. Multi-step cascade of tumor cell metastasis. *In Vivo*, 6: 309–316.

Suh, K.S., Chon, S., Oh, S., et al. 2010. Prooxidative effects of green tea polyphenol (–)-epigallocatechin-3-gallate on the HIT-T15 pancreatic beta cell line. *Cell Biol Toxicol*, 26: 189–199.

Tachibana, H. 2009. Molecular basis for cancer chemoprevention by green tea polyphenol EGCG. *Forum Nutr*, 61: 156–169.

Tang, G.Q., Yan, T.Q. Guo, W., et al. 2010. (–)-Epigallocatechin-3-gallate induces apoptosis and suppresses proliferation by inhibiting the human Indian Hedgehog pathway in human chondrosarcoma cells. *J Cancer Res Clin Oncol*, 136: 1179–1185.

Thakur, V.S., Ruhul Amin, A.R., Paul, R.K. et al. 2010. p53-Dependent p21-mediated growth arrest pre-empts and protects HCT116 cells from PUMA-mediated apoptosis induced by EGCG. *Cancer Lett*, 296: 225–232.

Thangapazham, R.L., Passi, N., and Maheshwari, R.K. 2007. Green tea polyphenol and epigallocatechin gallate induce apoptosis and inhibit invasion in human breast cancer cells. *Cancer Biol Ther*, 6:1938–1943.

Theunissen, J.W., and de Sauvage, F.J. 2009. Paracrine Hedgehog signaling in cancer. *Cancer Res*, 69: 6007–6010.

Tonks, N.K. 2006. Protein tyrosine phosphatases: From genes, to function, to disease. *Nat Rev Mol Cell Biol*, 7: 833–846.

Tsang, W., and Kwok, T.T. 2010. Epigallocatechin gallate up-regulation of miR-16 and induction of apoptosis in human cancer cells. *J Nutr Biochem*, 21: 140–146.

Valcic, S., Burr, J.A., Timmermann, B.N., et al. 2000. Antioxidant chemistry of green tea catechins. New oxidation products of (–)-epigallocatechin gallate and (–)-epigallocatechin from their reactions with peroxyl radicals. *Chem Res Toxicol*, 13: 801–810.

van Acker, S.A., van den Berg, D.J., Tromp, D.H., et al. 1996. Structural aspects of antioxidant activity of flavonoids. *Free Radic Biol Med*, 20: 331–342.

Van der Heide, L.P., Hoekman, M.F.M., and Smidt, M.P. 2004. The ins and outs of FoxO shuttling: Mechanisms of FoxO translocation and transcriptional regulation. *Biochem J*, 380: 297–309.

Vayalil, P.K., and Katiyar, S.K. 2004. Treatment of epigallocatechin-3-gallate inhibits matrix metalloproteinases-2 and -9 via inhibition of activation of mitogen-activated protein kinases, c-jun and NF-kappaB in human prostate carcinoma DU-145 cells. *Prostate*, 59: 33–42.

Vermeulen, K., Van Bockstaele, D.R., and Berneman, Z.N. 2003. The cell cycle: A review of regulation, deregulation and therapeutic targets in cancer. *Cell Prolif,* 36: 131–149.

Vittal, R., Selvanayagam, Z.E., Sun, Y., et al. 2004. Gene expression changes induced by green tea polyphenol (–)-epigallocatechin-3-gallate in human bronchial epithelial 21BES cells analyzed by DNA microarray. *Mol Cancer Ther,* 3: 1091–1099.

Volate, S.R., Muga, S.J., Issa, A.Y., et al. 2009. Epigenetic modulation of the retinoid X receptor alpha by green tea in the azoxymethane-Apc Min/+ mouse model of intestinal cancer. *Mol Carcinog,* 48: 920–933.

Wang, X., Wang, R., Hao, M.W., et al. 2008. The BH3-only protein PUMA is involved in green tea polyphenol-induced apoptosis in colorectal cancer cell lines. *Cancer Biol Ther,* 7: 902–908.

Wiseman, S.A., Balentine, D.A., and Frei, B. 1997. Antioxidants in tea. *Crit Rev Food Sci Nutr,* 37: 705–718.

Woodgett, J.R. 2005. Recent advances in the protein kinase B signaling pathway. *Curr Opin Cell Biol,* 17:150–157.

Wu, P.P., Kuo, S.C., Huang, W.W., et al. 2009. (–)-Epigallocatechin gallate induced apoptosis in human adrenal cancer NCI-H295 cells through caspase-dependent and caspase-independent pathway. *Anticancer Res,* 29: 1435–1442.

Wu, Y., and Zhou, B.P. 2008. New insights of epithelial-mesenchymal transition in cancer metastasis. *Acta Biochim Biophys Sin (Shanghai),* 40: 643–650.

Wu, Y., Zu, K., and Warren, M.A. 2006. Delineating the mechanism by which selenium deactivates Akt in prostate cancer cells. *Mol Cancer Ther,* 5: 246–252.

Xiao, G.S., Jin, Y.S., Lu, Q.Y., et al. 2007. Annexin-I as a potential target for green tea extract induced actin remodeling. *Int J Cancer,* 120: 111–120.

Yamakawa, S., Asai, T., Uchida, T., et al. 2004. (–)-Epigallocatechin gallate inhibits membrane type 1 matrix metalloproteinase, MT1-MMP, and tumor angiogenesis. *Cancer Lett,* 210: 47–55.

Yang, G.Y., Liao, J., Li, C., et al. 2000. Effect of black and green tea polyphenols on c-jun phosphorylation and $H(2)O(2)$ production in transformed and non-transformed human bronchial cell lines: Possible mechanisms of cell growth inhibition and apoptosis induction. *Carcinogenesis,* 21: 2035–2039.

Yang, G.Y., Liao, J., Kim, K., et al. 1998. Inhibition of growth and induction of apoptosis in human cancer cell lines by tea polyphenols. *Carcinogenesis,* 19: 611–616.

Yokoyama, S., Hirano, H., Wakimaru, N., et al. 2001. Inhibitory effect of epigallocatechin-gallate on brain tumor cell lines *in vitro. Neuro Oncol,* 3: 22–28.

Yuasa, Y., Nagasaki, H., Akiyama, Y., et al. 2005. Relationship between CDX2 gene methylation and dietary factors in gastric cancer patients. *Carcinogenesis,* 26: 193–200.

Zaveri, N.T. 2006. Green tea and its polyphenolic catechins: Medicinal uses in cancer and noncancer applications. *Life Sci,* 78: 2073–2080.

Zhang, Q.X., Tang, Q., Lu, Z., et al. 2006. Green tea extract and (–)-epigallocatechin-3-gallate inhibit hypoxia- and serum-induced HIF-1alpha protein accumulation and VEGF expression in human cervical carcinoma and hepatoma cells. *Mol Cancer,* 5: 1227–1238.

6

Green Tea Prevents Ultraviolet Radiation-Induced Skin Cancer through Rapid Repair of DNA Damage

Santosh K. Katiyar

Contents

6.1   Introduction

Cutaneous malignancies, including melanoma and nonmelanoma, represent a major public health problem, as the incidence of skin cancer is equivalent to the incidence of cancers in all other organs combined (Housman et al., 2003). It has been estimated that the annual cost of treating nonmelanoma and melanoma skin cancers in the United States is in the billions of dollars (www.cancer.org/statistics). The overexposure of the skin to solar ultraviolet (UV) radiation is a major etiologic factor for initiation of this disease. The depletion of the ozone layer that allows more solar UV radiation to reach the surface of the earth, the continuing increase in life expectancy, and the changing dietary habits and lifestyle appear to be contributing factors for the increasing risk of cutaneous malignancies. Effective chemopreventive and chemotherapeutic agents and strategies are urgently needed to address this health problem. One such strategy includes the regular consumption of those dietary phytochemicals that possess anticarcinogenic properties, and that can protect the skin from the risk of cutaneous malignancies. Naturally occurring phytochemicals, specifically phenolics, are widely distributed in plant foods, including fruits, vegetables, seeds, nuts, flowers, and bark. Important dietary sources of polyphenols, which are extensively studied, include grape seeds, grape skin, tea, apples, red wine, onions, and cacao (Nichols and Katiyar, 2010). These polyphenols possess beneficial health effects of dietary sources. Among others, green tea is better known for its health benefits in many organs, as well as in general health. In this chapter, we will particularly discuss the prevention of UV-induced skin cancer by green tea polyphenols, with particular emphasis on their DNA repair abilities in UV-exposed skin.

6.2   Skin and Skin Cells

The skin is the largest organ of the body, comprising a surface area of approximately 1.5 to 2.0 m^2, which protects the internal organs of the body against the detrimental effects of the external environment. Thus, skin provides a protective covering at this crucial interface between the inside and outside, and is a physical barrier between the external environment and internal tissues. The skin is made up of primarily the epidermis and dermis. The major cell type of the epidermis is the keratinocytes. Keratinocytes comprise >90% of the cells of the epidermal layer. Other cell types are Langerhans cells, melanocytes, and $\gamma\delta$ T-cells. The dermal components of the skin, including dermal fibroblast, microvasculature endothelial cells, dermal dendritic cells, mast cells, and resident perivascular T-cells, also participate in the cutaneous functions. These epidermal and dermal components have evolved into a dynamic network of interacting cells capable of sensing a variety of perturbations, including trauma, ultraviolet irradiation, toxic environmental chemicals, and pathogenic microorganisms in the cutaneous environment. Among many environmental and xenobiotic factors, exposure of the skin to solar UV radiation is the key factor in the initiation of various skin disorders, such as wrinkling, hypopigmentation and hyperpigmentation, and skin cancer (de Gruijl and van der Leun, 1994; Ichihashi et al., 2003; Mukhtar and Elmets, 1996). Statistically, the average annual UV dose that an average American typically receives in a year is about 2500 to 3300 mJ/cm^2 (Godar, 2001; Godar et al., 2001).

6.3 Solar UV Spectrum

It has been recognized that several environmental and genetic factors contribute to the development of skin diseases; however, the chronic exposure of the skin to solar UV radiation is a major etiological factor for initiation of cutaneous malignancies. The solar UV spectrum can be divided into three segments based on the wavelengths of the radiation as detailed below.

6.3.1 UVA (320 to 400 nm), Long-Wave Spectrum

This UVA comprises the largest spectrum of solar UV radiation (90 to 95%) and is considered the "aging ray." UVA can penetrate deeper into the epidermis and dermis of the skin. It can penetrate the skin to a depth of approximately 1000 μm. It has been shown that extensive UVA exposure can lead to benign tumor formation as well as malignant cancers (Bachelor and Bowden, 2004; Wang et al., 2001). The exposure of the skin to UVA radiation induces the generation of reactive oxygen species, such as singlet oxygen and hydroxyl free radicals, which can cause damage to cellular macromolecules, such as proteins, lipids, and DNA (DiGiovanni, 1992). In contrast to UVC or UVB, UVA is barely able to excite the DNA molecule directly and produces only a small number of pyrimidine dimers in the skin; therefore, it is assumed that much of the mutagenic and carcinogenic action of UVA radiation is mediated through reactive oxygen species (de Gruijl, 2000; Runger, 1999). This, however, is still a matter of debate. UVA is a significant source of oxidative stress in human skin, which causes photoaging in the form of skin sagging rather than wrinkling (Krutmann, 2001) and can suppress some immune functions (Ullrich, 1995).

6.3.2 UVB (280 to 320 nm), Midwave Spectrum

This solar UVB radiation constitutes approximately 5% of the total solar UV radiation and is mainly responsible for nonmelanoma and melanoma skin cancers. UVB radiation can penetrate the skin to a depth of approximately 160 to 180 μm. It can cross the whole epidermis layer and penetrate the dermis compartment of human skin. UVB radiation can induce both direct and indirect adverse biologic effects, including induction of oxidative stress, DNA damage, premature aging of the skin (de Gruijl and van der Leun, 1994; Katiyar, 2003; Mukhtar and Elmets, 1996), and multiple effects on the immune system (Kripke, 1990; Meunier et al., 1998; Parrish, 1983), which together play important roles in the generation and maintenance of UV-induced neoplasms (Hruza and Pentland, 1993; Katiyar and Mukhtar, 2001; Taylor et al., 1990). UVB can act as a tumor initiator (Kligman et al., 1980), tumor promoter (Katiyar et al., 1997) and cocarcinogen (Donawho and Kripke, 1991; Ziegler et al., 1994). Although skin possesses an elaborate defense system consisting of enzymatic and nonenzymatic components to protect the skin from these adverse biological effects, excessive exposure to UV radiation overwhelms and depletes the cutaneous defense system, leading to the development of various skin disorders, including skin cancer (Katiyar and Mukhtar, 2001; Katiyar et al., 1997, 2001).

6.3.3 UVC (200 to 280 nm), Short-Wave Spectrum

This UVC radiation can penetrate the skin to a depth of approximately 60 to 80 μm, and can damage DNA molecules. These wavelengths have enormous energy and are mutagenic in nature. However, UVC radiation is largely absorbed by the atmospheric ozone layer and normally does not reach the surface of the earth.

6.4 Solar UV Radiation: An Environmental Skin Carcinogen

Exposure of the skin to UVB radiation induces a variety of biological effects, including induction of inflammation and inflammatory mediators, generation of oxidative stress, suppression of immune system, and DNA damage. All together, these effects play important roles in the development of skin cancers (Hruza and Pentland, 1993; Katiyar and Mukhtar, 2001; Taylor et al., 1990). Although skin possesses an elaborate defense system, the excessive exposure to UV radiation overwhelms and depletes the cutaneous defense system and its ability to protect the skin from deleterious effects (Katiyar and Mukhtar, 2001; Katiyar et al., 1997; Mittal et al., 2003a). It has been recognized that chronic exposure to solar UV radiation is the major etiologic agent for over 1 million new nonmelanoma skin cancers in the United States each year and is an important factor in the pathogenesis of melanoma and immunosuppression. Nonmelanoma skin cancer is by far the most common cutaneous malignancy. Melanoma develops from melanocytes, a kind of skin cells that contain a pigment called melanin. Depending on the type of cells, nonmelanoma skin cancers are further classified as squamous cell and basal cell carcinomas. Nonmelanoma skin cancers can metastasize, but much less often than melanoma; therefore, mortality is rare. Chronic exposure of the skin to UV radiation leads to the development of actinic keratoses in human skin, and these actinic keratoses can lead to nonmelanoma skin cancers. Most of the mortality occurs due to melanoma because of its ability to metastasize to other organs. For the prevention of skin cancers, efforts can be made to educate people to minimize sun exposure specifically during midday (11:00 A.M. to 4:00 P.M.), at which time the intensity of UV radiation is high, wear protective clothing, including full-sleeve shirt and hat, and use adequate broad-spectrum sunscreen before going outdoors for recreational purposes.

There is ample clinical and experimental evidence to suggest that immune factors contribute to the pathogenesis of solar UV-induced skin cancer in mice and probably in humans as well (Urbach, 1991; Yoshikawa et al., 1990). Chronically immunosuppressed patients living in regions of intense sun exposure experience an exceptionally high rate of skin cancer (Kinlen et al., 1979). This observation is consistent with the hypothesis that immune surveillance is an important mechanism designed to prevent the generation and maintenance of neoplastic cells. Further, the incidence of skin cancers, especially squamous cell carcinoma, is also increased among organ transplant recipients (Cowen and Billingsley, 1999; Fortina et al., 2000; Otley and Pittelkow, 2000). The increased frequencies of squamous cell carcinoma, especially in transplant patients, are presumably attributable to a long-term immunosuppressive therapy (DiGiovanna, 1998); however, nonimmune mechanisms may also play a role (Hojo et al., 1999). These studies provide evidence in support of the concept that UV-induced immune suppression promotes skin cancer risk.

6.5 Green Tea Polyphenols and Cancer

In vitro and *in vivo* studies have shown that green tea polyphenols possess antioxidant, anti-inflammatory, and anticarcinogenic activities (Baliga and Katiyar, 2006; Katiyar et al., 2007). Of these major polyphenols, EGCG is the major component and the most effective molecule. EGCG has been extensively studied in several disease models, including its chemopreventive effects against solar ultraviolet radiation (Katiyar and Elmets, 2001; Katiyar et al., 2007). The polyphenols present in black tea are less well defined and known as thearubigens and theaflavins. However, extensive studies have been conducted with green tea polyphenols (GTPs). It may be because of their better health benefits than those of black tea. Importantly, the chemical composition of GTPs in water is not significantly changed at least for 3 days. Here, we will summarize and discuss the recent developments in the area of photoprotective potential of green tea polyphenols with emphasis on pre-vention of photocarcinogenesis through rapid repair of UVB-induced DNA damage in the skin.

6.6 Green Tea Inhibits UV Radiation-Induced Skin Cancer

Clinical and epidemiological studies suggest that chronic exposure of the skin to solar UV radiation is a major etiological agent for the development of skin cancers (Brash et al., 1991; Miller and Weinstock, 1994; Scotto and Fears, 1978; Strom, 1996). UV-induced nonmelanoma skin cancers, includ-ing basal cell and squamous cell carcinomas, represent the most common neoplasms in humans (Miller and Weinstock, 1994; Strom, 1996; Urbach, 1991). In laboratory studies, a well-known SKH-1 hairless mouse model is used for the determination of the chemopreventive effect of green tea polyphenols against UV-induced skin tumor development. Following standard photocarcinogenesis protocols, it has been found that oral administration of GTPs (a mixture of green tea polyphenols) in the drinking water of mice resulted in significant protection against skin tumor development in terms of tumor incidence, tumor multiplicity, and tumor size in the treated mice compared to non-GTP-treated UVB-irradiated mice (Mantena et al., 2005). Water extract of green tea leaves, which primarily contained a mixture of polyphenolic ingredients, when given as the sole source of drinking water to mice, also afforded significant protection against UVB radiation-induced skin tumorigenesis (Wang et al., 1992a). Wang et al. (1992b) have shown that drinking green tea also promoted partial regression of established skin papillomas in mice. Topical treatment of SKH-1 hairless mouse skin with GTPs or EGCG in a hydrophilic cream-based topical formulation also significantly inhibited UVB-induced skin tumor development in terms of tumor multiplicity and the growth of tumors (Mittal et al., 2003b). To understand the specific molecular targets of GTPs in prevention of UV-induced skin cancer, mechanism-based studies were performed using both *in vitro* cell culture and *in vivo* genetically modified animal models.

6.7 Molecular Mechanism: DNA as a Molecular Target

6.7.1 UV Irradiation Induces DNA Damage in Skin Cells

Exposure of the skin to UV radiation resulted in the formation of DNA photoproducts. The DNA photoproducts are altered DNA structures that activate a cascade of responses, beginning with the initiation of cell cycle arrest and activation of DNA repair mechanisms. The biologically harmful effects associated with UV irradiation are largely the result of errors in DNA repair, which can lead to oncogenic mutations (Timares et al., 2008). UV-induced DNA damage in the form of thymine dimers or cyclobutane pyrimidine dimers (CPDs) is a molecular trigger for the induction of immune suppression and initiation of nonmelanoma skin cancers (Kripke et al., 1992; Yarosh et al., 1992). It has been shown that exposure of the skin to UV radiation results in the formation of CPDs in skin cells (Katiyar et al., 2000a). The CPDs form immediately after the interaction of photons with the DNA molecule. It has been found that UV exposure of less than one minimal erythema dose is sufficient to damage the DNA in target cells. The CPD formation depends on the penetration ability of the UV radiation inside the skin. The DNA of the majority of epidermal cells got damaged after UVB irradiation. Thus, inhibition of DNA photodamage or its rapid repair may be a mechanism by which UV-induced immunosuppression and skin tumor development can be prevented.

6.7.2 GTPs Stimulate Repair of UVB-Induced DNA Damage: Role of Interleukin-12

Investigations by Katiyar et al. (2000b) revealed that topical treatment of human skin with GTPs prior to UV exposure resulted in a dose-dependent inhibition of CPD formation. Topical treatment of mouse skin with GTPs significantly inhibited UVB-induced DNA damage as assessed using a ^{32}P-postlabeling technique (Chatterjee et al., 1996). Camouse et al. (2009) found that topical application of green tea or white tea extracts provided human skin protection from solar-simulated ultraviolet light. These tea extracts were shown to provide protection against the detrimental effects of UV light on cutaneous immunity. This study also concluded that these protective effects were not due to direct UV absorption or sunscreen effects. Investigations on the effects of green tea polyphenols on the DNA repair kinetics and repair mechanisms of UV-induced CPDs have been carried out using *in vitro* cell culture and *in vivo* animal models as well as human skin. Studies showed that topical treatment of skin with EGCG does not prevent UVB-induced formation of CPDs immediately after UVB irradiation, which indicated that EGCG does not have a significant sunscreen effect. However, in skin samples obtained at 24 or 48 h after UVB exposure, the numbers of CPD-positive cells were significantly reduced (or repaired) in the EGCG-treated mouse skin compared to the mice that were not treated with EGCG (Meeran et al., 2006a,b). Studies of the DNA repair mechanisms suggested that the rapid repair of UV-induced CPDs by EGCG was mediated through stimulation of interleukin-12 (IL-12) on application of the EGCG onto the mouse skin (Meeran et al., 2006a,b). IL-12 is a 70 kDa heterodimeric cytokine composed of two disulfide-bonded protein chains, the p40 and p35 subunits (Katiyar, 2007b; Wolf et al., 1991). IL-12 has been shown to possess antitumor activity in a wide variety of murine tumor models (Brunda et al., 1993; Colombo et al., 1996; Robertson and Ritz, 1996), and also has been shown to have the capacity to induce DNA repair (Meeran et al., 2006b,c; Schwarz et al., 2002, 2005). This concept was verified by testing the effect of EGCG on UV-induced CPD formation in an IL-12 knockout mouse model. EGCG does not remove or repair UV-induced CPDs in the skin of IL-12 knockout mice but repaired in their wild-type counterparts, further confirming the role of IL-12 in repair of DNA damage by this polyphenol (Meeran et al., 2006a,b). Studies on the effects of oral administration of GTPs in the drinking water of mice with UVB-induced DNA damage were also carried out, and it was found that UV-induced CPDs were resolved rapidly in the GTP-treated mice compared to non-GTP-treated mice (Meeran et al., 2009). Further, the DNA repairing effect of GTPs was less pronounced in the skin of IL-12 knockout mice, as was observed in the case of EGCG treatment. Schwarz et al. (2008) observed that treatment of normal human keratinocytes and human skin equivalent with GTPs reduced UVB-induced DNA damage in the form of CPDs, and that this effect was mediated through the stimulation of IL-12 production. These investigations suggest that the difference in the GTP-associated DNA repair capacity between IL-12 knockout mice and their wild-type counterparts may be due to the absence of IL-12 in the IL-12 knockout mice. Zhao et al. (1999) demonstrated that application of green tea extract to Epiderm, a reconstituted human skin equivalent, also inhibited psoralen-UVA-induced formation of 8-methoxypsoralen-DNA adducts (Zhao et al., 1999). Wei et al. (1998) have found that water extract of green tea scavenges H_2O_2 and inhibits UV-induced oxidative DNA damage in an *in vitro* system. Treatment of skin with a green tea extract significantly inhibited DNA damage induced by solar simulator radiation when assessed using a ^{32}P-postlabeling technique (Chatterjee et al., 1996). These observations demonstrate the potential chemopreventive effects of green tea polyphenols against UVB-induced DNA damage in the skin cells.

6.7.3 GTPs Enhance the Levels of Nucleotide Excision Repair Genes

Further studies were conducted in the research laboratory of Dr. Katiyar to identify the DNA repair mechanism in UVB-exposed skin by green tea polyphenols. It was proposed that the nuclear excision repair (NER) mechanism is involved in the repair of UVB-induced DNA damage in the form of CPDs by green tea polyphenols, and that IL-12 has a role in this repairing process (Meeran et al., 2006a, 2006b). To determine whether the NER mechanism is required for the EGCG-induced IL-12-mediated repair of UVB-induced CPDs, NER-deficient human fibroblasts from a person suffering from xeroderma pigmentosum complementation group A (*XPA*) and NER-proficient fibroblasts from a healthy person (*XPA* proficient) were exposed to UVB with or without prior treatment with EGCG. The data analysis revealed that the numbers of CPD-positive cells were significantly lower in the EGCG-treated group at 24 h after UVB exposure in the *XPA*-proficient cells than in the non-EGCG-treated cells. In contrast, EGCG did not significantly remove or repair UVB-induced CPDs in *XPA*-deficient cells. Similar studies were also conducted with GTPs using *XPA*-proficient and *XPA*-deficient human fibroblasts. Treatment of human fibroblasts with GTPs was able to repair UVB-induced DNA damage in *XPA*-proficient fibroblasts but did not repair it in *XPA*-deficient fibroblasts (Katiyar et al., 2010). These *in vitro* observations indicated that EGCG- or GTP-induced DNA repair in the form of CPDs is mediated through a functional NER mechanism.

6.7.4 Stimulation of DNA Repair by GTPs Leads to a Reduction in Skin Inflammation

Skin exposure to UVB resulted in inflammation, and there is increasing evidence that chronic and sustained inflammation promotes the initiation of various skin diseases, including the development of skin cancers (Hussein, 2005; Mukhtar and Elmets, 1996). Both UV-induced inflammatory responses and UV-induced skin tumor development are causally related to UV-induced DNA damage. Therefore, it was of interest to explore the effects of green tea polyphenols on DNA repair and their relationship with inflammatory responses in the UVB-exposed skin. It is known that CPDs are formed immediately after the exposure of the skin to UV radiation, and inflammation develops at later stages. Following UV exposure, UV-induced DNA damage in the form of CPDs was repaired or removed more rapidly in the skin of mice that had been treated with either topical application of EGCG or administration of GTPs in drinking water of mice. Subsequently, the levels of UVB-induced inflammation were lower in the skin of treated mice than the nontreated mice. The levels of inflammation in the mouse skin were assessed through analysis of standard biomarkers of inflammation, such as cyclooxygenase-2 (COX-2) expression, prostaglandin E2 (PGE$_2$) production, and the levels of pro-inflammatory cytokines, such as tumor necrosis factor-α, IL-6, and IL-1β. In contrast, this effect of EGCG or GTPs was not observed in IL-12-deficient mice. This may be due to the fact that the treatment of mice with EGCG or GTPs was not able to repair UV-induced DNA damage significantly in the IL-12 knockout mice (Meeran et al., 2009). This new information supports the concept that UV-induced CPDs and inflammatory responses are causally related with the increased risk of UV carcinogenesis. This *in vivo* experimental evidence also suggests that the prevention of UVB-induced skin cancer by GTPs or EGCG is mediated through inhibition of UVB-induced inflammation, which in turn is mediated, at least in part, through rapid repair of damaged DNA. These new investigations suggest that regular consumption of green tea may be considered an effective strategy for the prevention of UV-induced inflammation-associated skin diseases, including UV radiation-induced melanoma and nonmelanoma skin cancers.

6.7.5 UVB-Induced DNA Damage Suppresses Immune System

Epidemiological as well as clinical studies suggest that UV radiation-induced immunosuppression is linked with the development of skin cancers, and UVB-induced DNA damage in the form of CPDs has been considered a molecular trigger for the induction of immunosuppression and initiation of UV carcinogenesis (Katiyar, 2007a,b; Meeran et al., 2006b, 2009). The UV-induced DNA damage also impairs the antigen-presenting capacity of Langerhans cells, which results in a lack of sensitization and the induction of tolerance to contact sensitizers (Vink et al., 1996, 1997). There is evidence suggesting that alterations in the immune system contribute to the pathogenesis of UV carcinogenesis in mice and probably in humans as well (Meunier et al., 1998; Yoshikawa et al., 1990). Chronically immunosuppressed patients living in regions of intense sun exposure experience an exceptionally high rate of skin cancer (Katiyar, 2007a,b; Meunier et al., 1998). This observation is consistent with the notion that immune surveillance is an important mechanism designed to prevent the generation and maintenance of neoplastic cells in the skin.

6.7.6 GTPs Prevent UVB-Induced Immunosuppression through Stimulation of DNA Repair

The administration of GTPs in the drinking water of mice inhibited UVB-induced suppression of contact hypersensitivity response to contact sensitizer, and simultaneously the migration of CPD-positive epidermal antigen presenting cells to draining lymph nodes of mice was reduced. This indicates that treatment of mice with GTPs might be able to repair UV-induced CPDs in the skin cells of mice. It is speculated that as the migrating antigen presenting cells in the epidermis were either not damaged or were repaired in mice, they were able to present Ag to T-cells in the draining lymph nodes, resulting in induction of sensitization to contact sensitizer after challenge in contact hypersensitivity reactions. The numbers of CPD-positive cells were found to be significantly higher in the non-GTP-treated mice in the subcapsular sinus to the paracortical region of the lymph nodes, including the interfollicular areas, which are the sites of T-cell localization (Katiyar et al., 2010). Thus, the damaged DNA in the lymph nodes of non-GTP-treated mice may adversely affect the ability of the antigen presenting cells to present

Ag to T-cells, thus abrogating sensitization after sensitizer treatment. In contrast, the reverse was observed in GTP-treated mice, and that may be one of the mechanisms, that intake of GTPs in drinking water of mice prevents UVB-induced immunosuppression.

6.7.7 GTPs Prevent UVB-Induced Immunosuppression through NER Mechanism

It has been suggested that nucleotide excision repair is the main mechanism of DNA repair in mammalian cells for the removal of UV radiation-induced DNA damage. As the treatment of GTPs enhances the removal or repair of UVB-induced DNA damage, it was of interest to examine whether the removal or repair of UV-induced CPDs by GTPs is mediated via induction of NER genes. Experimental data analysis reveals that treatment of mice with GTPs in the drinking water of mice increases the levels of some NER genes (e.g., *XPA*, *XPC*, and *RPA1*) in UVB-exposed skin compared to non-GTP-treated UVB-exposed mice, and that may have contributed to the enhanced repair of damaged DNA in mouse skin (Katiyar et al., 2010). The role of NER in DNA repair was further confirmed by assessing the effect of GTPs on UVB-induced immunosuppression in $XPA^{-/-}$ mice, and data were compared with the *XPA*-proficient ($XPA^{+/+}$) mice. Treatment of mice with GTPs in drinking water prevents UVB-induced suppression of contact hypersensitivity response in $XPA^{+/+}$ mice, but does not prevent in $XPA^{-/-}$ mice, further supporting the observations that inhibition of UVB-induced immunosuppression by GTPs requires functional NER genes. This observation was important, as the treatment of GTPs does not remove or repair UVB-induced DNA damage in $XPA^{-/-}$ (also called NER-deficient) mice but repairs in $XPA^{+/+}$ (NER-proficient or wild-type) mice that were exposed to UVB. These observations were further confirmed by using NER-deficient cells from *XPA* patients and NER-proficient cells from healthy persons. Cells derived from patients suffering from xeroderma pigmentosum either lack or have reduced DNA repair capacity due to genetic mutations in several components of the NER. The *XPA* complementation type represents the most severe phenotype, because the *XPA* gene is the most crucial component in the repair process, and thus cells lacking the *XPA* gene are completely deficient in NER (Carreau et al., 1995; Muotri et al., 2002). Following these experiments, GTPs were able to remove UV-induced CPDs in NER-proficient cells ($XPA^{+/+}$) but were not able to remove or repair in NER-deficient ($XPA^{-/-}$) human fibroblast cells. These observations indicate that repair of

UV-induced DNA damage by GTPs is mediated through the NER mechanism. These findings have important implications for the chemoprevention of skin cancer by GTPs, and identify a new mechanism by which GTPs prevent UV-induced immunosuppression.

In summary, the studies conducted with green tea polyphenols indicate that the prevention of UV radiation-induced immunosuppression, and subsequently the prevention of UV carcinogenesis by GTPs either through topical application or in the drinking water of mice, is mediated through rapid repair of UVB-induced DNA damage, as summarized in Figure 6.1. As UV-induced DNA damage and immunosuppression play important roles in melanoma and nonmelanoma skin cancers, it is suggested that drinking green tea should be further investigated as a chemopreventive agent for the prevention of skin cancers in humans, and its possible use in the future practice of medicine.

FIGURE 6.1
Schematic diagram depicting the chemopreventive mechanism of UV radiation-induced skin cancer by green tea polyphenols (GTPs). GTPs inhibit UVB-induced immunosuppression and initiation of skin cancer through rapid repair of UVB-induced DNA damage in the form of cyclobutane pyrimidine dimers (CPDs) or thymine dimers in the skin.

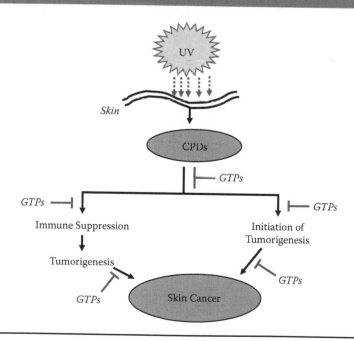

6.8 Bioavailability of Green Tea Catechins/Polyphenols

The potential health benefits of green tea depend not only on the amount consumed but on their bioavailability, which seems to be variable. The considerable structural diversity among the catechin derivatives/polyphenols can influence the bioavailability of the individual components. Small molecules, like catechin monomers in green tea, can be easily absorbed through the gut barrier, whereas the large molecular weight polyphenols, such as (–)-epigallocatechin-3-gallate or theaflavin and thearubigins from black tea, are poorly absorbed. Once absorbed, polyphenols are conjugated to glucuronide, sulfate, and methyl groups in the gut mucosa and inner tissues. Nonconjugated polyphenols are virtually not found in plasma. Bioavailability of green tea catechins appears to vary on an animal species basis. In rats given 0.6% GTPs in their drinking water over a period of 28 days, plasma concentrations of EGCG were much lower than those of EGC or EC, even though the ratio of EGCG to EGC was 5:1 in the GTP final product. When the same GTPs were given to mice in drinking water, the plasma levels of EGCG were much higher than those of EGC and EC (Kim et al., 2000). In humans, EGCG seems to be less bioavailable than other green tea catechins. Catechin levels in human plasma reach their peak 2 to 4 h after ingestion (Yang et al., 1998). ECG has been found to be more highly methylated than EGC and EGCG, and EGCG has been found to be less conjugated than EGC and EC (Chow et al., 2001). In case of topical application of the polyphenols, the penetration of polyphenols into the skin is limited and successful delivery of plant polyphenols requires cream-based, organic solvent-based, or lipid-soluble topical formulations that can enhance the penetration of the polyphenols, and that will result in more efficient skin photoprotection.

6.9 Translation of Animal Studies to Human Health

The intake of green tea polyphenols has shown significant DNA repair effects in UVB-irradiated mouse skin. These protective effects of polyphenols on UVB-induced DNA damage contribute to their antiphotocarcinogenic effects and act to abrogate the biochemical processes induced or mediated by UV radiation. Based on the epidemiological evidence and the results from laboratory studies conducted in *in vitro* and *in vivo* systems, it is suggested that routine consumption or topical treatment of green tea polyphenols may provide efficient protection against the adverse biological effects of solar ultraviolet radiation in humans. Based on the information obtained in animal models with the use of green tea polyphenols, it can be suggested that the consumption of 5 or 6 cups (1 cup = 150 ml) of green tea (1 g green tea leaves/150 ml) per day by humans may provide the same level of photoprotective effect in the human system as was observed in animal models. However, the magnitude of the photoprotective effects or UVB-induced immunosuppression and DNA repair by green tea may differ from person to person based on differences in race, and frequency and intensity and exposure time of UV radiation. For appropriate conversion of the doses of chemopreventive agent from animal studies to the human system, the body surface area normalization method has been prescribed (Reagan-Shaw et al., 2007). Based on this reference, the human equivalent dose of any chemopreventive agent can be calculated using the following formula:

$$\text{Human equivalent dose (mg/kg)} = \text{Animal dose (mg/kg)} \times \frac{\text{Animal } K_m \text{ factor}}{\text{Human } K_m \text{ factor}}$$

K_m factor for mouse = 3, and K_m factor for adult human = 37.

Moreover, the use of green tea polyphenols in combination with sunscreens or skin care lotions may provide an effective strategy for the protection of UV radiation-induced skin disorders, including the risk of melanoma and nonmelanoma skin cancers in humans.

Acknowledgments

The work reported from Dr. Katiyar's laboratory was supported by funds from the National Institutes of Health (CA104428, AT002536) and the Veteran Administration Merit Review Award. The content of this article does not necessarily reflect the views or policies of the funding sources.

References

Bachelor, M.A., and Bowden, G.T. 2004. UVA-mediated activation of signaling pathways involved in skin tumor promotion and progression. *Semin Cancer Biol*, 14: 131–138.

Baliga, M.S., and Katiyar, S.K. 2006. Chemoprevention of photocarcinogenesis by selected dietary botanicals. *Photochem Photobiol Sci*, 5: 243–253.

Brash, D.E., Rudolph, J.A., Simon, J.A., Lin, A., McKenna, G.J., Baden, H.P., Halperin, A.J., and Pontén, J. 1991. A role for sunlight in skin cancer: UV-induced p53 mutations in squamous cell carcinoma. *Proc Natl Acad Sci USA*, 88: 10124–10128.

Brunda, M.J., Luistro, L., Warrier, R.R., Wright, R.B., Hubbard, B.R., Murphy, M., Wolf, S.F., and Gately, M.K. 1993. Antitumor and antimetastatic activity of interleukin-12 against murine tumors. *J Exp Med*, 178: 1223–1230.

Camouse, M.M., Domingo, D.S., Swain, F.R., Conrad, E.P., Matsui, M.S., Maes, D., Declercq, L., Cooper, K.D., Stevens, S.R., and Baron, E.D. 2009. Topical application of green and white tea extracts provides protection from solar-simulated ultraviolet light in human skin. *Exp Dermatol*, 18: 522–526.

Carreau, M., Eveno, E., Quilliet, X., Chevalier-Lagente, O., Benoit, A., Tanganelli, B., Stefanini, M., Vermeulen, W., Hoeijmakers, J.H., and Sarasin, A. 1995. Development of a new easy complementation assay for DNA repair deficient human syndromes using cloned repair genes. *Carcinogenesis*, 16: 1003–1009.

Chatterjee, M.L., Agarwal, R., and Mukhtar, H. 1996. Ultraviolet B radiation-induced DNA lesions in mouse epidermis: An assessment using a novel ^{32}P-postlabelling technique. *Biochem Biophys Res Commun*, 229: 590–595.

Chow, H.H., Cai, Y., Alberts, D.S., Hakim, I., Dorr, R., Shahi, F., Crowell, J.A., Yang, C.S., and Hara, Y. 2001. Phase I pharmacokinetic study of tea polyphenols following single-dose administration of epigallocatechin gallate and Polyphenon E. *Cancer Epidemiol Biomarkers Prev*, 10: 53–58.

Colombo, M.P., Vagliani, M., Spreafico, F., Parenza, M., Chiodoni, C., Melani, C., and Stoppacciaro, A. 1996. Amount of interleukin 12 available at the tumor site is critical for tumor regression. *Cancer Res*, 56: 2531–2534.

Cowen, E.W., and Billingsley, E.M. 1999. Awareness of skin cancer by kidney transplant patients. *J Am Acad Dermatol*, 40: 697–701.

de Gruijl, F.R. 2000. Photocarcinogenesis: UVA vs UVB. Singlet oxygen, UVA, and ozone. *Methods Enzymol*, 319: 359–366.

de Gruijl, F.R., and van der Leun, J.C. 1994. Estimate of the wavelength dependency of ultraviolet carcinogenesis in humans and its relevance to the risk assessment of stratospheric ozone depletion. *Health Phys*, 67: 319–359.

DiGiovanna, J.J. 1998. Posttransplantation skin cancer: Scope of the problem, management and role for systemic retinoid chemoprevention. *Transplant Proc*, 30: 2771–2775.

DiGiovanni, J. 1992. Multistage carcinogenesis in mouse skin. *Pharmacol Ther*, 54: 63–128.

Donawho, C.K., and Kripke, M.L. 1991. Evidence that the local effect of ultraviolet radiation on the growth of murine melanomas is immunologically mediated. *Cancer Res*, 51: 4176–4181.

Fortina, A.B., Caforio, A.L., and Piaserico, S. 2000. Skin cancer in heart transplant recipients: Frequency and risk factor analysis. *J Heart Lung Transplant*, 19: 249–55.

Godar, D.E. 2001. UV doses of American children and adolescents. *Photochem Photobiol*, 74: 787–793.

Godar, D.E., Wengraitis, S.P., Shreffler, J., and Sliney, D.H. 2001. UV doses of Americans. *Photochem Photobiol*, 73: 621–629.

Hojo, M., Morimoto, T., and Maluccio, M. 1999. Cyclosporin induces cancer progression by a cell-autonomous mechanism. *Nature*, 397: 530–534.

Housman, T.S., Feldman, S.R., Williford, P.M., Fleischer Jr., A.B., Goldman, N.D., Acostamadiedo, J.M., and Chen, G.J. 2003. Skin cancer is among the most costly of all cancers to treat for the Medicare population. *J Am Acad Dermatol*, 48: 425–429.

Hruza, L.L., and Pentland, A.P. 1993. Mechanisms of UV-induced inflammation. *J Invest Dermatol*, 100: 35S–41S.

Hussein, M.R. 2005. Ultraviolet radiation and skin cancer: Molecular mechanisms. *J Cutan Pathol*, 32: 191–205.

Ichihashi, M., Ueda, M., Budiyanto, A., Bito, T., Oka, M., Fukunaga, M., Tsuru, K., and Horikawa, T. 2003. UV-induced skin damage. *Toxicology*, 189: 21–39.

Katiyar, S.K. 2003. Skin photoprotection by green tea: Antioxidant and immuno-modulatory effects. *Current Drug Targets Immune Endocr Metab Disord*, 3: 234–242.

Katiyar, S.K. 2007a. UV-induced immune suppression and photocarcinogenesis: Chemoprevention by dietary botanical agents. *Cancer Lett*, 255: 1–11.

Katiyar, S.K. 2007b. Interleukin-12 and photocarcinogenesis. *Toxicol Appl Pharmacol*, 224: 220–227.

Katiyar, S.K., Afaq, F., Perez, A., and Mukhtar, H. 2001. Green tea polyphenol (–)-epigallocatechin-3-gallate treatment of human skin inhibits ultraviolet radiation-induced oxidative stress. *Carcinogenesis*, 22: 287–294.

Katiyar, S.K., and Elmets, C.A. 2001. Green tea polyphenolic antioxidants and skin photoprotection. *Int J Oncol*, 18: 1307–1313.

Katiyar, S., Elmets, C.A., and Katiyar, S.K. 2007. Green tea and skin cancer: Photoimmunology, angiogenesis and DNA repair. *J Nutr Biochem*, 18: 287–296.

Katiyar, S.K., Korman, N.J., Mukhtar, H., and Agarwal, R. 1997. Protective effects of silymarin against photocarcinogenesis in a mouse skin model. *J Natl Cancer Inst*, 89: 556–566.

Katiyar, S.K., Matsui, M.S., and Mukhtar, H. 2000a. Kinetics of UV light-induced cyclobutane pyrimidine dimers in human skin *in vivo*: An immunohistochemical analysis of both epidermis and dermis. *Photochem Photobiol*, 72: 788–793.

Katiyar, S.K., and Mukhtar, H. 2001. Green tea polyphenol (–)-epigallocatechin-3-gallate treatment to mouse skin prevents UVB-induced infiltration of leukocytes, depletion of antigen presenting cells and oxidative stress. *J Leukoc Biol*, 69: 719–726.

Katiyar, S.K., Perez, A., and Mukhtar, H. 2000b. Green tea polyphenol treatment to human skin prevents formation of ultraviolet light B-induced pyrimidine dimers in DNA. *Clinical Cancer Res*, 6: 3864–3869.

Katiyar, S.K., Vaid, M., van Steeg, H., and Meeran, S.M. 2010. Green tea poly-phenols prevent UV-induced immunosuppression by rapid repair of DNA damage and enhancement of nucleotide excision repair genes. *Cancer Prev Res*, 3: 179–189.

Kim, S., Lee, M.J., Hong, J., Li, C., Smith, T.J., Yang, G.Y., Seril, D.N., and Yang, C.S. 2000. Plasma and tissue levels of tea catechins in rats and mice during chronic consumption of green tea polyphenols. *Nutr Cancer*, 37: 41–48.

Kinlen, L., Sheil, A., and Peta, J. 1979. Collaborative United Kingdom–Australia study of cancer in patients treated with immunosuppressive drugs. *Br J Med*, 2: 1461–1466.

Kligman, L.H., Akin, F.J., and Kligman, A.M. 1980. Sunscreens prevent ultraviolet photocarcinogenesis. *J Am Acad Dermatol*, 3: 30–35.

Kripke, M.L. 1990. Photoimmunology. *Photochem Photobiol*, 52: 919–923.

Kripke, M.L., Cox, P.A., Alas, L.G., and Yarosh, D.B. 1992. Pyrimidine dimers in DNA initiated systemic immunosuppression in UV-irradiated mice. *Proc Natl Acad Sci USA*, 89: 7516–7517.

Krutmann, J. 2001. The role of UVA rays in skin aging. *Eur J Dermatol*, 11: 170–171.

Mantena, S.K., Meeran, S.M., Elmets, C.A., and Katiyar, S.K. 2005. Orally administered green tea polyphenols prevent ultraviolet radiation-induced skin cancer in mice through activation of cytotoxic T cells and inhibition of angiogenesis in tumors. *J Nutr*, 135: 2871–2877.

Meeran, S.M., Akhtar, S., and Katiyar, S.K. 2009. Inhibition of UVB-induced skin tumor development by drinking green tea polyphenols is mediated through DNA repair and subsequent inhibition of inflammation. *J Invest Dermatol*, 129: 1258–1270.

Meeran, S.M., Mantena, S.K., Elmets, C.A., and Katiyar, S.K. 2006a. (–)-Epigallocatechin-3-gallate prevents photocarcinogenesis in mice through interleukin-12-dependent DNA repair. *Cancer Res*, 66: 5512–5520.

Meeran, S.M., Mantena, S.K., and Katiyar, S.K. 2006b. Prevention of ultraviolet radiation-induced immunosuppression by (–)-epigallocatechin-3-gallate in mice is mediated through interleukin 12-dependent DNA repair. *Clin Cancer Res*, 12: 2272–2280.

Meeran, S.M., Mantena, S.K., Meleth, S., and Katiyar, S.K. 2006c. Interleukin-12-deficient mice are at greater risk of ultraviolet radiation-induced skin tumors and malignant transformation of papillomas to carcinomas. *Mol Cancer Ther*, 5: 825–832.

Meunier, L., Raison-Peyron, N., and Meynadier, J. 1998. UV-induced immunosuppression and skin cancers. *Rev Med Interne*, 19: 247–254.

Miller, D.L., and Weinstock, M.A. 1994. Nonmelanoma skin cancer in the United States: Incidence. *J Am Acad Dermatol*, 30: 774–778.

Mittal, A., Elmets, C.A., and Katiyar, S.K. 2003a. Dietary feeding of proanthocyanidins from grape seeds prevents photocarcinogenesis in SKH-1 hairless mice: Relationship to decreased fat and lipid peroxidation. *Carcinogenesis*, 24: 1379–1388.

Mittal, A., Piyathilake, C., Hara, Y., and Katiyar, S.K. 2003b. Exceptionally high protection of photocarcinogenesis by topical application of (–)-epigallocatechin-3-gallate in hydrophilic cream in SKH-1 hairless mouse model: Relationship to inhibition of UVB-induced global DNA hypomethylation. *Neoplasia*, 5: 555–565.

Mukhtar, H., and Elmets, C.A. 1996. Photocarcinogenesis: Mechanisms, models and human health implications. *Photochem Photobiol*, 63: 355–447.

Muotri, A.R., Marchetto, M.C., Zerbini, L.F., Libermann, T.A., Ventura, A.M., Sarasin, A., and Menck, C.F. 2002. Complementation of the DNA repair deficiency in human xeroderma pigmentosum group A and C cells by recombinant adenovirus-mediated gene transfer. *Hum Gene Ther*, 13: 1833–1844.

Nichols, J.A., and Katiyar, S.K. 2010. Skin photoprotection by natural polyphenols: Anti-inflammatory, anti-oxidant and DNA repair mechanisms. *Arch Dermatol Res*, 302: 71–83.

Otley, C.C., and Pittelkow, M.R. 2000. Skin cancer in liver transplant recipients. *Liver Transpl*, 6: 253–262.

Parrish, J.A. 1983. Photobiologic considerations in photoradiation therapy. *Adv Exp Med Biol*, 160: 91–108.

Reagan-Shaw, S., Nihal, M., and Ahmad, N. 2007. Dose translation from animal to human studies revisited. *FASEB J*, 22: 659–661.

Robertson, M.J., and Ritz, J. 1996. Interleukin-12: Basic biology and potential applications in cancer treatment. *Oncologist*, 1: 88–97.

Runger, T.M. 1999. Role of UVA in the pathogenesis of melanoma and non-melanoma skin cancer. A short review. *Photodermatol Photoimmunol Photomed*, 15: 212–216.

Schwarz, A., Maeda, A., Gan, D., Mammone, T., Matsui, M.S., and Schwarz, T. 2008. Green tea phenol extracts reduce UVB-induced DNA damage in human cells via interleukin-12. *Photochem Photobiol*, 84: 350–355.

Schwarz, A., Maeda, A., Kernebeck, K., vanSteeg, H., Beissert, S., and Schwarz, T. 2005. Prevention of UV radiation-induced immunosuppression by IL-12 is dependent on DNA repair. *J Exp Med*, 201: 173–179.

Schwarz, A., Stander, S., Berneburg, M., Böhm, M., Kulms, D., van Steeg, H., Grosse-Heitmeyer, K., Krutmann, J., and Schwarz, T. 2002. Interleukin-12 suppresses ultraviolet radiation-induced apoptosis by inducing DNA repair. *Nat Cell Biol*, 4: 26–31.

Scotto, J., and Fears, T.R. 1978. Skin cancer epidemiology: Research needs. *Natl Cancer Inst Monogr*, 50: 169–177.

Strom, S. 1996. Epidemiology of basal and squamous cell carcinomas of the skin. In *Basal and squamous cell skin cancers of the head and neck*, ed. M.J. Miller, R.S. Weber, and H. Goepfert, 1–7. Baltimore: Williams and Wilkins.

Taylor, C.R., Stern, R.S., Leyden, J.J., and Gilchrest, B.A. 1990. Photoaging/photodamage and photoprotection. *J Am Acad Dermatol*, 22: 1–15.

Timares, L., Katiyar, S.K., and Elmets, C.A. 2008. DNA damage, apoptosis and Langerhans cells-activators of UV-induced immune tolerance. *Photochem Photobiol*, 84: 422–436.

Ullrich, S.E. 1995. Potential for immunotoxicity due to environmental exposure to ultraviolet radiation. *Hum Exp Toxicol*, 14: 89–91.

Urbach, F. 1991. Incidences of nonmelanoma skin cancer. *Dermatol Clin*, 9: 751–755.

Vink, A.A., Moodycliffe, A.M., Shreedhar, V., Ullrich, S.E., Roza, L., Yarosh, D.B., and Kripke, M.L. 1997. The inhibition of antigen-presenting activity of dendritic cells resulting from UV irradiation of murine skin is restored by *in vitro* photorepair of cyclobutane pyrimidine dimers. *Proc Natl Acad Sci USA*, 94: 5255–5260.

Vink, A.A., Strickland, F.M., Bucana, C., Cox, P.A., Roza, L., Yarosh, D.B., and Kripke, M.L. 1996. Localization of DNA damage and its role in altered antigen-presenting cell function in ultraviolet-irradiated mice. *J Exp Med*, 183: 1491–1500.

Wang, S.Q., Setlow, R., and Berwick, M. 2001. Ultraviolet A and melanoma: A review. *J Am Acad Dermatol*, 44: 837–846.

Wang, Z.Y., Huang, M.T., Ferraro, T., Wong, C.Q., Lou, Y.R., Reuhl, K., Iatropoulos, M., Yang, C.S., and Conney, A.H. 1992a. Inhibitory effect of green tea in the drinking water on tumorigenesis by ultraviolet light and 12-O-tetradecanoylphorbol-13-acetate in the skin of SKH-1 mice. *Cancer Res*, 52: 1162–1170.

Wang, Z.Y., Huang, M.T., Ho, C.T., Chang, R., Ma, W., Ferraro, T., Reuhl, K.R., Yang, C.S., and Conney, A.H. 1992b. Inhibitory effect of green tea on the growth of established skin papillomas in mice. *Cancer Res*, 52: 6657–6665.

Wei, H., Ca, Q., Rahn, R., Zhang, X., Wang, Y., and Lebwohl, M. 1998. DNA structural integrity and base composition affect ultraviolet light-induced oxidative DNA damage. *Biochemistry*, 37: 6485–6490.

Wolf, S.F., Temple, P.A., Kobayashi, M., Young, D., Dicig, M., Lowe, L., Dzialo, R., Fitz, L., Ferenz, C., and Hewick, R.M. 1991. Cloning of cDNA for natural killer cell stimulatory factor, a heterodimeric cytokine with multiple biologic effects on T and natural killer cells. Cloning of cDNA for natural killer cell stimulatory factor, a heterodimeric cytokine with multiple biologic effects on T and natural killer cells. *J Immunol*, 146: 3074–3081.

Yang, C.S., Chen, L., Lee, M.J., Balentine, D., Kuo, M.C., and Schantz, S.P. 1998. Blood and urine levels of tea catechins after ingestion of different amounts of green tea by human volunteers. *Cancer Epidemiol Biomarkers Prev*, 7: 351–354.

Yarosh, D., Alas, L.G., Yee, V., Oberyszyn, A., Kibitel, J.T., Mitchell, D., Rosenstein, R., Spinowitz, A., and Citron, M. 1992. Pyrimidine dimer removal enhanced by DNA repair liposomes reduces the incidence of UV skin cancer in mice. *Cancer Res*, 52: 4227–4231.

Yoshikawa, T., Rae, V., Bruins-Slot, W., vand-den-Berg, J.W., Taylor, J.R., and Streilein, J.W. 1990. Susceptibility to effects of UVB radiation on induction of contact hypersensitivity as a risk factor for skin cancer in humans. *J Invest Dermatol*, 95: 530–536.

Zhao, J.F., Zhang, Y.J., Jin, X.H., Athar, M., Santella, R.M., Bickers, D.R., and Wang, Z.Y. 1999. Green tea protects against psoralen plus ultraviolet A-induced photochemical damage to skin. *J Invest Dermatol*, 113: 1070–1075.

Ziegler, A., Jonason, A.S., Leffell, D.J., Simon, J.A., Sharma, H.W., Kimmelman, J., Remington, L., Jacks, T., and Brash, D.E. 1994. Sunburn and p[53] in the onset of skin cancer. *Nature*, 372: 773–776.

7

Green Tea Polyphenols in Cardiovascular Diseases

Hla-Hla Htay, Mahendra P. Kapoor, Theertham P. Rao, Tsutomu Okubo, and Lekh R. Juneja

Contents

7.1 Introduction

Cardiovascular disease (CVD) is defined as a group of diseases that include dysfunctional conditions of the heart and blood vessel system and other disorders, like atherosclerosis, deep vain thrombosis and pulmonary embolism, cerebrovascular accident (stroke), congestive heart failure, coronary artery disease (CAD), coronary heart disease (CHD), and myocardial infarction (MI) (heart attack). The mortality and morbidity rates due to cardiovascular disease are increasing around the globe without any discrimination of gender, ethnicity, and prosperity. The associated causes of cardiovascular disease include diabetes mellitus, hypertension, hypercholesterolemia, atherosclerosis, and hyperhomocysteinemia (Venardos and Kaye, 2007). Oxidative stress is one of the leading factors for cause of such chronic diseases. Consuming a number of fruits and vegetables having antioxidant capabilities might reduce the risk of cardiovascular disease (Misra et al., 2009).

Plant polyphenols like rutin, quercetin, catechins, pyrocatechol, and chlorogenic acid have been widely acknowledged as health tonics (Graham et al., 1992; Sajilata et al., 2008). Green tea polyphenols, rich in catechins, are well-known antioxidants that protect against several kinds of chronic diseases. Green tea has been heavily researched, wherein the multiple substances found in green tea, theaflavins, caffeine, theophylline, phenolic acids, theobromine, and theanine, have been shown to contribute to its health benefits (Cabrera et al., 2006). In general, green tea seems to be a negligible risk complementary therapy and may play a beneficial role in various conditions, including cancer, diabetes, genital warts, obesity and weight management, metabolic health, and cardiovascular disease (Bettuzzi et al., 2006; Gross et al., 2007; Kurahashi et al., 2008; Kuriyama et al., 2006; Nagao et al., 2007; Westerterp-Plantenga et al., 2005). In Japan it has been well postulated that green tea consumption effectively prevents considerable risk to mortality. This chapter focuses on the potential benefits that green tea polyphenols can play on cardiovascular disease by compiling the available laboratory, epidemiological, and clinical studies performed on cardiovascular conditions.

7.2 Prevalence of Cardiovascular Diseases

Behavioral risk factors are responsible for more than 80% of coronary heart disease and cerebrovascular disease. The major factors, also known as metabolic risk factors, are an unhealthy diet, physical inactivity, and harmful use of alcohol and tobacco. An unhealthy diet and physical inactivity may raise blood pressure, blood glucose, and blood lipids, and lead to overweight and obesity (Nagao et al., 2007; Westerterp-Plantenga et al., 2005). Secondary factors include poverty, stress, and heredity. Other underlying determinants include the major social driving forces, such as the economic and cultural changes of globalization, urbanization, and population aging. Cardiovascular diseases are the world's largest killers, claiming 17.3 million lives in 2008, representing 30% of all global deaths. Of these deaths, 7.3 million were due to coronary heart disease and 6.2 million were due to stroke, as per a report of the WHO in 2012. As the magnitude of cardiovascular diseases continues to accelerate globally, the pressing need for increased awareness and for stronger and more focused international responses is being increasingly recognized.

Hence, CHD remains the greatest killer in the Western world, and although the death rate from CHD has been falling, the current increase in major risk factors suggests it is likely that CHD incidence will increase over the next 20 years. It is estimated that by 2030, almost 23.6 million people will die from cardiovascular disease, mainly from heart disease and stroke, and it is projected to remain the single leading causes of death. The cost of stroke estimated in the United States alone is expected to exceed US$2 trillion between 2005 and 2050 (Flynn et al., 2008). In conjunction with preventive strategies, major advances in the treatment of acute coronary syndromes and myocardial infarction have occurred over the past two decades.

7.3 Role of Oxidative Stress and Cardiovascular Diseases

In general, oxidative stress is an injury of cardiac and vascular myocyte cells resulting from an increased formation of reactive oxygen species (ROS) or decreased antioxidant reserves. Thus, it represents an imbalance between production of ROS and detoxification intermediates (ROS scavengers) caused by biochemical or physiological processes in the human body (Misra et al., 2009). An increase in ROS generation seems to be due to impaired mitochondrial reduction of molecular oxygen, secretion of ROS by white blood cells, endothelial dysfunction, auto-oxidation of catecholamines, and exposure to radiation and air pollution. An imbalanced condition occurs when a normal redox state of tissue or organ has been destroyed by the toxic effect of peroxides and free radicals that damage all components of cells, including proteins, lipids, and DNA. ROS formed during oxidative stress can initiate lipid peroxidation, protein oxidation to active states, and cause DNA strand breaks, all potentially damaging to normal cellular functions. Also, the severe degrees of oxidative stress may cause apoptosis and necrosis. In normal physiological conditions, ROS production is usually homeostatically regulated by endogenous free radical scavengers such as superoxide dismutase (SOD), catalase, and the glutathione peroxide (GPX) and thioredoxin reductase (TxnRed) systems (Venardos and Kaye, 2007). Thus, the aforementioned cellular antioxidants play an important role to protect the cell from oxidative stress.

Generally, oxidative stress occurs in several diseases, such as atherosclerosis, thrombosis, heart failure, and stroke, due to increased production of oxidizing species. The heart is one of the major organs affected by ROS. Recent evidence suggests that oxidative stress is a common denominator in many aspects of cardiovascular disease (Victor et al., 2009). Although the cause-effect relationship of oxidative stress with any CVD still remains to be established, increased formation of ROS indicating the presence of oxidative stress has been observed in a wide variety of experimental and clinical conditions reported in *in vitro* and *in vivo* studies (Leeuwenburgh et al., 1997; Velayutham et al., 2008). Moreover, antioxidant therapy has been revealed to exert beneficial effects on hypertension, ischemic heart disease, atherosclerosis, congestive heart failure, and cardiomyopathies. During myocardial oxidative stress, the generation of ROS is enhanced and defense mechanisms of myocytes are altered, while the source of ROS in cardiac myocytes could be the mitochondrial electron transport chain, nitric oxide synthase (NOS), NADPH oxidase, xanthine oxidase, lipooxygenase/cyclooxygenase, and out-oxidation of catecholamines. Overproduction of ROS under pathophysiological conditions forms an integral part of the development of CVD, and in particular atherosclerosis (Dhalla et al., 2000). Endothelial dysfunction, usually characterized by a loss of nitric oxide bioactivity, occurs early in the development of atherosclerosis, and determines future vascular complications. Vascular endothelium plays an important role in regulation of vascular homeostasis, and alteration or dysfunction of vascular endothelium expresses the pathogenesis of cardiovascular diseases. Thus, the loss of nitric oxide derived from endothelium leads to endothelial dysfunction, and increased oxidative stress could be linked to impaired endothelial vasomotor function in atherosclerosis (Vita, 2005). Another feature of atherosclerosis development is due to an accumulation of oxidized low-density lipoprotein (ox-LDL) that inhibits NO release from endothelial cells (Duffy et al., 1999). LDL oxidation in vascular endothelium, which is a precursor to plaque formation, is also involved in ischemic cascade due to oxygen reperfusion injury followed by hypoxia (Li and Jackson, 2002). Gokce et al. (1999) studied a sustained beneficial effect of endothelial-derived nitric oxide (EDNO) in antioxidant treatment (ascorbic acid; 500 mg/day) in patients of coronary artery disease. In acute myocardial infarction (AMI), two distinct types of damage occur to the heart, including ischemic injury and reperfusion injury, which lead to mitochondrial dysfunction in heart cells. During ischemic and reperfusion injuries, ROS could be produced by both endothelial cells and circulating phagocytes. Although ischemia also causes alteration in the defense mechanism against ROS, some heat shock proteins are overexpressed in conditions of ischemic/reperfusion and can protect from cardiac injury.

7.4 Tea Polyphenols' Effect on Cardiovascular Diseases

Epidemiological studies have clearly indicated that tea polyphenols from green tea are beneficial to protect against cardiovascular diseases (Hodgson et al., 2008; Riemersma et al., 2001; Wolfram, 2007). Clinical trials have revealed its effectiveness on risk reduction, although they have shown inconsistent results on lipid levels, blood pressure, and coronary artery disease (Cooper et al., 2005; Zaveri, 2006). Oxidative stress is involved in many diseases, such as metabolic syndrome, obesity, diabetes, aging, and cardiovascular disease (Finkel and Holbrook, 2000). Interest is focused on green tea polyphenols as a dietary supplement in the prevention of CVD, as it has scavenging effects on reactive oxygen species, which plays a pivotal role in oxidative stress. Green tea polyphenols and its catechins may reduce the risk of coronary heart disease by lowering the plasma levels of cholesterol and triglycerides. Tea polyphenols having a galloyl moiety suppress postprandial hypertriacylglycerolenia by delaying lymphatic transport of dietary fat in rats. It has also been shown that regular consumption of green tea polyphenols from an early age prevents the development of spontaneous stroke in malignant stroke-prone spontaneously hypertensive rats, probably by inhibiting development of high blood pressure systems upon aging (Negishi et al., 2004). Green tea polyphenols also attenuated blood pressure increase in spontaneously hypersensitive rats, an effect attributed to its antioxidant properties (Ikeda et al., 2007). Another study indicated that green tea polyphenols, especially epicatechin gallate (ECG), interfere with the emulsification, digestion, and micellar solubilization of lipids as a critical step involved in intestinal absorption (Koo et al., 2004).

Han et al. (2003) studied the oxidative stress in the human saphenous veins, wherein oxidative stress was induced exogenously in the vein segments using xanthine/xanthine oxidase and 0.8 or 1.6 M H_2O_2 that affect the endothelium cells' viability and severe morphological and structural changes in veins. Another study was done on vein segments that were incubated with or without 1.0 of mg/ml green tea polyphenols for 1 to 14 days under physiological conditions. After incubation, the endothelium cell viability, vein histology, and endothelial nitric oxide synthase (eNOS) expression were observed. Cell viability decreased and structural and histological changes were found in nontreatment, but in green tea polyphenol-treated veins, cell viability was nearly 64% after 7 days and eNOS expression was maintained up to 40% compared to fresh veins. From these results green tea polyphenols may be effective to maintain eNOS, which is one of the relative biomarkers of CVDs. While these studies suggest that green tea polyphenols may act as biological antioxidants and protect veins from oxidative stress-induced toxicity, it is also reported that green tea is a better NO and peroxynitrite scavenger than black tea, with EGCg being the important contributor to both the NO and peroxynitrite scavenging characteristics (Heijnen et al., 2000).

Sano et al. (2004) demonstrated that green tea consumption is associated with a lower incidence of coronary artery disease in Japanese populations. Although a number of epidemiological and cellular studies that addressed the control conditions, differences in geographical regions and ethnic groups, have been conducted on the effects that green tea has on cardiovascular disease (Ahn et al., 2010; Akhlaghi and Bandy, 2010; Aneja et al., 2004; Bedoui et al., 2005; Sachinidis et al., 2002), a prospective cohort study among a large number of the Japanese population (40,530 subjects) revealed that consumption of five or more cups of green tea per day considerably reduced mortality rate and cardiovascular disease (Kuriyama et al., 2006). Women who consumed green tea had a 31% lower risk of dying from cardiovascular disease, while participants of the study acknowledge a significant reduction in incidence of stroke (~37%). Peters et al. (2001) carried out a meta-analysis of tea consumption in relation to myocardial infraction, which included 10 cohort and 7 case control studies.

Zhang et al. (2010) investigated green tea polyphenols' effect on cerebral ischemia established by middle cerebral artery occlusion (MCAO) in rats. After ischemia, triphenyltetrazolium chloride staining and Longa's score were used to determine the infarct volume and neurological deficit. The blood-brain barrier permeability was measured by Evans blue (EB) content in the brain tissue, while reverse transcription polymerase chain reaction (RT-PCR), immunohistochemistry, and Western blot assessment were used to detect expression of caveolin-1 in microvessel fragments of cerebral ischemic tissue and extracellular signal-regulated kinase (ERK) 1/2. The results showed that green tea polyphenols significantly reduced the infarct volume and ameliorated the neurological deficit as well as reduced the permeability of the blood-brain barrier. It was also suggested that the protective effect of green tea polyphenols in cerebral ischemia is probably due to a reduction of caveolin-1 mRNA and phosphorylated ERK1/2 in microvessel fragments of cerebral ischemic tissues.

Atherosclerosis risk factors are increasing blood pressure, dysfunction of kidney, obesity, diabetes, stress, lipid disorders, adrenal and thyroid gland problems, and excessive drinking of alcohol. Endothelial damage is a known key factor in the initiation of atherosclerosis. Concerning the above risk factors, tea polyphenols offer effective antioxidant properties, such as the enhancement of the total plasma antioxidant activities, an increased serum SOD level, and a decrease in plasma nitric oxide concentration (Negishi et al., 2004; Skrzydlewska et al., 2002; Yokozawa et al., 2002). In a study of regulatory effects of EGCg on the expression of C-reactive protein and other inflammatory markers in an experimental model of atherosclerosis rats, the result showed EGCg significantly decreased the serum C-reactive protein (CRP)-level erythrocyte sedimentation rate (ESR), total white blood cell (WBC), and platelet and differential leukocyte counts compared to the control group. This evidently supports that tea polyphenols can reduce the risk of CVDs through reducing inflammatory markers in rats fed an atherogenic diet (Ramesh et al., 2010). Green tea polyphenols have the ability to prevent free radical damage to cells and prevent the oxidation of low-density lipoprotein (LDL) cholesterol, both of which would be expected to inhibit the formation of atherosclerotic plaques (Chyu et al., 2004; Riemersma et al., 2001; Sasazuki et al., 2000). Therefore, atherogenesis could be slowed by reducing the oxidative modification of LDL cholesterol and associated events, such as foam cell formation, endothelial cytotoxicity, and induction of pro-inflammatory cytokines. Several studies have also looked at the effect of green tea polyphenols in the expression of adhesion molecules generally involved in the early stage of atherogenesis. The effect of green tea polyphenols in preventing the formation of atherosclerosis is promising, as apolipoprotein E is important in prevention because it is capable of removing excess cholesterol (Miura et al., 2000). Further, a mixture of green tea polyphenols with other types of antioxidants has shown enhanced modulation in inflammation of adipose tissues, improved endothelial dysfunction along with reduced oxidative stress, as well as increased liver fatty acid oxidation (Bakker et al., 2010). In a recent randomized controlled trial in obese subjects with metabolic syndrome, 35 subjects ((mean ± SE) age, 42.5 ± 1.7 years; body mass index, 36.1 ± 1.3 kg/m^2) completed the 8-week study. Participants were randomly assigned to receive a green tea beverage (4 cups/day), a green tea extract (two capsules and 4 cups water/day), or no treatment (4 cups water/day), wherein the EGCg content was the same in the green tea beverage and the green tea extract. It was noticed that both green tea beverage and extract significantly reduced the cardiovascular risk factor of serum amyloid alpha (Basu et al., 2011).

Green tea polyphenols also improve endothelial function and insulin sensitivity, reduce blood pressure, and protect against myocardial ischemia-reperfusion (I/R) injury in spontaneously hypertensive rats (Potenza et al., 2007). Myocardial ischemia-reperfusion injury is one of the heart diseases, wherein phase 2 enzymes (e.g., glutamate cysteine ligase and quinone reductase) are involved. Green tea polyphenols' ability to support the elevation of phase 2 enzymes and protect the heart against I/R injury was found in animal models as well as in culture cells (Akhlaghi and Bandy, 2010; Na and Surh, 2008; Potenza et al., 2007; Townsend et al., 2004; Yamazaki et al., 2008). The apoptosis and necrosis of myocardial cells are also well known to be involved in ischemia-reperfusion injury of the heart. Short-term dietary intake of green tea polyphenols showed an inhibition effect on apoptosis in an isolated heart through two indicators, such as myocardial caspase-3 activity and DNA fragmentation. The dosage of *ex vivo* study at green tea protection of I/R injury in rats was equivalent to 6 or 7 cups (200 ml) per day for humans.

Although some human studies have not demonstrated protective effects of tea consumption on serum lipid profiles and coronary heart disease morbidity and mortality, Mineharu et al. (2011) recently studied the effect of coffee, green tea, and oolong tea consumption and risk of mortality from cardiovascular disease in Japanese men and women. The study was designed on a daily consumption of beverages wherein an assessment was conducted through questionnaires among 76,979 individuals aged between 40 and 79 years, who declared themselves free from stroke and coronary heart disease as well as cancer. After 1,010,787 person-years of follow-up, 1362 deaths were documented from stroke and 650 deaths from coronary heart diseases. From the results, green tea intake was associated with a reduced risk of mortality from CVD.

Qin et al. (2010) studied green tea polyphenols' effect for reduced risk of coronary artery disease, wherein the underlying mechanism of action at the molecular level in the cardiac muscle of insulin-resistant rats was explored. In a high-fructose diet fed to rats for 6 weeks, supplemented with green tea polyphenols (200 mg/kg BW daily; dissolved in distilled water), rats showed a reduced systemic blood glucose, plasma insulin, retino-binding protein 4, soluble CD36, cholesterol, triglycerides, free fatty acids, tumor necrosis factor alpha and IL6, and LDL-C-level and pro-inflammatory cytokines. It also reported decreased inflammatory factors, TNF Il1b, and Il6nRNA levels, and enhanced anti-inflammatory protein, zinc-finger protein, proteins, and mRNA expressions. These results also suggested that tea polyphenols ameliorate the detrimental effects of a high-fructose diet on insulin-resistant rats in inflammation of cardiac muscle. Arab et al. (2009) presented a meta-analysis of green and black tea consumption based on 10 studies from six different countries, China, Japan, Finland, the Netherlands, Australia, and the United States of America, focusing on the risks of stroke among the diverse populations. It was demonstrated that subjects who consumed more than 3 cups of green tea per day had an approximately 21% reduced risk of fatal and nonfatal strokes compared to nondrinkers of tea. No significant difference was observed in type of tea (black or green tea) and ethnic group (Asian or non- Asian population). Another meta-analysis of tea consumption in relation to myocardial infarction was performed (Peters et al., 2001), which included 10 cohort and 7 case control studies. In this population study, the incidence rate of myocardial infarction was estimated to decrease by 11%, with an increase in tea consumption up to 3 cups a day. Since flavonoid intake has been reported to lower the risk of cardiovascular disease, a meta-analysis of seven prospective studies of flavonoids in relation to coronary heart disease revealed that the maximum tertile of flavonoids consumption was attributed to a 20% reduction in the risk of fatal coronary heart diseases (Huxley and Neil, 2003).

7.5 Mechanism and Bioavailability

The mechanism including interaction of green tea polyphenols on the molecular target of cardiovascular diseases had been postulated on the basis of its antioxidative properties, anti-inflammatory behaviors, antiproliferative power, preventive effect on thrombosis, antiplatelets, and antiangiogenic vasorelaxation, as well as improvement of lipid metabolism (Stangl et al., 2007). Green tea polyphenols exert both direct and indirect antioxidant effects on the cardiovascular system by decreasing the ox-LDL and oxidative enzyme concentrations, such as inducible nitric oxide synthase (iNOS) and xanthine oxidase (XO), cyclooxygenase (COX), and lipoxygenase (LOX). It also increases the activities of antioxidant enzymes like catalase (CAT), heme oxygenase-1 (HO-1), glutathione peroxidase (GP), and superoxide dismutase (SOD). Also, green tea polyphenols could play an important role in the scavenging of reactive oxygen species (ROS) as well as reactive nitrogen species (RNS). The induced oxidative stress and affected vascular damages due to progression of various vascular diseases (atherosclerosis, hypertension, ischemic heart disease, and congestive heart failure) and the interaction of green tea polyphenols with related genetic factors have been looked at. Kaul et al. (2004) observed the dose-dependent increase in transcriptional expression of genes coding for PPAR-α and LDL-R, and decreased PPAR-γ, LXR-α, and CD 36 gene expression. While the underlying mechanism for the development of atherosclerosis, which is also known as a chronic inflammatory disease that is caused by adhesion of leukocytes to endothelium cells, is usually regulated by both chemotactic cytokines and vascular adhesion molecules, the exact role and effect of green tea polyphenols needs to be explored further. However, it is believed that various adhesion molecules, such as endothelial leukocyte adhesion molecule-1 (E-selectin), intercellular adhesion molecule-1 (ICAM-1), and vascular adhesion molecule-1 (VCAM-1), are involved in such expressions, and nuclear factor (NF)-kB plays a pivotal role in attracting binding and transmigration of leukocytes into the inflammatory site. It has been demonstrated in *in vitro* studies using EGCg and ECG, wherein there were a reduced cytokine-induced VCAM-1 expression and monocyte adhesion to endothelium cells (Ecs) in a dose-dependent manner. In addition to the above, EGCg also showed an inhibitory effect on phorbol 12-myristate 13-acetate-induced monocyte chemoattractant protein-1 (MCP-1) mRNA and protein expression in human Ecs, and thus reduced the migration of monocytes that was mediated through the suppression of p38 mitogen-activated protein kinase (p38) and NF-kB activation (Hong et al., 2007; Ludwig et al., 2004). Recently, an EGCg-attenuated cardiac gap junction was studied on high-glucose-induced neonatal rat cardiomycytes, wherein EGCg (40 µM) is reported to activate the time-dependent phosphorylated Erk, JNK, and p38 mitogen-activated protein kinase (MAPK) (Yu et al., 2010). In another study by Tadano et al. (2010), the EGCg effect on cardiac muscle fibers was reported to decrease Ca^{2+} sensitivity of cardiac myofilaments, probably through its interaction with cardiac troponin C in a transgenic mouse model.

Mechanistically, a number of oxidative stresses are often involved in vascular inflammation and pathogenesis. At this stage, the redox-sensitive transcription nuclear factors NF-kB and activator protein-1 (AP-1) play a major role in cellular stimulations. Green tea polyphenols and EGCg have shown a reducing AP-1 activity via diminished phosphorylation of c-Jun (Aneja et al., 2004), and decreased NF-kB and tumor necrosis factor alpha (TNF-α), which are mediators in the inflammatory process. Since several adhesion molecules and pro-inflammatory cytokines and chemokines are regulated by NF-kB, this suggests that green tea polyphenols' suppression of NF-kB activity might reduce the pro-inflammatory molecules, while the matrix metalloproteinases (MMPs), which are a family of zinc-dependent endopeptitases, are responsible for degrading extracellular matrix protein and play an essential role in cell migration as well as tissue remodeling. MMP activity is elevated concomitant with increased infiltration of inflammatory cells, vascular smooth muscle cell (VSMC) hypertrophy migration, and proliferation in the atherosclerosis processes. In angiogenesis, which is an octapeptide involved in hypertrophy and proliferation of VSMC, the three main groups of MMPs, MMP-2 (type IV collagenase) and MMP-9 (type V collagenase and gelatinase B), are involved. MMP-2 is also involved in myocardial injury (Wang et al., 2002), and instability of MMPs may cause a number of effects during progression and maturation of atherosclerosis. EGC and

EGCg inhibit matrix protein degradation and VSMC invasion via reducing the gelatinolytic activity of MMP-2 (Bedoui et al., 2005; Maeda et al., 2003). In addition, EGCg also inhibits the MMP-9 expression, which is involved in the progression of atherosclerosis lesions (Kim and Moon, 2005). This could be regarded as an important mechanism in the prevention of atherosclerosis, which is stimulated by VSMC injury to the blood vessels.

Platelet activation and aggregation are generally associated with vascular endothelial injury, wherein rapid platelet aggregation forms the hemostatic plugs and arterial thrombi, which could trigger acute vascular events such as myocardial infarctions. Green tea polyphenols' effect on platelet aggregation has also been studied both *in vitro* and *in vivo* (Kang et al., 2001; Sachinidis et al., 2002). It is suggested that the antithrombotic activities of polyphenols might be due to antiplatelet activities rather than anticoagulation. Furthermore, an intracellular calcium concentration could be one of the factors in platelet aggregation and activation. Thus, green tea polyphenols inhibit the intracellular calcium mobilization via activation of $Ca2+$ -ATPase. Also, ECG and EGCg might reduce the platelet activation factors (PAFs) by inhibition of acetyl-CoA. lyso PAFacetyltransferase (Sugatani et al., 2004). In the inflammatory condition, arachidonic acid, a plasma membrane-bound fatty acid, could be released and readily metabolized by platelet generating prostaglandin, endoperoxides, and thromboxane A2, which are well-recognized key molecules for platelet aggregation and activation. Moreover, some studies reported (Jin et al., 2008; Son et al., 2004) that green tea polyphenols are capable of suppressing the PAFs, prostaglandin D2 (PGD2), and thromboxane A2 (TXA2). Also, a number of human studies have investigated the effect of flavonoid-rich foods and beverages on platelet function (Holt et al., 2006). Apart from the above, an abnormality of lipid metabolism is one of the risk factors of CVDs. Development of atherosclerotic plaque is related to elevated plasma lipids such as cholesterol, phospholipids, fatty acid, and triglycerides.

Green tea polyphenols are reported to reduce blood cholesterol levels and prevent an accumulation of cholesterol in various liver and heart tissues. *In vitro* study has expressed that EC, EGC, ECG, or EGCG at 5 µM could suppress intracellular lipid accumulation (Furuyashiki et al., 2004). The cholesterol-lowering effect of green tea polyphenols suggested its influence on intestinal lipid absorption in animal studies. The structural difference of green tea polyphenols on the inhibition of lipid absorption is well known, showing that the gallate ester component of green tea polyphenols, such as in EGCG and ECG, is comparatively much more effective than in EC and EGC (Ikeda et al., 1992). Furthermore, its absorption in the digestive tract is effectively involved in prevention of CVDs, wherein reduced oxidative damage and cholesterol reduction are in the true list of important mechanisms for cardiovascular protection.

7.6 Conclusion

Dietary foods containing an abundance of flavonoids and flavonoid polymers are capable of altering metabolic processes and thus provide positive health benefits. Although the worldwide consumption of tea beverage is rather large, it has been recorded that dietary intake of flavonoids varies from country to country, among various populations, as well as by differences in habitual food intakes. Many studies on health benefits of green tea have been linked to their polyphenol content. Flavanols, a subclass of flavonoids, contained in green tea provide a number of physiological benefits owing to their antioxidant properties. Cabrera et al. (2003, 2006) have demonstrated that green tea polyphenol content also varies among different types of commercial teas due to geographical and agricultural practice. Available information suggests that green tea polyphenols are likely to provide modest protection against cardiovascular diseases. Significance of processing parameters (infusion and temperature) to get the suitable functionality with controlled content of green tea polyphenols consistently is a key factor to observe the beneficial effect on CVD risk factors (Nantz et al., 2009). The effect of tea flavonoids on endothelial function is a major example of its effect on risk of cardiovascular disease. It is noteworthy to mention that the role of green tea polyphenols in CVD prevention has been hypothesized from a large number of studies in cell culture as well as animal models of cardiovascular/metabolic diseases carried out over more than a decade. Cellular mechanism, associated to CVDs, is recognized mostly on its antioxidative properties for indirect risk reduction. Such an oxidative stress is one of the risk factors of atherosclerosis. In addition to antioxidant properties, green tea polyphenols are capable of reducing the caveolin-1 mRNA and phosphorylated ERK1/2 in microvessel fragments of cerebral ischemic tissue to prevent cerebral ischemia. Also, green tea polyphenols help elevate the phase 2 enzymes in myocardial ischemia-reperfusion (I/R) injury and significantly reduce the risk of heart damage. In general, green tea polyphenols provide promising health benefits against cardiovascular disease, as they act as oxidants and prevent tissue damage by free radicals, thus lowering the risk of CVDs. The abundance of evidence related to the nutraceutical characteristics of green tea polyphenols has stimulated the food and supplement industries to develop and promote green tea polyphenol-rich products to heal impairments to cardiovascular system.

References

Ahn, H.Y., Kim, C.H., and Sandra, T.D. 2010. Epigallocatechin-3-gallate regulates NADPH oxidase expression in human umbilical vein endothelial cells. *Kor J Physiol Pharmacol*, 14: 325–329.

Akhlaghi, M., and Bandy, B. 2010. Dietary green tea extract increases phase 2 enzyme activities in protecting. *Nutr Res*, 30: 32–39.

Aneja, R., Hake, P.W., Burroughs, T.J., Denenberg, A.G., Wong, H.R., and Zingarelli, B. 2004. Epigallocatechin, a green tea polyphenol, attenuates myocardial ischemia reperfusion injury in rats. *Mol Med*, 10: 55–62.

Arab, L., Liu, W., and Elashoff, D. 2009. Green and black tea consumption and risk of stroke: A meta-analysis. *Stroke*, 40: 1786–1792.

Bakker, G.C.M., Erk, M.I.V., Pellis, L., Wopereis, S., Rubingh, C.M., Cnubben, N.H.P., Kooistra, T., Ommen, B.V., and Hendriks, H.F. 2010. An anti-inflammatory dietary mix modulates inflammation and oxidative and metabolic stress in overweight men: A nutrigenomics approach. *Am J Clin Nutr*, 91: 1044–1059.

Basu, A., Du, M., Sanchez, K., Leyva, M.J., Betts, N.M., Blevins, S., Wu, M., Aston, C.E., and Lyons, T.J. 2011. Green tea minimally affects biomarkers of inflammation in obese subjects with metabolic syndrome. *Nutrition*, 27: 206–213.

Bedoui, J.E., Oak, M.H., Anglard, P., and Schini-Kerth, V.B. 2005. Catechins prevent vascular smooth muscle cell invasion by inhibiting MT1-MMP activity and MMP-2 expression. *Cardiovasc Res*, 67: 317–325.

Bettuzzi, S., Brausi, M., Rizzi, F., Castagnetti, G., Peracchia, G., and Corti, A. 2006. Chemoprevention of human prostate cancer by oral administration of green tea catechins in volunteers with high-grade prostate intraepithelial neoplasia: A preliminary report from a one-year proof-of-principle study. *Cancer Res*, 66: 1234–1240.

Cabrera, C., Gimenez, R., and Lopez, C.M. 2003. Determination of tea components with antioxidant activity. *J Agric Food Chem*, 51: 4427–4235.

Cabrera, C., Artacho, R., and Giménez, R. 2006. Beneficial effects of green tea; a review. *J Am Coll Nutr*, 25: 79–99.

Chyu, K.Y., Babbidge, S.M., Zhao, X., Dandillaya, R., Reitveld, A.G., Yano, J., Dimayuga, P., Cercek, B., and Shah, P.K. 2004. Differential effect of green tea derived catechin on developing versus established atherosclerosis in apolipoprotein E-null mice. *Circulation*, 109: 2448–2453.

Cooper, R., Morré, D.J., and Morré, D.M. 2005. Medicinal benefits of green tea: Part I. Review of noncancer health benefits. *J Altern Complement Med*, 11: 521–528.

Dhalla, N.S., Temsah, R.M., and Netricadan, T. 2000. Role of oxidative stress in cardiovascular diseases. *J Hypertens*, 18: 655–673.

Duffy, S.J., Vita, J.A., and Keaney Jr., J.F. 1999. Antioxidants and endothelial function. *Heart Failure*, 15: 135–152.

Finkel, T., and Holbrook, N.J. 2000. Oxidants, oxidative stress and the biology of ageing. *Nature*, 408: 239–247.

Flynn, R.W.V., MacWalter, R.S.M., and Donney, A.S.F. 2008. The cost of cerebral ischaemia. *Neuropharmacology*, 55: 250–256.

Furuyashiki, T., Nagayasu, H., Aoki, Y., Bessho, H., Hashimoto, T., Kanazawa, K., and Ashida, H. 2004. Tea catechin suppresses adipocyte differentiation accompanied by down-regulation of PPARgamma2 and C/EBPalpha in 3T3-L1 cells. *Biosci Biotechnol Biochem*, 68: 2353–2359.

Gokce, N., Keaney Jr., J.F., Frei, B., Holbrook, M., Olesiak, M., Zachariah, B.J., Leewenburgh, C., Heinecke, J.W., and Vita, J.A. 1999. Long-term ascorbic acid administration reverses endothelial vasomotor dysfunction in patients with coronary artery disease. *Circulation*, 99: 3234–3240.

Graham, D.Y., Lew, G.M., and Klein, P.D.1992. Effect of treatment of *Helicobacter pylori* infection on the long-term recurrence of gastric or duodenal ulcer: Randomized controlled study. *Ann Intern Med*, 116: 705–708.

Gross, G., Meyer, K.G., Press, H., Thielert, C., Tawfik, H., and Mescheder, A. 2007. A randomized, double-blind, four-arm parallel-group, placebo-controlled phase II/III study to investigate the clinical efficacy of two galenic formulations of Polyphenon E in the treatment of external genital warts. *J Eur Acad Dermatol Venereol*, 21: 1404–1412.

Han, D.W., Suh, H., Park, Y.H., Cho, B.K., Hyon, S.H., and Park, Y.H. 2003. Preservation of human saphenous vein against reactive oxygen species-induced oxidative stress by green tea polyphenol pretreatment. *Artif Organs*, 27: 1137–1142.

Heijnen, C.G.M., Haenen, G.R.M., Wiseman, S.A., Tijburg, L.B.M., and Bast, A. 2000. The interaction of tea flavonoids with the NO-system: Discrimination between good and bad NO. *Food Chem*, 70: 365–370.

Hodgson, J.M. 2008. Tea flavonoids and cardiovascular disease. *Asia Pac J Clin Nutr*, 17(S1): 288–290.

Holt, R.R., Actis-Goretta, L., Momma, T.Y., and Keen, C.L. 2006. Dietary flavanols and platelet reactivity. *J Cardio Pharmacol*, 47: S187–S196.

Hong, M.H., Kim, M.H., Chang, H.J., Kim, N.H., Shin, B.A., Ahn, B.W., and Jung, Y.D. 2007. (–)-Epigallocatechin-3-gallate inhibits monocyte chemotactic protein-1 expression in endothelial cells via blocking NF-kappaB signaling. *Life Sci*, 80: 1957–1965.

Huxley, R.R., and Neil, H.A. 2003. The relation between dietary flavonol intake and coronary heart disease mortality: A meta-analysis of prospective cohort studies. *Eur J Clin Nutr*, 57: 904–908.

Ikeda, I., Imasato, Y., Sasaki, E., Nakayama, M., Nagao, H., Takeo, T., Yayabe, F., and Sugano, M. 1992. Tea catechins decrease micellar solubility and intestinal absorption of cholesterol in rats. *Biochim Biophys Acta*, 1127: 141–146.

Ikeda, M., Suzuki, C., Umegaki, K., Saito, K., Tabuchi, M., and Tomita, T. 2007. Preventive effects of green tea catechins on spontaneous stroke in rats. *Med Sci Monit*, 13: BR 40–BR 45.

Jin, Y.R., Im, J.H., Park, E.S., Cho, M.R., Han, X.H., Lee, J.J., Lim, Y., Kim, T.J., and Yun, Y.P. 2008. Antiplatelet activity of epigallocatechin gallate is mediated by the inhibition of PLCgamma2 phosphorylation, elevation of PGD2 production, and maintaining calcium-ATPase activity. *J Cardio Pharmacol*, 51: 45–54.

Kang, W.S., Chung, K.H., Chung, J.H., Lee, J.Y., Park, J.B., Zhang, Y.H., Yoo, H.S., and Yun, Y.P. 2001. Antiplatelet activity of green tea catechins is mediated by inhibition of cytoplasmic calcium increase. *J Cardio Pharmacol*, 38: 875–884.

Kaul, D., Sikand, K., and Shukla, A.R. 2004. Effect of green tea polyphenols on the genes with atherosclerotic potential. *Phytother Res*, 18: 177–179.

Kim, C.H., and Moon, S.K. 2005. Epigallocatechin-3-gallate causes the p21/WAF1-mediated G(1)-phase arrest of cell cycle and inhibits matrix metalloproteinase-9 expression in TNF-alpha-induced vascular smooth muscle cells. *Arch Biochem Biophys*, 435: 264–272.

Koo, M.W.L., and Cho, C.H. 2004. Pharmacological effects of green tea on the gastrointestinal system. *Eur J Pharmacol*, 500: 177–185.

Kurahashi, N., Sasazuki, S., Iwasaki, M., Inoue, M., and Tsugane, S. 2008. Green tea consumption and prostate cancer risk in Japanese men: A prospective study for the JPHC Study Group. *Am J Epidemiol*, 167: 71–77.

Kuriyama, S., Shimazu, T., Ohmori, K., Kikuchina, N., Nakaya, N., Nishino, Y., Tsubono, Y., and Tsuji, I. 2006. Green tea consumption and mortality due to cardiovascular disease, cancer, and all causes in Japan: The Ohsaki study. *JAMA*, 296: 1255–1265.

Leeuwenburgh, C., Hardy, M.M., Hazen, S.L., Wagner, P., Oh-ishi, S., Steinbrecher, U.P., and Heinecke, J.W. 1997. Reactive nitrogen intermediates promote low density lipoprotein oxidation in human atherosclerotic intima. *J Biol Chem*, 272: 1433–1436.

Li, C., and Jackson, R.M. 2002. Reactive species mechanisms of cellular hypoxia-reoxygenation injury. *Am J Physiol Cell Physiol*, 282: C227–C241.

Ludwig, A., Lorenz, M., Grimbo, N., Steinle, F., Meiners, S., Bartsch, C., Stangl, K., Baumann, G., and Stangl, V. 2004. The tea flavonoid epigallocatechin-3-gallate reduces cytokine-induced VCAM-1 expression and monocyte adhesion to endothelial cells. *Biochem Biophys Res Commun*, 316: 659–665.

Maeda, K., Kuzuya, M., Cheng, X.W., Asai, T., Kanda, S., Tamaya-Mori, N., Sasaki, T., Shibata,T., and Iguchi, A. 2003. Green tea catechins inhibit the cultured smooth muscle cell invasion through the basement barrier. *Atherosclerosis*, 166: 23–30.

Mineharu, Y., Koizumi, A., Wada, Y., Iso, H., Watanabe, Y., Date, C., Yamamoto, A., Kikuchi, S., Inaba, Y., Toyoshima, H., Kondo, T., and Tamakoshi, A. 2011. Coffee, green tea, black tea and oolong tea consumption and risk of mortality from cardiovascular disease in Japanese men and women. *J Epidemiol Commun Health*, 65: 230–240.

Misra, M.K., Sarwat, M., Bhakuni, P., Tuteja, R., and Tuteja, N. 2009. Oxidative stress and ischemic myocardial syndromes. *Med Sci Monit*, 5: RA209–RA219.

Miura, Y., Chiba, T., Miura, S., Tomita, I., Umegaki, K., Ikeda, M., and Tomita, T. 2000. Green tea polyphenols (flavan 3-ols) prevent oxidative modification of low density lipoproteins: An *ex vivo* study in humans. *J Nutr Biochem*, 11: 216–222.

Na, H.K., and Surh, Y.J. 2008. Modulation of Nrf2-mediated antioxidant and detoxifying enzyme induction by the green tea polyphenol EGCG. *Food Chem Toxicol*, 46: 1271–1278.

Nagao, T., Hase, T., and Tokimitsu, I. 2007. A green tea extract high in catechins reduces body fat and cardiovascular risks in humans. *Obesity (Silver Spring)*, 15: 1473–1483.

Nantz, M.P., Rowe, C.A., Bukowski, J.F., Susan, S., and Percival, S.S. 2009. Standardized capsule of *Camellia sinensis* lowers cardiovascular risk factors in a randomized, double-blind, placebo-controlled study. *Nutrition*, 25: 147–154.

Negishi, H., Xu, J.W., Ikeda, K., Njelekela, M., Nara, Y., and Yamori, Y. 2004. Black and green tea polyphenols attenuate blood pressure increases in stroke-prone spontaneously hypertensive rats. *J Nutr*, 134: 38–42.

Peters, U., Poole, C., and Arab, L. 2001. Does tea affect cardiovascular disease? A meta-analysis. *Am J Epidemiol*,154: 495–503.

Potenza, M.A., Marasciulo, F.L, Tarquinio, M., Tiravanti, E., Colantuono G., Federici, A., Kim, J.A., Quon, M.J., and Montagnani, M. 2007. EGCG, a green tea polyphenol, improves endothelial function and insulin sensitivity, reduces blood pressure, and protects against myocardial I/R injury in SHR. *Am J Physiol Endocrinol Met*, 292: E1378–E1387.

Qin, B., Polansky, M.M., Harry, D., and Anderson, R.A. 2010. Green tea polyphenols improve cardiac muscle mRNA and protein levels of signal pathways related to insulin and lipid metabolism and inflammation in insulin-resistant rats. *Mol Nutr Food Res*, 54: S14–S23.

Ramesh, E., Geraldine, P., and Thomas, P.A. 2010. Regulatory effect of epigallo-catechin gallate on the expression of C-reactive protein and other inflammatory markers in an experimental model of atherosclerosis. *Chemicobiol Interactions*, 183: 125–132.

Riemersma, R.A., Rice-Evans, C.A., Tyrrell, R.M., and Clifford, M.N. 2001. Tea flavonoids and cardiovascular health. *Q J Med*, 94: 277–282.

Sachinidis, A., Skach, R.A., Seul, C., Ko, Y., Hescheler, J., Ahn, H.Y., and Fingerle, J. 2002. Inhibition of the PDGF beta-receptor tyrosine phosphorylation and its downstream intracellular signal transduction pathway in rat and human vascular smooth muscle cells by different catechins. *FASEB J*, 16: 893–895.

Sajilata, M.G., Bajaj, P.R., and Singhal, R.S. 2008. Tea polyphenols as nutraceuticals. *Comp Rev Food Sci Food Safety*, 7: 229–254.

Sano, J., Inami, S., Seimiya, K., Ohba, T., Sakai, S., Takano, T., and Mizuno, K. 2004. Effects of green tea intake on the development of coronary artery disease. *Circ J*, 68: 665–670.

Sasazuki, S., Kodama, H., Yoshimasu, K., Liu Y., Mohri, M., and Takeshita, A. 2000. Relation between green tea consumption and the severity of coronary atherosclerosis among Japanese men and woman. *Ann Epidemiol*, 10: 401–408.

Skrzydlewska, E., Ostrowska, J., Farbiszewski, R., and Michalak, K. 2002. Protective effect of green tea against lipid peroxidation in the rat liver, blood serum and brain. *Phytomedicine*, 9: 232–238.

Son, D.J., Cho, M.R., Jin, Y.R., Kim, S.Y., Park, Y.H., Lee, S.H., Akiba, S., Sato, T., and Yun, Y.P. 2004. Antiplatelet effect of green tea catechins: A possible mechanism through arachidonic acid pathway. *Prostaglandins Leukotrines Essential Fatty Acids*, 71: 25–31.

Stangl, V., Dreger, H., Stangl, K., and Lorenz, M. 2007. Molecular targets of tea polyphenols in cardiovascular system. *Cardiovas Res*, 73: 348–358.

Sugatani, J., Fukujawi, N., Ujihara, K., Yoshinari, K., Abe, I., Noguchi, H., and Miwa, M. 2004. Tea polyphenols inhibit acetyl-CoA: 1-alkyl-sn-glycero-3-phosphocholine acetyltransferase (a key enzyme in platelet-activating factor biosynthesis) and platelet-activating factor-induced platelet aggregation. *Int Arch Allergy Immunol*, 134: 17–28.

Tadano, N., Du, C.K., Yumoto, F., Morimoto, S., Ohta, M., Xie, M.F., Nagata, K., Zhan, D.Y., Lu, Q.W., Miwa, Y., Takahashi-Yanaga, F., Tanokura, M., Ohtsuki, I., and Sasaguri, T. 2010. Biological actions of green tea catechins on cardiac troponin C. *Br J Pharmacol*, 161: 1034–1043.

Townsend, P.A., Scarabelli, T.M., Pasini, E., Gitti, G., Menegazzi, M., and Suzuki, H. 2004. Epigallocatechin-3-gallate inhibits STAT-1 activation and protects cardiac myocytes from ischemia/reperfusion-induced apoptosis. *FASEB J*, 18: 1621–1623.

Velayutham, P., Babu, A., and Liu, D. 2008. Green tea catechins and cardiovascular health: An update. *Curr Med Chem*, 15: 1840–1850.

Venardos, K.M., and Kaye, D.M. 2007. Myocardial ischemia-reperfusion injury, antioxidant enzyme systems, and selenium: A review. *Curr Med Chem*, 14: 1539–1549.

Victor, V.M., Milagros, R., Eva, S., Celia, B., Katherine, G.M., and Antonio, H.M. 2009. Oxidative stress, endothelial dysfunction and atherosclerosis. *Curr Pharm Design*, 15: 2988–3002.

Vita, J.A. 2005. Polyphenols and cardiovascular disease: Effects on endothelial and platelet function. *Am J Clin Nutr*, 81(Suppl.): 292S–297S.

Wang, W., Sawicki, G., and Schulz, R. 2002. Peroxynitrite-induced myocardial injury is mediated through matrix metalloproteinase-2. *Card Res*, 53: 165–174.

Westerterp-Plantenga, M.S., Lejeune, M.P., and Kovacs, E.M. 2005. Body weight loss and weight maintenance in relation to habitual caffeine intake and green tea supplementation. *Obes Res*, 13: 1195–1204.

WHO. 2012. Global atlas on cardiovascular disease prevention and control. http://www.who.int/cardiovascular_diseases/.

Wolfram, S. 2007. Effects of green tea and EGCG on cardiovascular and metabolic health. *J Am Coll Nutr*, 26: 373S–388S.

Yamazaki, K.G., Romero-Perez, D., Barraza-Hidalgo, M., Cruz, M., Rivas, M., Cortez-Gomez, B., Ceballos, G., and Villarreal, F. 2008. Short- and long-term effects of (–)-epicatechin on myocardial ischemia-reperfusion injury. *Am J Physiol Heart Circul Physiol*, 295: H761–H767.

Yokozawa T., Nakagawa, T., and Kitani, K. 2002. antioxidative activity of green tea polyphenol in cholesterol-fed rats. *J Agric Food Chem*, 50: 3549–3552.

Yu, L., Zhao, Y., Fan, Y., Wang, M., Xu, S., and Fu, G. 2010. Epigallocatechin-3 gallate, a green tea catechin, attenuated the downregulation of the cardiac gap junction induced by high glucose in neonatal rat cardiomyocytes. *Cell Physiol Biochem*, 26: 403–412.

Zaveri, N.T. 2006. Green tea and its polyphenolic catechins: Medicinal uses in cancer and noncancer applications. *Life Sci*, 78: 2073–2080.

Zhang, S., Liu, Y., Zhao, Z., and Xue, Y. 2010. Effects of green tea polyphenols on caveolin-1 of microvessel fragments in rats with cerebral ischemia. *Neurol Res*, 32: 963–970.

8

Green Tea Polyphenols in Weight Management (Obesity) and Diabetes

Tadashi Sakuma, Hideto Takase, Tadashi Hase, and Ichiro Tokimitsu

Contents

8.1 Introduction

The Japanese lifestyle has changed markedly over the last 50 years: the era of nutritional deficiency has shifted to the age of plenty. Together with the changes in the social environment, such as the development of transportation, the evolving lifestyle has led to an increase in health problems, such as obesity and lifestyle-related diseases. Westernized diets and a sedentary lifestyle are reported to be the main causes of the increase. Visceral fat-type obesity was recently clarified to be most closely associated with lifestyle-related diseases such as hyperlipidemia, hypertension, and hyperglycemia. Visceral fat-type obesity concomitant with other symptoms, such as hypertension and hyperlipidemia, is defined as metabolic syndrome (Examination Committee of Criteria for "Metabolic Syndrome" in Japan, 2005), and the syndrome is considered to increase the risk of arteriosclerotic disease (Matsuzawa et al., 1995). The latest National Health and Nutrition Survey (2008) (Ministry of Heath, Labour, and Welfare, Japan, 2009) reported that one in two men and one in five women aged 40 to 74 years are strongly suspected of having or considered to develop metabolic syndrome (visceral fat-type obesity). Improvement of individual lifestyles (dietary therapy and combination of dietary and exercise therapies) is very important to treat and prevent obesity and lifestyle-related diseases. Changing lifestyles markedly, however, can be very difficult. Therefore, the development of an effective treatment against obesity that can be readily incorporated into individual daily life is needed.

8.2 Evidence of Effects of Tea Catechins on Reducing Body Fat Levels in Studies

Tea catechins are polyphenols contained in green tea, a beverage long consumed by the Japanese, and various physiologic actions, such as antioxidative and anticancer effects, have been reported (Katiyar and Mukhtar, 1996; Kurahashi et al., 2008; Nozawa et al., 2002; Yoshino et al., 1994). Recent epidemiologic studies suggest that tea catechins also reduce the risk of circulatory disease (Kuriyama et al., 2006b; Mukhtar and Ahmad, 2000) and prevent cognitive disorders (Kuriyama et al., 2006a). Tea catechins have been shown to reduce body fat levels (visceral fat) in clinical studies, and continuous ingestion of beverages with a high-concentration of tea increases energy expenditure in humans and animals. One mechanism by which tea catechins reduce body fat is activation of lipid metabolism in the liver and muscle, allowing for ready utilization of fat as energy. Recent findings are introduced below.

8.2.1 Body Fat-Reducing Effects (Animal Studies)

The C57BL/6J mouse is an animal model of dietary-induced obesity of humans (Murase et al., 2002). In this model, obesity is not observed when mice are given a standard diet containing 5% fat. A high-fat, high-sucrose diet containing 30% fat and 13% sucrose, however, leads to a significant increase in body weight and body fat (visceral fat tissue) weight compared to a standard diet. The addition of caffeine-free tea catechins (0.1, 0.2, and 0.5%) to a high-fat, high-sucrose diet over a period of 11 months significantly inhibits body weight and body fat weight gains in a dose-dependent manner (Figures 8.1 and 8.2) (Murase et al., 2002) compared to a high-fat, high-sucrose diet alone.

FIGURE 8.1
Tea catechin-induced inhibition of weight gain in a mouse obesity model (Murase et al., 2002). LD: Standard diet containing 5% lipids. HD: High-fat, high-sucrose diet containing 30% lipids and 13% sucrose. Values are means ± SD ($n = 5$). Significantly different from HD (tea catechins = 0 mg) group by t-test (unpaired; *, $p < 0.05$; **, $p < 0.01$; ***, $p < 0.001$).

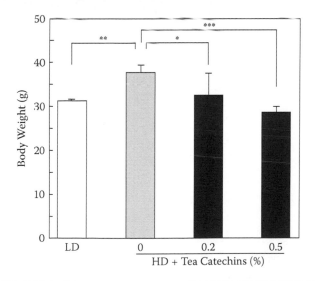

FIGURE 8.2
Tea catechin-induced inhibition of an increase in body fat weight in a mouse obesity model (Murase et al., 2002). LD: Standard diet containing 5% lipids. HD: High-fat, high-sucrose diet containing 30% lipids and 13% sucrose. Values are means ± SD ($n = 5$). Significantly different from HD (tea catechins = 0 mg) group by t-test (unpaired, *, $p < 0.05$; **, $p < 0.01$; ***, $p < 0.001$).

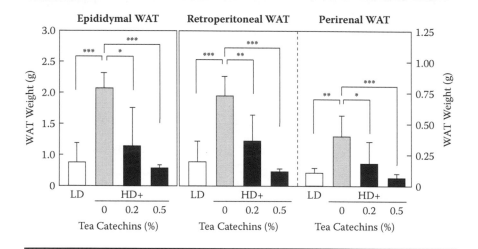

8.2.2 Body Fat-Reducing Effects (Human Studies)

Many studies have reported the body fat-reducing effects of tea catechins in humans. The following findings have been reported:

1. The effective dose is 500 to 600 mg/day (Nagao et al., 2001, Takase et al., 2008a).
2. Tea catechins are effective in people inclined to be obese (with higher body mass index (BMI)) (Hase et al., 2001; Kozuma et al., 2005; Maki et al., 2009; Nagao et al., 2005, 2007; Otsuka et al., 2002; Takase et al., 2008b; Tsuchida et al., 2002).
3. Tea catechins have similar effects between sexes (Kozuma et al., 2005; Tsuchida et al., 2002).
4. Rebound phenomena are not observed after discontinuation (Kozuma et al., 2005; Tsuchida et al., 2002).

5. Physical and hematologic findings are normal. A total of more than 1000 subjects have been investigated. (Kozuma et al., 2005; Tsuchida et al., 2002).

In one study (Nagao et al., 2007), 240 adults with a mean age of 41.7 ± 9.9 years and a mean BMI of 26.8 ± 2.0 kg/m² were divided into two groups in a randomized double-blind, controlled parallel multicenter trial, a control group (n = 117, 49 women and 68 men; tea catechins = 96.3 mg/day), and a catechin group (n = 123, 51 women and 72 men; tea catechins = 582.8 mg/day). Test beverages (340 ml) containing tea catechins were given for 12 weeks. After 12 weeks, the BMI value, abdominal fat areas (visceral and subcutaneous fat areas) on abdominal CT images, and total fat area (TFA) (calculated by adding the visceral fat area (VFA) to the subcutaneous fat area (SFA)) in the catechin group were significantly lower than those in the control group (rates of decrease: BMI, 0.6 kg/m²; body weight, 1.6 kg; VFA, 6.4 cm²; SFA, 9.7 cm²; and TFA, 16.1 cm², respectively) (Table 8.1 and Figure 8.3A) (Nagao et al., 2007). In our subclass analysis with respect to sex, the effects were similar between men and women.

Continuous ingestion of the high-concentration tea catechin beverage significantly reduced not only the body weight and body fat level, but also waist circumference, systolic blood pressure (initial value, 130 or higher), and serum LDL cholesterol levels, suggesting that tea catechins contribute to reducing the risks of obesity and cardiovascular diseases (Table 8.1 and Figure 8.3B and C) (Nagao et al., 2007).

Published randomized control studies with groups having a daily intake of more than 539.7 mg of tea catechins in a beverage were selected for a meta-analysis. All participants were overweight/obese Japanese adults without diabetes or any medication. The total number of participants was 902 (519 men and 383 women, mean BMI = 26.3kg/m²) (Kataoka et al., 2004; Kozuma et al., 2005; Nagao et al., 2005, 2007; Otsuka et al., 2002; Takase et al., 2008b; Tsuchida et al., 2002). Participants were classified into three groups based on the daily intake of tea catechins: a low-catechin group (LC) (0.0 to 277.9 mg, n = 457), a high-catechin 1 group (HC1) (539.7 to 587.5 mg, n = 403), and a high-catechin 2 group (HC2) (689.9 to 844.7 mg, n = 42). The VFA reduction determined by computed tomography significantly correlated with the total catechin intake during the intervention (p = 0.006) and was significantly greater in the HC group than in the LC group (Figure 8.4) (Takase et al., 2008a). BMI, waist circumference, and blood pressure were also significantly reduced in the HC group. In addition, the prevalence of metabolic syndrome was reduced in the HC group (Table 8.2) (Takase et al., 2008a, 2009).

TABLE 8.1
Effects of Tea Catechin Ingestion for 12 Weeks in Humans

	Catechin/ Control[a]	n	Week 0	Week 12	Δ Value at Week 12[b]
Body weight (kg)	Catechin	123	73.3 ± 9.7	71.6 ± 9.8	−1.7 ± 1.5]*
	Control	117	72.1 ± 10.0	72.1 ± 10.3	−0.1 ± 1.7
Body mass index (kg/m²)	Catechin	123	26.9 ± 1.9	26.2 ± 1.9	−0.6 ± 0.6]*
	Control	117	26.7 ± 2.1	26.6 ± 2.2	−0.0 ± 0.6
Waist circumference (cm)	Catechin	123	87.2 ± 5.2	84.7 ± 5.5	−2.5 ± 2.2]*
	Control	117	86.5 ± 6.1	86.5 ± 6.7	0.0 ± 2.5
Total fat area (cm²)	Catechin	123	324.3 ± 79.3	308.4 ± 79.9	−16.0 ± 46.6]*
	Control	117	315.8 ± 77.1	316.0 ± 79.0	0.1 ± 32.6
Visceral fat area (cm²)	Catechin	123	109.2 ± 42.3	98.9 ± 38.6	−10.3 ± 23.3]*
	Control	117	107.7 ± 44.0	103.8 ± 38.9	−3.9 ± 24.9
Subcutaneous fat area (cm²)	Catechin	123	215.2 ± 66.9	209.5 ± 66.3	−5.7 ± 38.5]*
	Control	117	208.1 ± 60.7	212.1 ± 64.9	4.0 ± 24.8
Systolic blood pressure (mmHg)	Catechin	56	139.8 ± 7.9	130.8 ± 13.0	−9.0 ± 12.1]*
Initial ≥ 130 min Hg	Control	61	139.3 ± 8.6	136.4 ± 12.7	−2.9 ± 12.0
Serum LDL cholesterol (mg/dl)	Catechin	123	131.9 ± 33.3	128.0 ± 34.4	−3.5 ± 18.9]*
	Control	117	129.2 ± 31.7	130.7 ± 33.3	1.6 ± 20.1

Source. Nagao, T., et al., *Obesity* 15: 1473–1483, 2007.

Note. Values are mean ± SD.

[a] Control group (tea catechins = 96.3 mg/day); catechin group (tea catechins = 582.8 mg/day).

[b] The value is the change from week 0 to week 12.

* Significant difference in Δ value at week 12 compared with controls (tea catechins = 96.3 mg/day) by unpaired *t*-test (two-sided, *, $p < 0.05$).

8.3 Effects of Tea Catechins on Energy and Fat Metabolism in Humans

Body fat changes depend on the balance between energy intake and expenditure. Body fat weight decreases when energy intake decreases or when energy expenditure increases. Humans expend 60 to 75%, approximately 20%, and approximately 10% of their daily energy as basal metabolic energy, physical activity energy, and diet-induced thermogenesis (DIT), respectively (Granata and Brandon, 2002). DIT is lower in obese patients than in normal-weight persons (Laville et al., 1993), and postprandial lipid oxidation is also reduced (Thomas et al., 1992) in obese patients.

FIGURE 8.3

Effects of tea catechins ingestion for 12 weeks in humans (Nagao et al., 2007). (A): Body fat area (☐: control group; n = 117, tea catechins = 96.3 mg/day; ■: catechin group; n = 123, tea catechins = 582.8 mg/day). (B): Systolic blood pressure (initial ≥ 130 mmHg; ☐: control group, n = 61, tea catechins = 96.3 mg/day; ■: catechin group, n = 56, tea catechins = 582.8 mg/day). (C): Serum LDL cholesterol (☐: control group, n = 117, tea catechins = 96.3 mg/day; ■: catechin group, n = 123, tea catechins = 582.8 mg/day). Values as means ± SEM are expressed as changes. Significantly different from control group by t-test (*, p < 0.05).

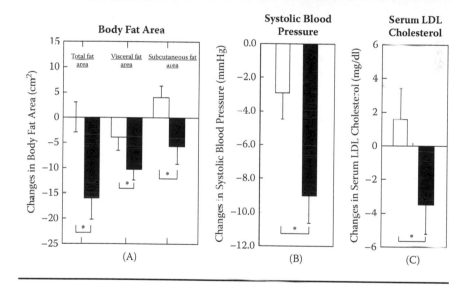

Studies of the effects of continuous consumption of high-concentration tea catechins on energy and fat metabolism in humans are summarized below.

8.3.1 Effects of Continuous Ingestion of High-Concentration Tea Catechins on Fat Metabolism

In a randomized controlled study (Harada et al., 2005), 12 subjects with either normal body weight or obesity (class I), according to the criteria set by

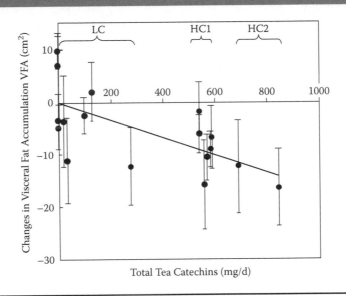

FIGURE 8.4
Effect of long-term ingestion of tea catechins on visceral fat accumulation (Takase et al., 2008a). Groups: LC, low-catechin group (*n* = 457, tea catechins = 0 ~ 277.9 mg/day); HC1, high-catechin group 1 (*n* = 403, tea catechins = 539.7 ~ 587.5 mg/day); HC2, high-catechin group 2 (*n* = 42, tea catechins = 689.9 ~ 844.7 mg/day). Changes in VFA were significantly correlated with total tea catechins (*p* = 0.006). (From Takase, H., et al., *Jpn Pharmacol Ther*, 36: 509–514, 2008. With permission.)

the Japanese Society of Obesity, consumed a beverage containing 77.7 mg of tea catechins (control group) or 592.9 mg of tea catechins (catechin group) for 12 weeks. DIT (energy expenditure) was measured based on oxygen consumption during the 8 h following ingestion of a standard diet (800 kcal) and before and after continuous ingestion. Dietary lipid oxidation based on the excretion of $^{13}CO_2$ after the injection of dietary ^{13}C-labeled lipids during the DIT measurement was also examined. The control group showed no increase in DIT after 12 weeks of ingestion, compared with the baseline levels (week 0). The catechin group, after 12 weeks of ingestion, showed a significantly ($p < 0.05$) higher DIT than the control group (Figure 8.5A) (Harada et al., 2005). The mean rate of increase in DIT over 12 weeks in the catechin group was 38.9 kcal. Furthermore, the increase was more marked in subjects with a higher VFA prior to the start of the study. Fat oxidation was

TABLE 8.2
Effects of Long-Term Ingestion of Tea Catechins on Metabolic Syndrome

	Group	Base	End	Change	t-Test vs. LC
			Measurement		
Body mass index (kg/m²)	LC	26.2 ± 2.6	26.2 ± 2.8	0.0 ± 0.0	
	HC1	26.5 ± 2.5	25.7 ± 2.5	−0.7 ± 0.0	0.000
	HC2	24.7 ± 2.9	24.1 ± 3.0	−0.6 ± 0.1	0.000
Waist (cm)	LC	85.8 ± 8.1	85.5 ± 8.4	−0.3 ± 0.1	
	HC1	85.8 ± 7.8	83.7 ± 7.9	−2.1 ± 0.1	0.000
	HC2	86.8 ± 8.0	84.6 ± 7.8	−2.2 ± 0.4	0.000
Visceral fat area (cm²)	LC	94.6 ± 45.1	95.8 ± 47.1	1.2 ± 0.9	
	HC1	103.8 ± 52.0	95.4 ± 47.5	−8.4 ± 1.0	0.000
	HC2	76.9 ± 25.9	65.3 ± 23.5	−11.6 ± 2.3	0.000
Systolic blood pressure (mmHg)	LC	129 ± 14	129 ± 14	1 ± 0	
	HC1	128 ± 14	126 ± 14	−2 ± 1	0.003
	HC2	127 ± 16	126 ± 16	−1 ± 2	0.437
Diastolic blood pressure (mmHg)	LC	80 ± 10	79 ± 10	0 ± 0	
	HC1	79 ± 10	78 ± 9	−2 ± 0	0.025
	HC2	80 ± 10	78 ± 10	−1 ± 1	0.443
Triglyceride (mg/dl)	LC	129 ± 75	127 ± 72	2 ± 3	
	HC1	144 ± 100	144 ± 105	0 ± 4	0.740
	HC2	123 ± 82	124 ± 81	1 ± 12	0.809
High-density lipoprotein (mg/dl)	LC	55 ± 14	55 ± 14	0 ± 0	
	HC1	55 ± 15	54 ± 14	−1 ± 0	0.364
	HC2	51 ± 11	52 ± 12	2 ± 1	0.078
Blood sugar (mg/dl)	LC	94 ± 16	93 ± 16	−1 ± 1	
	HC1	97 ± 21	94 ± 17	−2 ± 1	0.276
	HC2	94 ± 7	93 ± 7	−1 ± 1	0.930

Source. Data taken from Takase, H., et al., *Jpn Pharmacol Ther*, 36: 509–514, 2008; Takase, H., et al., Effects of Long-Term Ingestion of Tea Catechins on Metabolic Syndrome among Different Criteria: Meta-Analysis of 7 Randomized Controlled Trials. Paper presented at Obesity 2009, the 27th Annual Scientific Meeting of the Obesity Society, Washington, DC, October 24–28, 2009.
Note. Groups: LC, low-catechin group (n = 457, tea catechins = 0 – 277.9 mg/day); HC1, high-catechin group 1 (n = 403, tea catechins = 539.7 – 587.5 mg/day); HC2, high-catechin group 2 (n = 42, tea catechins = 689.9 – 844.7 mg/day).

also significantly enhanced only in the catechin group, and these parameters were significantly higher than those in the control group after 12 weeks of ingestion (Figure 8.5B) (Harada et al., 2005).

These findings suggest that continuous ingestion of catechins increases DIT and postprandial fat oxidation in obese subjects, in whom these measures are reported to be lower than those in normal-weight subjects. These

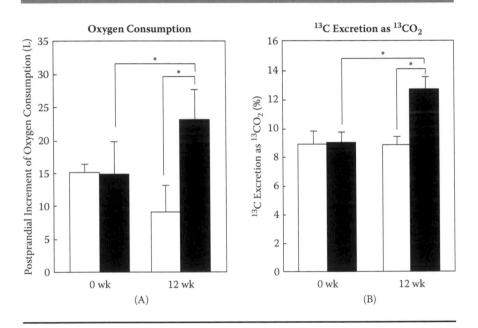

FIGURE 8.5
Effects of continuous ingestion of high-concentration tea catechins on fat metabolism (Harada et al., 2005). (A): DIT (oxygen consumption). (B): Lipid β-oxidation (rate of ^{13}C release). ■: Control group (n = 6, tea catechins = 77.7 mg/day). ■: Catechin group (n = 6, tea catechins = 592.9 mg/day). Values are expressed as the mean ± SEM. Significantly different from control group by t-test (*, p < 0.05). (From Harada, U., et al., *J Health Sci*, 51: 248–252, 2005. With permission.)

effects may contribute to body fat reduction during continuous ingestion of tea catechins.

8.3.2 Effects of High-Concentration Tea Catechins on Lipid Metabolism in Daily Physical Activities

In a recent study (Ota et al., 2005), 14 healthy men aged 26 to 42 years were divided into a high-concentration tea catechin beverage group (group A, 7 subjects) and a control group drinking a beverage containing no tea catechin (group B, 7 subjects). Groups A and B drank 500 ml of a beverage containing 570 and 0 mg of tea catechins daily for 8 weeks, respectively, and

FIGURE 8.6
Energy expenditure, substrate oxidation, and respiratory exchange ratio during treadmill exercise (Ota et al., 2005). ▢: Control group (n = 7, tea catechins = 0 mg/day) + treadmill exercise. ■: Catechin group (n = 7, tea catechins = 570 mg/day) + treadmill exercise (Treadmill exercise condition 5 km/h, 30 min × 3/week). Values are expressed as the mean ± SD. Significantly different from control group by t-test (*, p < 0.05). (From Ota, N., et al., J Health Sci, 51: 233–236, 2005. With permission.)

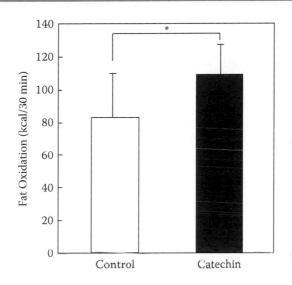

performed periodic 30 min treadmill exercise at 5 km/h three times a weak during this period. After 8 weeks, breath analysis was performed, and energy expenditure was measured in the resting state and during treadmill exercise. The fat combustion level during exercise was significantly increased when continuous tea catechin ingestion was combined with periodic exercise, compared to that without tea catechin ingestion (Figure 8.6) (Ota et al., 2005). Other human studies have also revealed that the combination of continuous tea catechin ingestion and habitual exercise effectively reduce body fat levels (Kataoka et al., 2004; Takashima et al., 2004).

The above findings suggest that continuous ingestion of a high-concentration tea catechin beverage during eating and daily living activities increases fat combustion.

8.4 Mechanisms of the Effects of Tea Catechins on Energy and Fat Metabolism

Body fat weight decreases with continuous ingestion of high-concentration tea catechins when energy intake decreases or when energy expenditure increases. In an experiment using an obesity animal model, continuous ingestion of high-concentration tea catechins did not inhibit lipid absorption, and the total energy intake was not altered (Meguro et al., 2001).

It was recently reported that tea catechin ingestion increases energy expenditure in humans and animals (Dulloo et al., 1999, 2000), and catechins ingested at a high concentration enter the circulation and reach the hepatocytes (Chen et al., 1997; Hackett and Griffiths, 1983). Furthermore, a recent study (Murase et al., 2002) using an animal obesity model reported that ingestion of high-concentration tea catechins elevates the gene expression level of fat-burning enzymes in the liver (β-oxidation-related enzymes: acyl-CoA oxidase (ACO) and medium-chain acyl-CoA dehydrogenase (MCAD)) by nearly 40% and the lipid β-oxidation activity level by about three times (Figure 8.7) (Murase et al., 2002).

These findings suggest that ingestion of high-concentration tea catechins activates lipid metabolism in the liver and subsequently increases energy expenditure by utilizing lipids (Figure 8.8) (Murase et al., 2002). Recent studies using animal obesity models also confirmed that the combination of periodic exercise and continuous tea catechin ingestion activates lipid metabolism in the liver and skeletal muscle (Murase et al., 2006; Shimotoyodome et al., 2005).

8.5 Clinical Application

As described above, the long-term effect on body fat accumulation and the mechanism of the effects on energy metabolism of high-concentration tea catechins have been elucidated in considerable detail. Several recent studies of high-concentration tea catechins in clinical use have been reported.

8.5.1 Tea Catechins in Combination with Nutritional Guidance

Healthy adult males (n = 134, mean BMI = 24.6kg/m^2) ingested a catechin-containing beverage (n = 77, 588 mg of tea catechins) or a control beverage (n = 57, 126 mg of tea catechins) for 1 year under nutritional guidance by a registered dietitian every 3 months (Yoneda et al., 2009).

FIGURE 8.7

Tea catechin-induced enhancement of mRNA expression of liver β-oxidation enzymes and liver lipid β-oxidation activity in a mouse obesity model (Murase et al., 2002). (A): mRNA expression of liver β-oxidation enzymes (ACO, acyl-CoA oxidase; MCAD, medium-chain acyl-CoA dehydrogenase). (B): Liver lipid β-oxidation activity. HD, high-fat, high-sucrose diet containing 30% lipids and 13% sucrose. ■: HD + tea catechins = 0% (n = 5). ■: HD + tea catechins = 0.5% (n = 5). Values are expressed as mean ± SD, n = 5. Significantly different from HD (tea catechins = 0 mg) group by t-test (unpaired, **, p < 0.01; ***, p < 0.001).

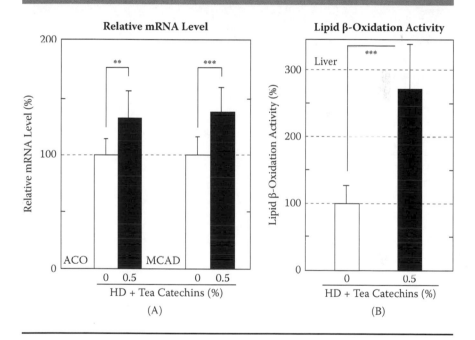

There was a greater reduction in body weight and body mass index ($p = 0.092$) in the catechin group compared to the control group. Under the guidance of a registered dietitian, subjects in the catechin group who showed a reduced dietary fat-derived energy percentage during the test period tended to show more reduced body weight than those with an increase in the fat-derived energy percentage, although the total energy intake was not different between groups (Figure 8.9) (Yoneda et al., 2009).

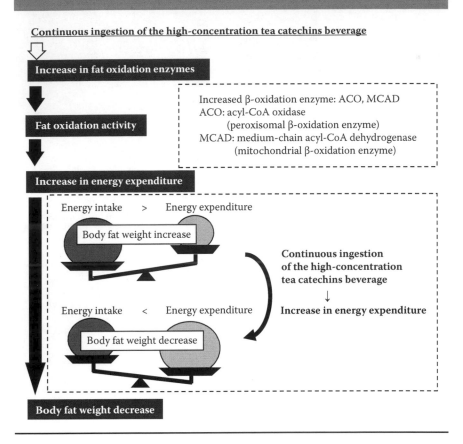

FIGURE 8.8
Mechanism of effects of tea catechins on energy and fat metabolism (Murase et al., 2002).

8.5.2 Tea Catechins in Combination with Exercise Intervention

Healthy elderly adults (n = 44, mean BMI = 26.2kg/m^2) ingested a catechin-containing beverage (n = 22, 637.5 mg of tea catechins) or a control beverage (n = 22, 0 mg of tea catechins) for 13 weeks with gait exercise intervention using a pedometer (Miyazaki et al., 2010).

Only in the catechin group, the body weight reduction significantly correlated with the change in number of steps; that is, catechins enhanced the impact of exercise intervention (Figure 8.10) (Miyazaki et al., 2010). This result is considered to be due primarily to the enhancement of fat utilization during exercise with continuous catechin ingestion, as described in Section 8.3.2.

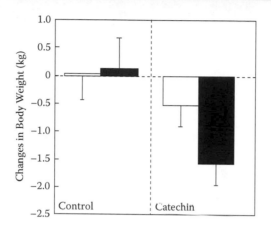

FIGURE 8.9
Effect of changes in fat-derived energy percentage during test period on tea catechins (Yoneda et al., 2009). ▢: Fat-derived energy (%) increase. ▇: Fat-derived energy (%) decrease. Values are expressed as mean ± SE. Difference between groups during study period was calculated by repeated-measure two-way analysis of variance (ANOVA) ($p = 0.092$).

8.6 Application for Weight and Glucose Control of Type 2 Diabetes

Patients with type 2 diabetes who were not receiving insulin therapy ingested a high-concentration catechin-containing beverage ($n = 23$, 582.8 mg of tea catechins) or a control beverage ($n = 20$, 96.3 mg of tea catechins) for 12 weeks (Nagao et al., 2009). The reduction in waist circumference was significantly greater in the catechin group than in the control group (Figure 8.11A) (Nagao et al., 2009). Adiponectin, which is negatively correlated with visceral adiposity, increased significantly only in the catechin group (Figure 8.11B) (Nagao et al., 2009). In patients treated with insulinotropic agents, the increase in the insulin level and reduction of the HbA1c level were significantly greater in the catechin group than in controls (Figure 8.11C and D) (Nagao et al., 2009).

These findings suggest that a catechin-rich beverage might be therapeutically beneficial in patients with type 2 diabetes who may not yet require insulin therapy.

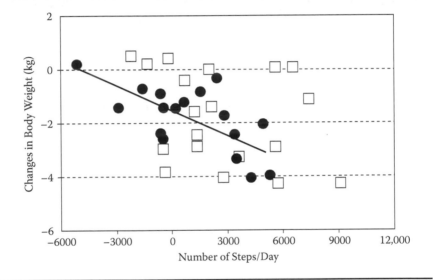

FIGURE 8.10
Tea catechin ingestion in combination with exercise intervention (Miyazaki et al., 2010). ■: Control group (n = 20, tea catechins = 0 mg); r = −0.149 (ns). ■: Catechin group (n = 18, tea catechins = 637.5mg); y = −0.00031x − 1.539; r = −0.685 (p < 0.01). Changes in body weight were significantly correlated with changes in number of steps only in the catechin group. (From Miyazaki, R., et al., *Jpn Soc Study Obes*, 16: 74–81, 2010. With permission.)

8.7 Conclusion

To prevent and relieve obesity and metabolic syndrome, it is crucial to improve lifestyle-related factors, such as habitual exercise and diet. Tea can be readily ingested every day and may be useful in initiating changes in lifestyle habits.

Tea has been scientifically demonstrated to prevent and relieve obesity. Habitual ingestion of green tea components, especially tea catechins, may decrease body fat, reduce risk of various lifestyle-related diseases such as diabetes, hyperlipidemia, and hypertension, and may contribute to the prevention of arteriosclerotic disorders.

FIGURE 8.11

Effects of tea catechin ingestion for 12 weeks by type 2 diabetes treatments (Nagao et al., 2009). (A): Waist circumference (□: control group, $n = 20$, tea catechins = 96.3 mg/day; ■: catechin group, $n = 23$, tea catechins = 582.8 mg/day; values are expressed as mean ± SE; significant difference between week 0 and week 12 values in catechin group by paired t-test, *, $p < 0.05$). (B): Adiponectin (□: control group, $n = 20$, tea catechins = 96.3 mg/day; ■: catechin group, $n = 23$, tea catechins = 582.8 mg/day; values are expressed as mean ± SE; significant difference between week 0 and week 12 values in catechin group by paired t-test, *, $p < 0.05$). (C): Insulin level (□: control group, $n = 17$, tea catechins = 96.3 mg/day; ■: catechin group, $n = 16$, tea catechins = 582.8 mg/day; values are expressed as the mean ± SE; significantly different from control group by t-test, *, $p < 0.05$). (D): Hemoglobin A1c (□: control group, $n = 17$, tea catechins = 96.3 mg/day; ■: catechin group, $n = 16$, tea catechins = 582.8 mg/day; values are expressed as the mean ± SE; significantly different from control group by t-test, *, $p < 0.05$).

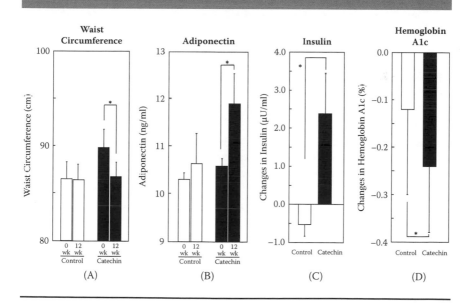

References

Chen, L., Lee, M.J., Li, H., and Yang, C.S. 1997. Absorption, distribution, and elimination of tea polyphenols in rats. *Drug Met Dispos*, 25: 1045–1050.

Dulloo, A.G., Duret, C., Rohrer, D., Girardier, L., Mensi, N., Fathi, M., Chantre, P., and Vandermander, J. 1999. Efficacy of a green tea extract rich in catechin polyphenols and caffeine in increasing 24-h energy expenditure and fat oxidation in humans. *Am J Clin Nutr*, 70: 1040–1045.

Dulloo, A.G., Seydoux, J., Girardier, L., Chantre, P., and Vandermander, J. 2000. Green tea and thermogenesis: Interactions between catechin-polyphenols, caffeine and sympathetic activity. *Int J Obes*, 24: 252–258.

Examination Committee of Criteria for "Metabolic Syndrome" in Japan. 2005. The Japanese Society of Internal Medicine. New criteria for "metabolic syndrome" in Japan. *Jpn Soc Int Med*, 94: 188–203.

Granata, G.P., and Brandon, L.J. 2002. The thermic effect of food and obesity: Discrepant results and methodological variations. *Nutr Rev*, 60: 223–233.

Hackett, A.M., and Griffiths, L.A. 1983. The disposition of 3-O-methyl-(+)-catechin in the rat and the marmoset following oral administration. *Eur J Drug Metab Pharmacol*, 8: 35–42.

Harada, U., Chikama, A., Saito, S., Takase, H., Nagao, T., Hase T., and Tokimitsu, I. 2005. Effects of the long-term ingestion of tea catechins on energy expenditure and dietary fat oxidation in healthy subjects. *J Health Sci*, 51: 248–252.

Hase, T., Komine, Y., Meguro, S., Takeda, Y., Takahashi, H., Matsui, Y., Inaoka, S., Katsuragi, Y., Tokimitsu, I., Shimasaki, H., and Itakura, H. 2001. Anti-obesity effects of tea catechins in humans. *J Oleo Sci*, 50: 599–605.

Kataoka, K., Takashima, S., Shibata, E., and Hoshino, E. 2004. Body fat reduction by the long term intake of catechins and the effects of physical activity. *Prog Med*, 24: 3358–3370.

Katiyar, S.K., and Mukhtar, H. 1996. Tea in chemoprevention of cancer: Epidemiologic and experimental studies (review). *Int J Oncol*, 8: 221–238.

Kozuma, K., Chikama, A., Hoshino, E., Kataoka, K., Mori, K., Hase, T., Katsuragi, Y., Tokimitsu, I., and Nakamura, H. 2005. Effect of intake of a beverage containing 540mg catechins on the body composition of obese women and men. *Prog Med*, 25:1945–1957.

Kurahashi, N., Sasazuki, S., Iwasaki, M., Inoue, M., and Tsugane, S., 2008. Green tea consumption and prostate cancer risk in Japanese men: A prospective study. JPHC Study Group. *Am J Epidemiol*, 167: 71–77.

Kuriyama, S., Hozawa, A., Ohmori, K., Shimazu, T., Matsui, T., Ebihara, S., Awata, S., Nagatomi, R., Arai, H., and Tsuji, I. 2006a. Green tea consumption and cognitive function: A cross-sectional study from the Tsurugaya Project. *Am J Clin Nutr*, 83: 355–361.

Kuriyama, S., Shimazu, T., Ohmori, K., Kikuchi, N., Nakaya, N., Nishino, Y., Tsubono, Y., and Tsuji, I. 2006b. Green tea consumption and mortality due to cardiovascular disease, cancer, and all causes in Japan. *JAMA*, 296:1255–1265.

Laville, M., Cornu, C., Normand, S., Mithieux, G., Beylot, M., and Riou, J.P. 1993. Decreased glucose-induced thermogenesis at the onset of obesity. *Am J Clin Nutr*, 57: 851–856.

Maki, K.C., Reeves, M.S., Farmer, M., Yasunaga, K., Matsuo, N., Katsuragi, Y., Komikado, M., Tokimitsu, I., Wilder, D., Jones, F., Blumberg, J.B., and Cartwright, Y. 2009. Green tea catechin consumption enhances exercise-induced abdominal fat loss in overweight and obese adults. *J Nutr*, 139: 264–270.

Matsuzawa, Y., Nakamura, T., Shimomura, I., and Kotani, K. 1995. Visceral fat accumulation and cardiovascular disease. *Obes Res*, 3: 645S–647S.

Meguro, S., Mizuno, T., Onizawa, K., Kawasaki, K., Nakagiri, H., Komine, Y., Suzuki, J., Matsui, Y., Hase, T., Tokimitsu, I., Shimasaki, H., and Itakura, H. 2001. Effects of tea catechins on diet-induced obesity in mice. *J Oleo Sci*, 50: 593–598.

Ministry of Heath, Labour, and Welfare, Japan. 2009. Summary of the National Health and Nutrition Survey in Japan 2008. http://www.mhlw.go.jp/houdou/2009/11/h1109-1.html.

Miyazaki, R., Takase, H., Harada, U., and Ishii, K. 2010. Effects of combining the long-term ingestion of tea catechins with a change in steps per day on items for a specific health checkup in men and women with metabolic syndrome. *Jpn Soc Study Obes*, 16: 74–81.

Mukhtar, H., and Ahmad, N. 2000. Tea polyphenols: Prevention of cancer and optimizing health. *Am J Clin Nutr*, 71:1698S–1702S.

Murase, T., Nagasawa, A., Suzuki, J., Hase, T., and Tokimitsu, I. 2002. Beneficial effects of tea catechins on diet-induced obesity: Stimulation of lipid catabolism in the liver. *Int J Obes*, 26: 1459–1464.

Murase, T., Haramizu, S., Shimotoyodome, A., and Tokimitsu, I. 2006. Reduction of diet-induced obesity by a combination of tea-catechin intake and regular swimming. *Int J Obes*, 30: 561–568.

Nagao, T., Hase, T., and Tokimitsu, I. 2007. A green tea extract high in catechins reduces body fat and cardiovascular risks in humans. *Obesity* 15: 1473–1483.

Nagao, T., Komine, Y., Soga, S., Meguro, S., Hase, T., Tanaka, Y., and Tokimitsu, I. 2005. Ingestion of a tea rich in catechins leads to a reduction in body fat and malondialdehyde-modified LDL in men. *Am J Clin Nutr*, 81: 122–129.

Nagao, T., Meguro, S., Hase, T., Otsuka, K., Komikado, M., Tokimitsu, I., Yamamoto, T., and Yamamoto, K. 2009. A catechin-rich beverage improves obesity and blood glucose control in patients with type 2 diabetes. *Obesity* 17: 310–317.

Nagao, T., Meguro, S., Soga, S., Otsuka, A., Tomonobu, K., Fumoto, S., Chikama, A., Mori, K., Yuzawa, M., Watanabe, H., Hase, T., Tanaka, Y., Tokimitsu, I., Shimasaki, H., and Itakura, H. 2001. Tea catechins suppress accumulation of body fat in humans. *J Oleo Sci*, 50:717–728.

Nozawa, A., Sugimoto, A., Nagata, K., Kakuda, T., and Horiguchi, T. 2002. The effects of a beverage containing tea catechins on serum cholesterol level. *J Nutr Food Sci*, 5: 1–9.

Ota, N., Soga, S., Shimotoyodome, A., Haramizu, S., Inaba, M., Murase, T., and Tokimitsu, I. 2005. Effects of combination of regular exercise and tea catechins intake on energy expenditure in humans. *J Health Sci*, 51: 233–236.

Otsuka, K., Uchida, H., Yuzawa, M., Fumoto, S., Tomonobu, K., Chikama, A., Hase, T., Watanabe, H., Tokimitsu, I., and Itakura, H. 2002. Effects of tea catechins on body fat metabolism in women. *Jpn J Nutr Assess*, 19: 365–376.

Shimotoyodome, A., Haramizu, S., Inaba, M., Murase, T., and Tokimitsu, I. 2005. Exercise and green tea extract stimulate fat oxidation and prevent obesity in mice. *Med Sci Sports Exer*, 37: 1884–1892.

Takase, H., Nagao, T., Otsuka, K., Kozuma, K., Kataoka, K., and Katashima, M. 2009. Effects of long-term ingestion of tea catechins on metabolic syndrome among different criteria: Meta-analysis of 7 randomized controlled trials. Paper presented at Obesity 2009, the 27th Annual Scientific Meeting of the Obesity Society, Washington, DC, October 10, 24–28.

Takase, H., Nagao, T., Otsuka, K., Kozuma, K., Kataoka, K., Meguro, S., Komikado, M., and Tokimitsu, I. 2008a. Effects of long-term ingestion of tea catechins on visceral fat accumulation and metabolic syndrome—Pooling-analysis of 7 randomized controlled trials. *Jpn Pharmacol Ther*, 36: 509–514.

Takase, H., Nagao, T., Otsuka, K., Meguro, S., Komikado, M., and Tokimitsu, I. 2008b. Effects of long-term ingestion of tea catechins on visceral fat accumulation and metabolic syndrome risk in women with abdominal obesity. *Jpn Pharmacol Ther*, 36: 237–245.

Takashima, S., Kataoka, K., Shibata, E., and Hoshino, E. 2004. The long term intake of catechins improves lipid catabolism during exercise. *Prog Med*, 24: 3371–3379.

Thomas, C.D., Peters, J.C., Reed, G.W., Abumrad, N.N., Sun, M., and Hill, J.O. 1992. Nutrient balance and energy expenditure during *ad libitum* feeding of high-fat and high-carbohydrate diets in humans. *Am J Clin Nutr*, 55: 934–942.

Tsuchida, T., Itakura, H., and Nakamura, H. 2002. Reduction of body fat in humans by long-term ingestion of catechins. *Prog Med*, 22: 2189–2203.

Yoneda, T., Shoji, K., Takase, H., Hibi, M., Hase, T., Meguro, S., Tokimitsu, I., and Kambe, H. 2009. Effectiveness and safety of 1-year *ad libitum* consumption of a high-catechin beverage under nutritional guidance. *Met Synd Rel Disord*, 7: 349–356.

Yoshino, K., Hara, Y., Sano, M., and Tomita, I. 1994. Antioxidative effects of black tea theaflavins and thearubigin on lipid peroxidation of rat liver homogenates induced by *t*-butyl hydroperoxide. *Biol Pharma Bull*, 17: 146–149.

9

Green Tea Polyphenols for the Protection of Internal Organs— Focus on Renal Damage Caused by Oxidative Stress

Takako Yokozawa, Jeong Sook Noh,
Chan Hum Park, and Jong Cheol Park

Contents

9.1 Introduction

Clinical and experimental studies have resulted in extensive discussions of the link between renal disease and oxidative stress, which is directly or indirectly derived from various pathological conditions, such as hyperglycemia, free radical-generating toxic substances, and inflammation. The free radicals are highly reactive and harmful to lipids, proteins, and nucleic acids, resulting in structural and functional impairment. Increased levels of end products mediated by the reactions between biomolecules and free radicals, such as malondialdehyde, 3-nitrotyrosine, and 8-hydroxy-2'-deoxyguanosine, were observed with various pathological phenomena, such as acute renal failure and hemodialysis (Fiorillo et al., 1998; Handelman et al., 2001; Kakimoto et al., 2002; Yokozawa et al., 2002). Inhibitors of free radicals and anti-oxidants have also been shown to protect against renal damage in a number of studies (Hahn et al., 1999).

Green tea polyphenols have been shown to act as metal chelators, preventing the metal-catalyzed formation of radical species, antioxidant enzyme modulators, and scavengers of free radicals, including the hydroxyl radical ($\bullet$OH), superoxide anion (O_2^-), nitric oxide (NO), and peroxynitrite ($ONOO^-$) (Chung et al., 1998; Nakagawa and Yokozawa, 2002; Yokozawa et al., 1998, 2000). These antioxidant activities are considered to be closely related to their protective effects against various diseases, including renal disease, arteriosclerosis, cancer, and inflammation caused by lipid peroxidation and excessive free radical production (Halliwell and Gutteridge, 1990). The polyphenolic compounds of green tea mainly comprise (–)-epigallocatechin 3-O-gallate, (–)-epicatechin 3-O-gallate, (–)-epigallocatechin, and (–)-epicatechin, which are classified as the flavan-3-ol class of flavonoids. This chapter gives a review of our recent findings, with emphasis on the therapeutic potential of the polyphenols of green tea in a useful experimental model of renal damage.

9.2 Green Tea Polyphenols and Arginine-Induced Renal Failure

Arginine is essential for ammonia detoxification via urea synthesis, which prevents metabolic derangements caused by elevated tissue ammonia levels. However, the administration of excess arginine causes an imbalance of amino acids and changes in protein metabolism. In addition, arginine is the key substance of guanidino compounds such as creatinine (Cr), methylguanidine (MG), and guanidinosuccinic acid (GSA), which are considered to be uremic toxins (Natelson and Sherwin, 1979; Natelson et al., 1978). Moreover, NO, an important mediator of diverse pathological damage because of its toxic effects, is formed from arginine by the NO synthase family of enzymes. Numerous uremic toxins such as Cr and MG, as well as NO produced from excessive arginine, are responsible for acute renal failure (Orita et al., 1978; Paller et al., 1984; Wills, 1985; Yokozawa et al., 1991).

To determine whether green tea polyphenols ameliorate the pathological conditions induced by excessive dietary arginine, they were administered to rats at a daily dose of 50 or 100 mg/kg body weight for 30 days with a 2% (w/w) arginine diet. The green tea polyphenol mixture was composed of (–)-epigallocatechin 3-O-gallate (18.0%), (–)-gallocatechin 3-O-gallate (11.6%), (–)-epicatechin 3-O-gallate (4.6%), (–)-epigallocatechin (15%), (+)-gallocatechin (14.8%), (–)-epicatechin (7.0%), and (+)-catechin (3.5%).

The arginine-feeding rats showed enhanced levels of urea, arginine, guanidinoacetic acid (GAA), Cr, and MG in the serum and urine, suggesting that the uremic toxin-eliminating activity of the kidney was impaired and renal failure developed (Tables 9.1 and 9.2). However, green tea polyphenols

TABLE 9.1
Effect of Green Tea Polyphenols on Guanidino Compounds and Urea Nitrogen in Serum

Group	Dose (mg/kg B.W./day)	Arginine (mg/dl)	GAA (μg/dl)	Cr (mg/dl)	Urea Nitrogen (mg/dl)
Casein-fed rats	—	2.99 ± 0.08	166 ± 5	0.30 ± 0.01	17.5 ± 2.9
Arginine-fed rats					
Control	—	5.70 ± 0.36[b]	206 ± 5[b]	0.39 ± 0.02[b]	21.6 ± 0.6[a]
Polyphenols	50	5.26 ± 0.27[b]	206 ± 5[b]	0.38 ± 0.03[b]	20.1 ± 0.6
Polyphenols	100	5.12 ± 0.39[b,c]	220 ± 7[b,d]	0.29 ± 0.04[e]	18.5 ± 0.7[c]

Significance: [a]$p < 0.01$, [b]$p < 0.001$ versus casein-fed rats; [c]$p < 0.05$, [d]$p < 0.01$, [e]$p < 0.001$ versus arginine-fed control rats

TABLE 9.2
Effect of Green Tea Polyphenols on Guanidino Compounds in Urine

Group	Dose (mg/kg B.W./day)	Arginine (mg/day)	GAA (μg/day)	Cr (mg/day)	MG (μg/day)
Casein-fed rats	—	0.32 ± 0.01	426 ± 45	7.14 ± 0.48	7.7 ± 1.4
Arginine-fed rats					
Control	—	0.92 ± 0.16[c]	633 ± 104[c]	10.14 ± 1.77[c]	10.7 ± 1.1[a]
Polyphenols	50	0.91 ± 0.15[c]	738 ± 28[c]	8.41 ± 0.55	11.4 ± 1.1[b]
Polyphenols	100	0.80 ± 0.12[c]	784 ± 58[c,d]	7.15 ± 0.72[e]	11.3 ± 1.9[b]

Significance: [a]$p < 0.05$, [b]$p < 0.01$, [c]$p < 0.001$ versus casein-fed rats; [d]$p < 0.01$, [e]$p < 0.001$ versus arginine-fed control rats.

TABLE 9.3
Effect of Green Tea Polyphenols on NO Metabolites in Urine

Group	Dose (mg/kg B.W./day)	NO_2^- (μmol/day)	NO_3^- (μmol/day)	$NO_2^- + NO_3^-$ (μmol/day)
Casein-fed rats	—	0.008 ± 0.008	4.579 ± 1.334	4.587 ± 1.342
Arginine-fed rats				
Control	—	0.035 ± 0.028	6.197 ± 0.857[a]	6.232 ± 0.881[a]
Polyphenols	50	0.035 ± 0.027	5.277 ± 0.417	5.312 ± 0.410
Polyphenols	100	0.030 ± 0.010	4.821 ± 0.328	4.851 ± 0.328

Significance: [a] $p < 0.05$ versus casein-fed rats.

reduced the levels of urea and Cr in the serum and urine; therefore, they would be expected to ameliorate renal failure by reducing the levels of uremic toxins. GSA formation is known to increase depending on the serum and urinary urea levels (Aoyagi et al., 1983; Cohen and Patel, 1982; Kopple et al., 1977). The decreases in the blood and urinary urea nitrogen levels were brought by green tea polyphenol administration, which suggests that GSA formation would also be lowered (Tables 9.1 and 9.2).

It has been well established that serum and urinary concentrations of nitrite (NO_2^-) and nitrate (NO_3^-) are elevated in patients with renal failure (Kone, 1997). In the present investigation, the level of NO_2^- plus NO_3^- was higher in arginine- than casein-fed rats (Table 9.3). It is believed that the increase in NO production is attributable to dietary arginine, and that it causes renal failure. However, green tea polyphenols reduced the production of NO derived from excessive arginine administration (Table 9.3).

Herein we describe the activities of the radical scavenging enzymes superoxide dismutase (SOD), catalase, and glutathione peroxidase (GSH-Px) in the kidney to elucidate whether free radicals participate in the process of arginine-induced renal failure. Arginine influenced the activities of radical-scavenging enzymes in the kidney, leading to a decrease in the activity of SOD, which catalyzes O_2^- to hydrogen peroxide (H_2O_2). Moreover, the activity of catalase, which specifically eliminates H_2O_2, was also suppressed in rats fed the 2% arginine diet (Table 9.4). These results indicate that arginine affected the activity of antioxidative enzymes in the renal peroxisomes, and the reductions in the SOD and catalase activities in the arginine-fed control group suggest that oxygen-derived free radicals were generated and the biological defense system was weakened. However, the administration of green tea polyphenols increased the activities of SOD and catalase (Table 9.4). The

TABLE 9.4
Effect of Green Tea Polyphenols on Oxygen Species-Scavenging Enzymes in Kidney

Group	Dose (mg/kg B.W./day)	SOD (U/mg protein)	Catalase (U/mg protein)	GSH-Px (U/mg protein)
Casein-fed rats	—	26.09 ± 1.66	177.0 ± 13.5	60.61 ± 1.99
Arginine-fed rats				
Control	—	17.69 ± 1.68[b]	76.0 ± 7.0[b]	64.76 ± 2.18[a]
Polyphenols	50	23.32 ± 1.75[c]	111.6 ± 6.6[b,e]	60.94 ± 1.24[c]
Polyphenols	100	25.04 ± 4.68[d]	114.9 ± 7.6[b,e]	60.04 ± 2.55[d]

Significance: [a]$p < 0.05$, [b]$p < 0.001$ versus casein-fed rats; [c]$p < 0.05$, [d]$p < 0.01$, [e]$p < 0.001$ versus arginine-fed control rats.

results suggest that excessive dietary arginine evokes renal failure through increasing the production of uremic toxins and NO, and through decreasing oxygen species-scavenging enzyme activity in the kidney. However, green tea polyphenols showed a protective effect against the renal injury induced by arginine. Therefore, green tea polyphenols have the potential to offer a promising new therapeutic approach to renal disorders.

9.3 (–)-Epicatechin 3-O-Gallate and ONOO$^-$-Mediated Renal Damage

Evidence on the role of reactive oxygen and nitrogen metabolites in the pathogenesis of renal diseases has accumulated. Also, ONOO$^-$ formed *in vivo* from O_2^- and NO has been suggested to be an important causative agent in the pathogenesis of cellular damage and renal dysfunction (Douki et al., 1996; Radi et al., 1991). The pathological effects of ONOO$^-$ and its decomposition product, •OH, contribute to the antioxidant depletion, alterations of the protein structure and function by tyrosine nitration, and oxidative damage observed in human diseases and animal models of diseases (Ceriello et al., 2001; Cuzzocrea and Reiter, 2001; Fukuyama et al., 1997; Ischiropoulos, 1998; Nakazawa et al., 2000). The protective effect of (–)-epicatechin 3-O-gallate against ONOO$^-$-mediated damage was examined using an animal model and cell culture system. This study was also carried out to elucidate whether the effect of (–)-epicatechin 3-O-gallate is distinct from the activity of several well-known free radical inhibitors, the ONOO$^-$ inhibitors ebselen and uric acid, O_2^- scavenger copper zinc SOD

TABLE 9.5
Effect of (–)-Epicatechin 3-O-Gallate and Free Radical Inhibitors on Plasma NO and O$_2^-$ Radicals in Rats

Group	NO (μM)	O$_2^-$ (O.D.)
Sham operation	1.71 ± 0.18	0.315 ± 0.013
LPS plus ischemia-reperfusion		
Control	15.33 ± 0.72[b]	0.371 ± 0.011[a]
(–)-Epicatechin 3-O-gallate (10 mg/kg B.W.)	15.02 ± 1.15[b]	0.377 ± 0.019[b]
(–)-Epicatechin 3-O-gallate (20 mg/kg B.W.)	14.24 ± 0.33[b]	0.401 ± 0.008[b]
Ebselen (5 mg/kg B.W.)	15.98 ± 1.35[b]	0.345 ± 0.007
Uric acid (62.5 mg/kg B.W.)	15.08 ± 1.15[b]	0.360 ± 0.026[a]
SOD (10,000 U/kg B.W.)	19.04 ± 1.72[b,d]	0.336 ± 0.016[c]
L-N^6-(1-iminoethyl)lysine hydrochloride (3 mg/kg B.W.)	3.39 ± 0.25[e]	0.363 ± 0.022[a]

Significance: [a] $p < 0.01$, [b] $p < 0.001$ versus sham operation values; [c] $p < 0.05$, [d] $p < 0.01$, [e] $p < 0.001$ versus LPS plus ischemia-reperfused control values.

(CuZnSOD), and the selective inducible NO synthase inhibitor L-N^6-(1-iminoethyl)lysine hydrochloride. To generate ONOO$^-$, male Wistar rats were subjected to ischemia-reperfusion (occlusion of the renal artery and vein with clamps) together with lipopolysaccharide (LPS) injection.

In this study, the significant stimulation of NO and O$_2^-$ generation in response to the LPS injection plus ischemia-reperfusion process declined markedly after treatment with L-N^6-(1-iminoethyl)lysine hydrochloride and CuZnSOD, respectively (Table 9.5). (–)-Epicatechin 3-O-gallate, however, did not reverse the elevations in the plasma NO and O$_2^-$ levels resulting from LPS plus ischemia-reperfusion. This suggests that (–)-epicatechin 3-O-gallate does not act as a scavenger of the ONOO$^-$ precursors NO and O$_2^-$. In light of these results, we hypothesized that the protective activity of (–)-epicatechin 3-O-gallate against ONOO$^-$ could be attributed to the direct scavenging of ONOO$^-$, and so we evaluated the levels of 3-nitrotyrosine and myeloperoxidase (MPO) activity as indicators of ONOO$^-$ formation.

The LPS plus ischemia-reperfusion process led to elevation of the plasma 3-nitrotyrosine level in rats, suggesting that oxidative damage due to the formation of ONOO$^-$ had occurred (Figure 9.1) and the cellular formation of ONOO$^-$ increased by 3-morpholinosydnonimine (SIN-1) treatment (Figure 9.2). However, (–)-epicatechin 3-O-gallate reduced nitrotyrosine formation markedly in a dose-dependent manner compared with ebselen and CuZnSOD. The activity of (–)-epicatechin 3-O-gallate was comparable with that of L-N^6-(1-iminoethyl)lysine hydrochloride, although (–)-epicatechin 3-O-gallate did not scavenge NO (Figure 9.1 and Table 9.5). The magnitudes

FIGURE 9.1
Effect of (–)-epicatechin 3-O-gallate and free radical inhibitors on plasma 3-nitrotyrosine level in rats. N, sham operation; C, LPS plus ischemia-reperfusion; E1, LPS plus ischemia-reperfusion after (–)-epicatechin 3-O-gallate (10 mg/kg body weight); E2, LPS plus ischemia-reperfusion after (–)-epicatechin 3-O-gallate (20 mg/kg body weight); EB, LPS plus ischemia-reperfusion after ebselen (5 mg/kg body weight); U, LPS plus ischemia-reperfusion after uric acid (62.5 mg/kg body weight); SOD, LPS plus ischemia-reperfusion after CuZnSOD (10,000 U/kg body weight); NIL, LPS plus ischemia-reperfusion after L-N^6-(1-iminoethyl)lysine hydrochloride (3 mg/kg body weight). N.D., not detectable. Significance: [a]$p < 0.001$ versus LPS plus ischemia-reperfused control values.

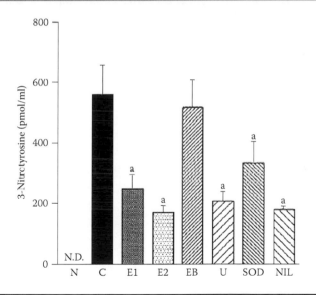

of the significant elevations of $ONOO^-$ production in the cellular system were decreased by (–)-epicatechin 3-O-gallate treatment (Figure 9.2). Taken together, these findings indicate that (–)-epicatechin 3-O-gallate scavenges $ONOO^-$ directly, but not its precursors NO and O_2^-. In addition, the elevation of MPO activity was reversed by the administration of (–)-epicatechin 3-O-gallate, uric acid, and SOD, but not by that of L-N^6-(1-iminoethyl)lysine hydrochloride (Figure 9.3). We consider that the reduction of MPO activity

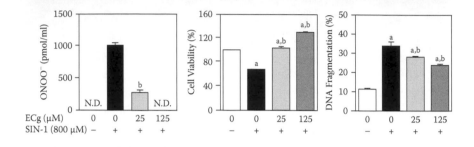

FIGURE 9.2
Effect of (–)-epicatechin 3-O-gallate on SIN-1-induced ONOO$^-$ formation, viability, and DNA fragmentation in renal epithelial cells, LLC-PK$_1$. N.D., not detectable. Significance: ^{a}$p < 0.001$ versus no treatment values; ^{b}$p < 0.001$ versus SIN-1 treatment values.

by (–)-epicatechin 3-O-gallate ameliorated ONOO$^-$-induced oxidative damage by inhibiting protein nitration and lipid peroxidation through a mechanism distinct from that of L-N^6-(1-iminoethyl)lysine hydrochloride, which actually increased MPO activity. In addition, uric acid acted in a similar way to (–)-epicatechin 3-O-gallate as a direct scavenger of ONOO$^-$ through the inhibition of 3-nitrotyrosine and MPO activity, and not as a scavenger of ONOO$^-$ precursors (Figures 9.1 and 9.3).

The antioxidative defense system was significantly suppressed by the excessive increase of ONOO$^-$ resulting from the LPS plus ischemia-reperfusion process. The administration of (–)-epicatechin 3-O-gallate resulted in concentration-dependent elevations of the activities of the antioxidative enzymes, SOD, catalase, and GSH-Px, and the cellular antioxidant reduced glutathione (GSH) (Tables 9.6 and 9.7). Furthermore, the excessive ONOO$^-$ increased lipid peroxidation of renal mitochondria (Table 9.7), and we confirmed the mitochondrial oxidative damage caused by ONOO$^-$. In contrast, the administration of (–)-epicatechin 3-O-gallate reduced the magnitude of the lipid peroxidation level elevation caused by the experimental process (Table 9.7).

Since ONOO$^-$ decomposes to form a strong and reactive oxidant, •OH, the effects of free radical scavengers and (–)-epicatechin 3-O-gallate on •OH also have to be evaluated to compare their protective actions against ONOO$^-$. Herein, the spin-trap method was used to determine the level of •OH in rat renal tissue formed by the Fenton reaction, and found that the magnitude of the increase in the height of the DMPO-OH peak of rats that

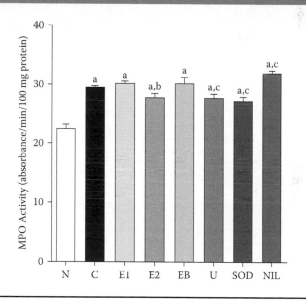

FIGURE 9.3

Effect of (–)-epicatechin 3-O-gallate and free radical inhibitors on renal MPO activity in rats. N, sham operation; C, LPS plus ischemia-reperfusion; E1, LPS plus ischemia-reperfusion after (–)-epicatechin 3-O-gallate (10 mg/kg body weight); E2, LPS plus ischemia-reperfusion after (–)-epicatechin 3-O-gallate (20 mg/kg body weight); EB, LPS plus ischemia-reperfusion after ebselen (5 mg/kg body weight); U, LPS plus ischemia-reperfusion after uric acid (62.5 mg/kg body weight); SOD, LPS plus ischemia-reperfusion after CuZnSOD (10,000 U/kg body weight); NIL, LPS plus ischemia-reperfusion after L-N^6-(1-iminoethyl)lysine hydrochloride (3 mg/kg body weight). Significance: [a]$p < 0.001$ versus sham operation values; [b]$p < 0.01$, [c]$p < 0.001$ versus LPS plus ischemia-reperfused control values.

underwent LPS plus ischemia-reperfusion was reduced by treatment with (–)-epicatechin 3-O-gallate, CuZnSOD, and L-N^6-(1-iminoethyl)lysine hydrochloride (Table 9.8). These findings indicate that the effect of (–)-epicatechin 3-O-gallate on the highly reactive radical •OH plays a crucial role in its protective action against ONOO⁻-induced oxidative damage. Furthermore, the effects of (–)-epicatechin 3-O-gallate on ONOO⁻ and •OH were stronger

TABLE 9.6
Effect of (–)-Epicatechin 3-O-Gallate on Oxygen Species-Scavenging Enzymes in Renal Tissue

Group	SOD (U/mg protein)	Catalase (U/mg protein)	GSH-Px (U/mg protein)
Sham operation	31.82 ± 2.29	255.3 ± 35.0	138.7 ± 10.3
LPS plus ischemia-reperfusion			
Control	16.67 ± 2.52[c]	146.8 ± 19.3[c]	79.5 ± 7.2[c]
(–)-Epicatechin 3-O-gallate (10 μmoles/kg B.W./day)	18.18 ± 1.70[c]	176.0 ± 15.3[c]	105.7 ± 8.0[c,e]
(–)-Epicatechin 3-O-gallate (20 μmoles/kg B.W./day)	21.45 ± 3.67[c,d]	194.4 ± 22.6[b,d]	118.7 ± 11.0[a,f]

Significance: [a]$p < 0.05$, [b]$p < 0.01$, [c]$p < 0.001$ versus sham operation values; [d]$p < 0.05$, [e]$p < 0.01$, [f]$p < 0.001$ versus LPS plus ischemia-reperfused control values.

TABLE 9.7
Effect of (–)-Epicatechin 3-O-Gallate on the Oxidative Damages of Renal Mitochondria

Group	GSH (nmol/mg protein)	TBA-Reactive Substance (nmol/mg protein)
Sham operation	4.42 ± 0.09	0.121 ± 0.001
LPS plus ischemia-reperfusion		
Control	2.75 ± 0.14[a]	0.165 ± 0.007[a]
(–)-Epicatechin 3-O-gallate (10 μmoles/kg B.W./day)	3.72 ± 0.18[a,b]	0.147 ± 0.003[a,b]
(–)-Epicatechin 3-O-gallate (20 μmoles/kg B.W./day)	3.77 ± 0.21[a,b]	0.144 ± 0.007[a,b]

Significance: [a]$p < 0.001$ versus sham operation values; [b]$p < 0.001$ versus LPS plus ischemia-reperfused control values.

than those of the other well-known free radical inhibitors tested, which can also be regarded as a mechanism distinct from that of the others.

The LPS plus ischemia-reperfusion process resulted in a significant elevation of the uric acid level, indicating that a pathological condition in the kidney had developed (Table 9.9). However, the administration of (–)-epicatechin 3-O-gallate reduced the uric acid level, while the other free radical inhibitors did not (Table 9.9). This effect of (–)-epicatechin 3-O-gallate on excessive uric acid levels is also considered to be a property distinct from the other free radical scavengers. The renal function parameters of serum urea nitrogen and Cr levels were elevated markedly by LPS plus ischemia-reperfusion, while the administration of (–)-epicatechin 3-O-gallate reduced these levels significantly,

TABLE 9.8
Effect of (–)-Epicatechin 3-O-Gallate and Free Radical Inhibitors on Renal Hydroxyl Radical in Rats

Group	Hydroxyl Radical (DMPO-OH)
Sham operation	0.29 ± 0.07
LPS plus ischemia-reperfusion	
Control	1.15 ± 0.15[a]
(–)-Epicatechin 3-O-gallate (10 mg/kg B.W.)	0.18 ± 0.01[b]
(–)-Epicatechin 3-O-gallate (20 mg/kg B.W.)	0.17 ± 0.01[b]
Ebselen (5 mg/kg B.W.)	1.10 ± 0.18[a]
Uric acid (62.5 mg/kg B.W.)	1.06 ± 0.07[a]
SOD (10,000 U/kg B.W.)	0.22 ± 0.01[b]
L-N^6-(1-iminoethyl)lysine hydrochloride (3 mg/kg B.W.)	0.20 ± 0.03[b]

Significance: [a]$p < 0.001$ versus sham operation values; [b]$p < 0.001$ versus LPS plus ischemia-reperfused control values.

TABLE 9.9
Effect of (–)-Epicatechin 3-O-Gallate and Free Radical Inhibitors on Plasma Uric Acid Level in Rats

Group	Uric Acid (mg/dl)
Sham operation	1.53 ± 0.18
LPS plus ischemia-reperfusion	
Control	1.95 ± 0.03[a]
(–) Epicatechin 3-O-gallate (10 mg/kg B.W.)	1.64 ± 0.24
(–)-Epicatechin 3-O-gallate (20 mg/kg B.W.)	1.12 ± 0.11[c]
Ebselen (5 mg/kg B.W.)	2.15 ± 0.37[b]
Uric acid (62.5 mg/kg B.W.)	1.96 ± 0.35
SOD (10,000 U/kg B.W.)	2.09 ± 0.09[a]
L-N^6-(1-iminoethyl)lysine hydrochloride (3 mg/kg B.W.)	1.57 ± 0.25

Significance: [a]$p < 0.05$, [b]$p < 0.01$ versus sham operation values; [c]$p < 0.001$ versus LPS plus ischemia-reperfused control values.

indicating the amelioration of renal dysfunction by (–)-epicatechin 3-O-gallate. In addition, uric acid and L-N^6-(1-iminoethyl)lysine hydrochloride protected against renal dysfunction induced by this process, although their activity was relatively low compared with (–)-epicatechin 3-O-gallate.

Also, results in rats showed that the LPS plus ischemia-reperfusion process led to proteinuria, demonstrated by the sodium dodecyl sulfate–polyacrylamide gel electrophoresis (SDS-PAGE) pattern with an abundance of low- and

FIGURE 9.4
Effect of (–)-epicatechin 3-*O*-gallate and free radical inhibitors on SDS-PAGE pattern of proteinuria in rats. 1, marker; 2, sham operation; 3, LPS plus ischemia-reperfusion; 4, LPS plus ischemia-reperfusion after ebselen (5 mg/kg body weight); 5, LPS plus ischemia-reperfusion after uric acid (62.5 mg/kg body weight); 6, LPS plus ischemia-reperfusion after CuZnSOD (10,000 U/kg body weight); 7, LPS plus ischemia-reperfusion after L-N^6-(1-iminoethyl)lysine hydrochloride (3 mg/kg body weight); 8, LPS plus ischemia-reperfusion after (–)-epicatechin 3-*O*-gallate (20 mg/kg body weight); 9, LPS plus ischemia-reperfusion after (–)-epicatechin 3-*O*-gallate (10 mg/kg body weight). Markers (kDa): 107, phosphorylase B; 76, bovine serum albumin; 52, ovalbumin; 37, carbonic anhydrase; 27, soybean trypsin inhibitor.

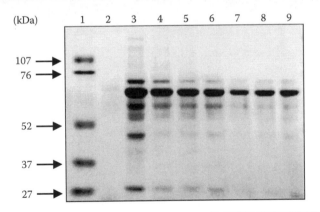

high-molecular-weight proteins relative to the marker albumin (76 kDa) (Figure 9.4). The administration of (–)-epicatechin 3-*O*-gallate and L-N^6-(1-iminoethyl)lysine hydrochloride reduced the intensity of the low- and high-molecular-weight protein bands to a greater extent than the other radical inhibitors, which suggests that (–)-epicatechin 3-*O*-gallate would ameliorate proteinuria due to renal failure caused by $ONOO^-$-induced oxidative damage.

In the LPS plus ischemia-reperfusion rat model, (–)-epicatechin 3-*O*-gallate, L-N^6-(1-iminoethyl)lysine hydrochloride, and uric acid showed a strong protective effect against ONOO$^-$-induced oxidative damage, while CuZnSOD and ebselen exerted relatively low activity. In light of the results of this study, we suggest that the activity of (–)-epicatechin 3-O-gallate is distinct from that of the other free radical inhibitors, especially L-N^6-(1-iminoethyl)lysine hydrochloride and uric acid. (–)-Epicatechin 3-O-gallate scavenged ONOO$^-$ directly, but it did not scavenge its precursors O_2^- and NO. Furthermore, (–)-epicatechin 3-O-gallate indirectly inhibits the generation of ONOO$^-$ through the enhancement of antioxidant enzyme activities. In addition, the inhibition of MPO activity by (–)-epicatechin 3-O-gallate would contribute to the effective inhibition of protein nitration and lipid peroxidation. (–)-Epicatechin 3-O-gallate was also a stronger scavenger of the ONOO$^-$ decomposition product •OH than any of the other free radical inhibitors tested. The improvement by (–)-epicatechin 3-O-gallate of the renal dysfunction caused by ONOO-related oxidative damage was marked and distinct from that induced by any of the other free radical inhibitors.

9.4 (–)-Epigallocatechin 3-O-Gallate and Adenine-Induced Renal Failure

MG is widely recognized as a strong uremic toxin. The •OH specifically plays an important role in the pathway of MG production from Cr. It is also investigated whether the oral administration of (–)-epigallocatechin 3-O-gallate suppresses MG production in rats with chronic renal failure after intraperitoneal Cr injection.

In normal rats, Cr was rapidly excreted into the urine after Cr loading, whereas in rats with renal failure, urinary Cr excretion was low, and high levels of Cr were present in the serum, muscle, kidneys, and liver, suggesting that the body was susceptible to oxidative alterations (Figure 9.5). After Cr loading, the MG levels in the serum, urine, muscle, liver, and kidneys of rats with renal failure were higher than those of normal rats, confirming that MG production from Cr was increased in rats with renal failure (Figure 9.6). The oral administration of (–)-epigallocatechin 3-O-gallate dose-dependently reduced the serum MG levels, showing that (–)-epigallocatechin 3-O-gallate effectively inhibited increased MG production in which oxidative reactions markedly participate. (–)-Epigallocatechin 3-O-gallate (20 mg/kg body weight) reduced the urinary and kidney MG levels, which were reduced further and significantly in the 100 and 500 mg treated groups. In the muscle and liver, a significant reduction was only observed in the high-dose-treated group (500 mg) (Figure 9.6).

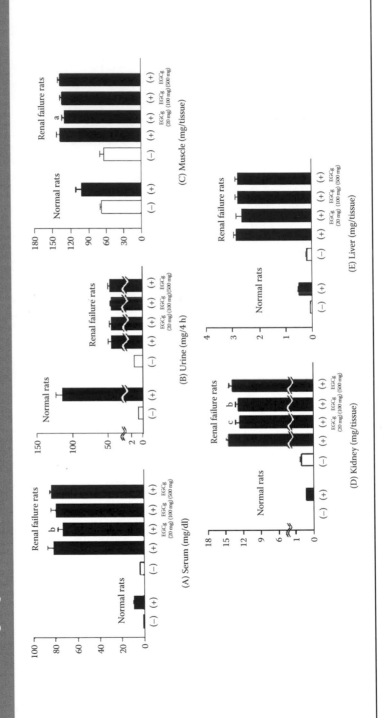

Figure 9.5

Cr levels in serum (A), urine (B), muscle (C), kidney (D), and liver (E). (−), without Cr loading; (+), with Cr loading. Significance: $^a p < 0.05$, $^b p < 0.01$, $^c p < 0.001$ versus renal failure control rats with Cr loading.

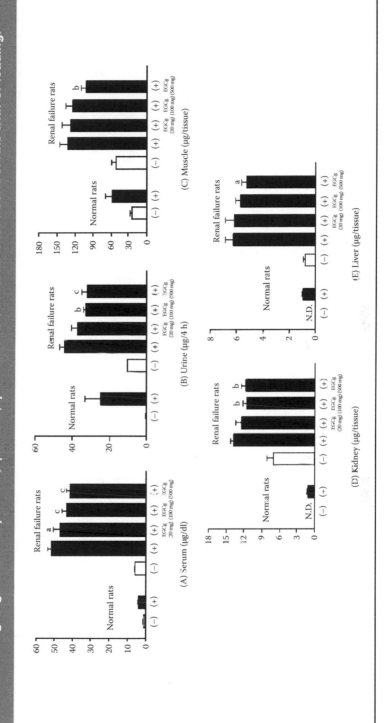

Figure 9.6

MG levels in serum (A), urine (B), muscle (C), kidney (D), and liver (E). (–), without Cr loading; (+), with Cr loading. Significance: $^a p < 0.05$, $^b p < 0.01$, $^c p < 0.001$ versus renal failure control rats with Cr loading.

We have already demonstrated that green tea polyphenols (daily dose, 400 mg) administered for 6 months to 50 patients on dialysis decreased the blood levels of MG (Yokozawa et al., 1996a), and that concomitant treatment with green tea polyphenols during 25-day adenine feeding periods produced a dose-dependent decrease in the serum MG level (Yokozawa et al., 1997). Furthermore, we reported that concomitant treatment with green tea polyphenols had protective effects against the increased serum Cr and urinary protein levels and the decreased creatinine clearance (Ccr) (Yokozawa et al., 1996b, 1998), indicating that green tea polyphenols can delay deterioration of the renal function. Taking the evidence from previous and present studies into consideration, we propose that green tea polyphenols exert an MG-lowering effect in dialysis patients and rats with adenine-induced renal failure through, at least in part, two actions: the improvement of renal dysfunction and inhibition of MG production from Cr due to their ability to scavenge •OH.

9.5 (–)-Epigallocatechin 3-O-Gallate and Diabetic Nephropathy

The pathogenesis of diabetic nephropathy has been extensively discussed for many years, and it has been accepted that oxidative stress is closely involved as a causative factor stemming from persistent hyperglycemia (Baynes and Thorpe, 1999; Ha and Kim, 1999). Within the diabetic kidney, glucose-dependent pathways such as increasing oxidative stress, polyol formation, and advanced glycation end product (AGE) accumulation, are activated.

To evaluate the effect of (–)-epigallocatechin 3-O-gallate as a representative polyphenol on diabetic nephropathy, rats with subtotal nephrectomy plus streptozotocin (STZ) injection were orally administered (–)-epigallocatechin 3-O-gallate at doses of 25, 50, and 100 mg/kg body weight/day for 50 days.

Hyperglycemia is the principle factor responsible for structural alterations at the renal level, and the Diabetes Control and Complications Trial Research Group (1993) has elucidated that hyperglycemia is directly linked to diabetic microvascular complications, particularly in the kidney; therefore, glycemic control remains the main target of therapy. In this study, the glucose level of diabetic nephropathy rats showed a significant and approximate threefold increase; however, (–)-epigallocatechin 3-O-gallate inhibited this increase dose-dependently (Table 9.10). In addition, the typical pattern

TABLE 9.10
Serum Constituents at 50 Days of Administration

Items	Normal	Control	(–)-Epigallocatechin 3-O-gallate 25 mg/kg B.W./day	50 mg/kg B.W./day	100 mg/kg B.W./day
Glucose (mg/dl)	193 ± 9	592 ± 38^c	497 ± 22^{c,e}	487 ± 22^{c,e}	460 ± 19^{c,e}
Total protein (g/dl)	4.75 ± 0.11	4.21 ± 0.08^c	4.20 ± 0.10^c	4.37 ± 0.07^{c,d}	4.44 ± 0.06^{c,e}
Albumin (g/dl)	2.88 ± 0.04	2.38 ± 0.08^c	2.43 ± 0.06^c	2.56 ± 0.06^{c,e}	2.62 ± 0.05^{c,e}
Total cholesterol (mg/dl)	46.4 ± 2.4	113.6 ± 12.7^c	102.3 ± 6.0^c	83.3 ± 6.4^{c,e}	77.7 ± 6.8^{c,e}
Triglyceride (mg/dl)	63.7 ± 6.3	143.1 ± 31.4^c	126.6 ± 15.7^a	120.9 ± 27.3^a	116.6 ± 26.3^a
TBA-reactive substance (nmol/ml)	1.56 ± 0.08	3.70 ± 0.39^c	2.48 ± 0.18^{b,e}	2.50 ± 0.34^{b,e}	2.16 ± 0.24^e

Significance: $^a p < 0.05$, $^b p < 0.01$, $^c p < 0.001$ versus normal values; $^d p < 0.05$, $^e p < 0.001$ versus diabetic nephropathy control values.

of serum constituents, that is, a decrease in total protein and albumin due to their excessive excretion via urine, and also an increase in lipids, for example, total cholesterol and triglyceride, whose abnormal metabolism has been proven to play a role in the pathogenesis of diabetic nephropathy (Sun et al., 2002) and to enhance lipid peroxidation, was improved by the administration of (–)-epigallocatechin 3-O-gallate (Table 9.10). Therefore, we suggest that (–)-epigallocatechin 3-O-gallate had a positive effect on serum glucose and lipid metabolic abnormalities.

The results of the study presented here demonstrate that diabetic nephropathy rats showed significant increases in the serum urea nitrogen, Cr, and urinary protein excretion rate, whereas the Ccr level showed a significant decrease compared with normal rats, representing a decline in the renal function (Table 9.11). However, the (–)-epigallocatechin 3-O-gallate treatment positively affected these parameters, especially in the group given 100 mg (Table 9.11). For further investigation, we performed pattern analysis of proteinuria using SDS-PAGE, and the (–)-epigallocatechin 3-O-gallate treatment led to a clear decrease at all parts of the molecule (Figure 9.7). These data suggest that not only the improvement of proteinuria, but also its individual fractions, may, at least in part, ameliorate the development of glomerular and tubulointerstitial injury.

In the state of diabetic nephropathy, there is increased glomerular basement membrane thickening and mesangial extracellular matrix (ECM) deposition, followed by mesangial hypertrophy and diffuse and nodular glomerular sclerosis, and these structural changes may be directly influenced by AGEs through

TABLE 9.11
Renal Functional Parameters at 50 Days of Administration

Items	Normal	Control	(–)-Epigallocatechin 3-O-gallate 25 mg/kg B.W./day	50 mg/kg B.W./day	100 mg/kg B.W./day
Serum urea nitrogen (mg/dl)	16.8 ± 0.5	44.5 ± 3.1[b]	37.9 ± 1.8[b,d]	38.0 ± 2.6[b,d]	28.8 ± 1.4[b,d]
Serum Cr (mg/dl)	0.38 ± 0.01	0.94 ± 0.09[b]	0.90 ± 0.08[b]	0.82 ± 0.06[b]	0.66 ± 0.05[b,d]
Ccr (ml/kg B.W./min)	7.20 ± 0.26	3.35 ± 0.43[b]	3.41 ± 0.32[b]	3.65 ± 0.37[b]	4.09 ± 0.35[b,c]
Urinary protein (mg/day)	19.1 ± 0.7	82.3 ± 13.3[b]	64.0 ± 11.9[b]	47.9 ± 14.6[a,d]	40.6 ± 6.4[d]

Significance: [a]$p < 0.05$, [b]$p < 0.001$ versus normal values; [c]$p < 0.05$, [d]$p < 0.001$ versus diabetic nephropathy control values.

FIGURE 9.7
SDS-PAGE pattern of proteinuria in normal rats (N) and diabetic nephrectomized rats treated with (–)-epigallocatechin 3-O-gallate at 25 mg/kg body weight/day (E25), 50 mg/kg body weight/day (E50), 100 mg/kg body weight/day (E100), or water (control, C) for 50 days. Lane M shows the molecular weight marker.

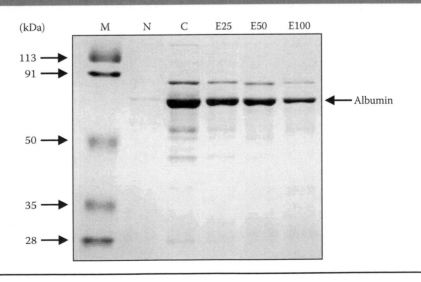

excessive cross-linking of the matrix molecules in a receptor-independent way (Vlassara et al., 1992, 1994). In this study, we demonstrated that renal AGE accumulation observed in diabetic nephropathy rats was decreased by (–)-epigallocatechin 3-O-gallate administration, although (–)-epigallocatechin 3-O-gallate showed only a slight tendency to reduce renal receptors for advanced glycation end product (RAGE) expression in diabetic nephropathy rats (Figure 9.8). However, a marked antioxidative activity of renal tissue was shown in the level of lipid peroxidation at 50 and 100 mg doses of (–)-epigallocatechin 3-O-gallate, resembling the results of inducible nitric oxide synthase (iNOS), cyclo-oxygenase (COX)-2, nuclear factor-κB (NF-κB), and phosphorylated inhibitor binding protein κB-α (IκB-α) (Figure 9.9), and the fibrogenic cytokine transforming growth factor (TGF)-β_1 and fibronectin protein expression in the renal cortex (Figure 9.8).

Moreover, diabetic nephropathy rats used in the present study showed significant glomerular hypertrophy and diffuse and exudative lesions. Longitudinal hyperfiltration is associated with renal enlargement such as an increase in the glomerular size, and diffuse lesion development is dependent on increased mesangial matrix and glomerular basement membrane thickening, because both are composed of ECM molecules, as in the case of the TGF-β system, and they also correlate with proteinuria. The other phenomenon, the exudative lesion called the capsular drop and fibrin cap, is suggested to consist of plasma components, such as IgM, fibrinogen, and AGEs. According to the results of histopathological evaluation, although diabetic nephropathy rats showed a 2.2-fold increase in the glomerular area, mild but significant increases in diffuse and exudative lesions, and a slight increase in the mesangial matrix, (–)-epigallocatechin 3-O-gallate could affect glomerular hypertrophy, and the lesions at 50 and 100 mg doses, reflecting the effects of AGEs, TGF-β_1, and fibronectin levels (Figures 9.8 and 9.10). Hence, we may hypothesize that (–)-epigallocatechin 3-O-gallate could be advantageous against diabetic kidney damage, which correlates with AGEs with or without a receptor-dependent pathway and their related inflammatory responses, and then (–)-epigallocatechin 3-O-gallate subsequently suppresses the induction of mesangial hypertrophy and fibronectin synthesis in diabetic nephropathy.

Our observations presented here suggest that (–)-epigallocatechin 3-O-gallate has a beneficial effect on diabetic nephropathy via suppressing hyperglycemia, AGEs, their related oxidative stress and cytokine activations, and also pathological states due to its synergistic effect. This study may provide original and strong supporting evidence for the efficacy of (–)-epigallocatechin 3-O-gallate in the early stage of diabetic nephropathy, suggesting that it would be a superior aid for the management of patients with diabetic nephropathy.

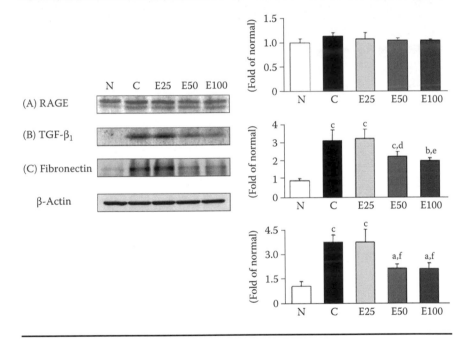

FIGURE 9.8
Western blot analyses of RAGE (A), TGF-β_1 (B), and fibronectin (C) protein expression in the renal cortex of normal rats (N) and diabetic nephrectomized rats treated with (–)-epigallocatechin 3-O-gallate at 25 mg/kg body weight/day (E25), 50 mg/kg body weight/day (E50), 100 mg/kg body weight/day (E100), or water (control, C) for 50 days. Significance: $^{a}p < 0.05$, $^{b}p < 0.01$, $^{c}p < 0.001$ versus normal values; $^{d}p < 0.05$, $^{e}p < 0.01$, $^{f}p < 0.001$ versus diabetic nephropathy control values.

9.6 Matcha (Green Tea Powder) and Type 2 Diabetic Renal Damage

The control of hyperglycemia and dyslipidemia such as hypertriglyceridemia and hypercholesterolemia in type 2 diabetes has been considered a therapeutic strategy along with body weight loss through lifestyle modification (Scheen, 2003). In the present study, we investigated the effects of matcha on Otsuka Long-Evans Tokushima Fatty (OLETF) rats, a model of human type 2 diabetes mellitus, to identify its effects on hyperglycemia,

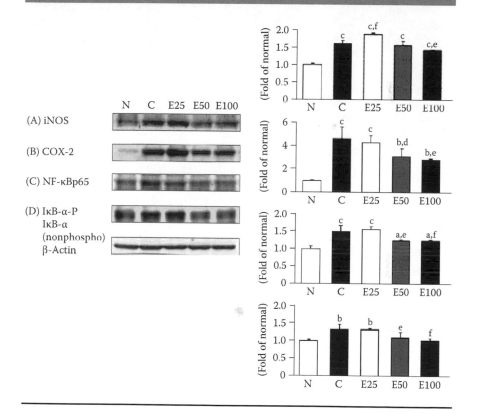

FIGURE 9.9
Western blot analyses of iNOS (A), COX-2 (B), NF-κBp65 (C), and IκB-α (phosphorylated and nonphosphorylated) (D) protein expression in the renal cortex of normal rats (N) and diabetic nephrectomized rats treated with (−)-epigallocatechin 3-*O*-gallate at 25 mg/kg body weight/day (E25), 50 mg/kg body weight/day (E50), 100 mg/kg body weight/day (E100), or water (control, C) for 50 days. Significance: [a]$p < 0.05$, [b]$p < 0.01$, [c]$p < 0.001$ versus normal values; [d]$p < 0.05$, [e]$p < 0.01$, [f]$p < 0.001$ versus diabetic nephropathy control values.

dyslipidemia, AGEs and their receptors, and oxidative stress. Matcha (50, 100, or 200 mg/kg body weight) was orally administered for 16 consecutive weeks to 22-week-old OLETF rats. Following matcha administration, serum glucose, triglyceride, and total cholesterol levels were significantly reduced in the OLETF rats (Figure 9.11 and Table 9.12). In addition, the elevated serum

FIGURE 9.10
Photomicrographs of the glomeruli in normal rats (A) and diabetic nephrectomized rats treated with (−)-epigallocatechin 3-*O*-gallate at 25 mg/kg body weight/day (D), 50 mg/kg body weight/day (E), 100 mg/kg body weight/day (F), or water (control, B and C) for 50 days. Scale bar, 100 μm.

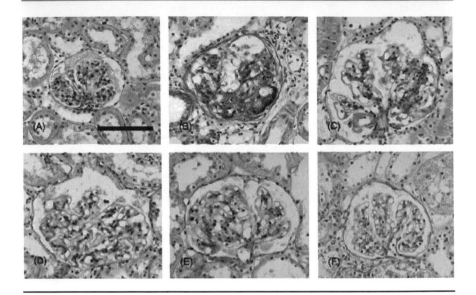

thiobarbituric acid (TBA)-reactive substance levels, used as an indicator of oxidative stress (Wright et al., 2006), in OLETF rats were also significantly decreased with 100 and 200 mg matcha administration (Table 9.12).

In this experiment, serum glycosylated protein in serum and AGE levels in the liver and kidney were increased in OLETF rats. Matcha administration (100 and 200 mg/kg body weight) markedly reduced the elevated AGE levels in the kidney, but not in the serum or liver (Tables 9.12 and 9.13). Then, we investigated the effects of matcha on the expression of RAGE as well as N^{ε}-(carboxylmethyl)lysine (CML) and N^{ε}-(carboxylethyl)lysine (CEL), two major products of AGEs in the kidney. In addition, RAGE is activated by CML, and the AGE-RAGE interaction in mesangial and endothelial cells increases reactive oxygen species formation, with the subsequent activation of NF-κB. Western blotting analyses revealed that matcha effectively reduced the elevated expression levels of CML, CEL, and RAGE in OLETF rats (Figure 9.12).

FIGURE 9.11
Profiles of (A) serum glucose and (B) total cholesterol levels every 4 weeks and (C) urinary protein and (D) Ccr levels every 8 weeks during the 16-week experimental period: normal (LETO) (■), control (■), matcha (50 mg/kg body weight/day) (■), matcha (100 mg/kg body weight/day) (■), and matcha (200 mg/kg body weight/day) (■). Significance: [a]p < 0.05, [b]p < 0.01, [c]p < 0.001 versus OLETF control rats.

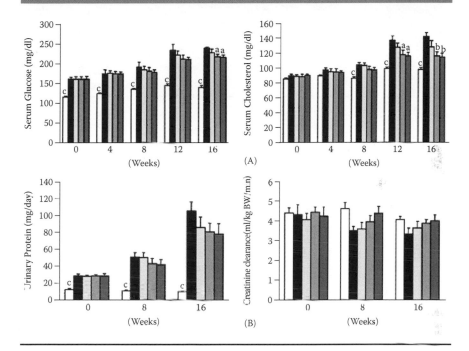

Sterol regulatory element binding proteins (SREBPs) are transcription factors that regulate fatty acid and cholesterol synthesis, as well as insulin sensitivity (Horton et al., 2002). SREBP-1 preferentially activates genes involved in fatty acid synthesis, whereas SREBP-2 preferentially activates those involved in cholesterol biosynthesis. Hepatic glucose, triglyceride, and total cholesterol levels are significantly reduced by matcha treatment (Table 9.13). In this experiment, matcha restored the decreased SREBP-2 expression to the values of normal rats, but not that of SREBP-1 (Figure 9.13).

In conclusion, we revealed that matcha has beneficial effects on renal and hepatic damage through the suppression of renal AGE accumulation and reductions in the hepatic glucose, triglyceride, and total cholesterol levels

TABLE 9.12
Serum Biochemical Levels in OLETF and LETO Rats

| | | | OLETF Rats | | |
| | | | Matcha | | |
Items	LETO Rats	Control	50 mg/kg B.W./day	100 mg/kg B.W./day	200 mg/kg B.W./day
Glycosylated protein (nmol/mg protein)	15.7 ± 0.8[b]	19.2 ± 0.4	18.8 ± 0.5	18.4 ± 0.3	18.2 ± 0.4
Albumin (g/dl)	3.57 ± 0.02	3.42 ± 0.05	3.49 ± 0.03	3.52 ± 0.05	3.55 ± 0.06
Total protein (g/dl)	5.55 ± 0.06	5.60 ± 0.07	5.64 ± 0.07	5.76 ± 0.06	5.86 ± 0.08[a]
Triglyceride (mg/dl)	65.0 ± 5.4[c]	247.0 ± 19.7	211.3 ± 16.3	190.3 ± 13.8[a]	188.8 ± 16.6[a]
TBA-reactive substance (nmol/ml)	2.31 ± 0.11[c]	3.05 ± 0.17	2.62 ± 0.20	2.21 ± 0.11[c]	2.13 ± 0.17[c]

Significance: [a]$p < 0.05$, [b]$p < 0.01$, [c]$p < 0.001$ versus OLETF control values.

TABLE 9.13
Renal and Hepatic Parameters in OLETF and LETO Rats

| | | | OLETF Rats | | |
| | | | Matcha | | |
Items	LETO Rats	Control	50 mg/kg B.W./day	100 mg/kg B.W./day	200 mg/kg B.W./day
Renal					
Glucose (mg/g tissue)	1.20 ± 0.04[c]	1.43 ± 0.05	1.44 ± 0.04	1.43 ± 0.04	1.47 ± 0.04
AGEs (AU)	0.85 ± 0.05[a]	0.98 ± 0.03	0.93 ± 0.02	0.88 ± 0.01[b]	0.88 ± 0.02[b]
Total cholesterol (mg/g tissue)	4.27 ± 0.02[c]	4.90 ± 0.07	4.67 ± 0.05[a]	4.72 ± 0.06[a]	4.71 ± 0.06[a]
Triglyceride (mg/g tissue)	4.47 ± 0.12[c]	6.13 ± 0.12	5.82 ± 0.15	6.14 ± 0.10	6.13 ± 0.21
Hepatic					
Glucose (mg/g tissue)	5.42 ± 0.27[c]	6.95 ± 0.24	6.38 ± 0.21[a]	6.27 ± 0.11[b]	6.10 ± 0.18[b]
AGEs (AU)	0.67 ± 0.04	0.81 ± 0.02	0.77 ± 0.02	0.76 ± 0.04	0.74 ± 0.03
Total cholesterol (mg/g tissue)	4.58 ± 0.37[c]	7.61 ± 0.26	6.83 ± 0.24	6.60 ± 0.36[b]	6.14 ± 0.32[c]
Triglyceride (mg/g tissue)	14.25 ± 0.76[c]	37.11 ± 1.35	29.49 ± 0.98[b]	28.90 ± 1.10[b]	28.68 ± 1.41[b]

Significance: [a]$p < 0.05$, [b]$p < 0.01$, [c]$p < 0.001$ versus OLETF control values.

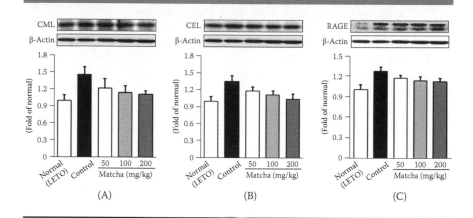

FIGURE 9.12
Effect of matcha on (A) CML, (B) CEL, and (C) RAGE levels in the renal cortex.

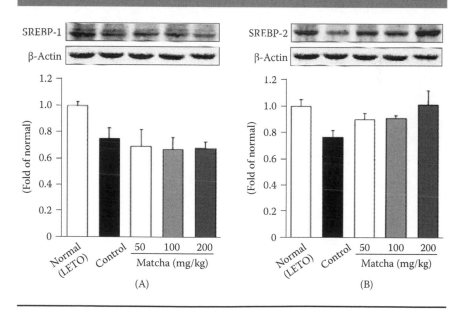

FIGURE 9.13
Effect of matcha on nuclear (A) SREBP-1 and (B) SREBP-2 levels in hepatic tissue.

and in antioxidant activities. Therefore, daily matcha consumption may have beneficial effects on the control of obesity and type 2 diabetes.

9.7 Conclusion and Future Prospects

Much attention regarding green tea's benefits has been focused on the role of antioxidant activity in relation to the aging process and degenerative diseases, like cancer, cardiovascular disease, and diabetes. This chapter shows that based on antioxidant activity, green tea polyphenols and their constituents exert protective effects on renal damage caused by various toxic situations, such as excessive arginine supply, strong oxidative radicals, renal toxin, diabetic nephropathy, and type 2 diabetes. Therefore, we expect that green tea polyphenols have the potential to prevent organ failure and, in particular, provide a promising therapeutic approach to renal disorders. As green tea is already one of the most popular beverages worldwide, its role should be understandably elucidated in the direct and indirect prevention of chronic diseases. In order to explain the potential mechanisms of green tea polyphenols for protection against organ damage concomitant with chronic disease, additional research is needed on the pharmacokinetics of tea constituents as well as exploration at the cellular level. Furthermore, well-designed observational epidemiological studies and intervention trials will generate clear and safe conclusions concerning the protective effects of tea.

References

Aoyagi, K., Ohba, S., Narita, M., and Tojo, S. 1983. Regulation of biosynthesis of guanidinosuccinic acid in isolated rat hepatocytes and *in vivo*. *Kidney Int Suppl*, 16: 224–228.

Baynes, J.W., and Thorpe, S.R. 1999. Role of oxidative stress in diabetic complications: A new perspective on an old paradigm. *Diabetes*, 48: 1–9.

Ceriello, A., Mercuri, F., Quagliaro, L., Assaloni, R., Motz, E., Tonutti, L., and Taboga, C. 2001. Detection of nitrotyrosine in the diabetic plasma: Evidence of oxidative stress. *Diabetologia*, 44: 834–838.

Chung, H.Y., Yokozawa, T., Soung, D.Y., Kye, I.S., No, J.K., and Baek, B.S. 1998. Peroxynitrite-scavenging activity of green tea tannin. *J Agric Food Chem*, 46: 4484–4486.

Cohen, B.D., and Patel, H. 1982. Guanidinosuccinic acid and the alternate urea cycle. *Adv Exp Med Biol*, 153: 435–441.

Cuzzocrea, S., and Reiter, R.J. 2001. Pharmacological action of melatonin in shock, inflammation and ischemia-reperfusion injury. *Eur J Pharmacol*, 426: 1–10.

Diabetes Control and Complications Trial Research Group. 1993. The effect of intensive treatment of diabetes on the development and progression of long-term complications in insulin-dependent diabetes mellitus. *N Engl J Med*, 329: 977–986.

Douki, T., Cadet, J., and Ames, B.N. 1996. An adduct between peroxynitrite and 2'-deoxyguanosine: 4,5-dihydro-5-hydroxy-4-(nitrosooxy)-2'-deoxyguanosine. *Chem Res Toxicol*, 9: 3–7.

Fiorillo, C., Oliviero, C., Rizzuti, G., Nediani, C., Pacini, A., and Nassi, P. 1998. Oxidative stress and antioxidant defenses in renal patients receiving regular haemodialysis. *Clin Chem Lab Med*, 36: 149–153.

Fukuyama, N., Takebayashi, Y., Hida, M., Ishida, H., Ichimori, K., and Nakazawa, H. 1997. Clinical evidence of peroxynitrite formation in chronic renal failure patients with septic shock. *Free Radic Biol Med*, 22: 771–774.

Ha, H., and Kim, K.H. 1999. Pathogenesis of diabetic nephropathy: The role of oxidative stress and protein kinase C. *Diabetes Res Clin Pract*, 45: 147–151.

Hahn, S., Krieg, R.J., Hisano, S., Kuemmerle, N.B., Saborio, P., and Chan, J.C. 1999. Vitamin E suppresses oxidative stress and glomerulosclerosis in rat remnant kidney. *Pediatr Nephrol*, 13: 195–198.

Halliwell, B., and Gutteridge, J.M.C. 1990. Role of free radicals and catalytic metal ions in human disease: An overview. *Methods Enzymol*, 186(B): 1–85.

Handelman, G.J., Walter, M.F., Adhikarla, R., Gross, J., Dalal, G.E., Levin, N.W., and Blumberg, J.B. 2001. Elevated plasma F2-isoprostanes in patients on long-term hemodialysis. *Kidney Int*, 59: 1960–1966.

Horton, J.D., Goldstein, J.L., and Brown, M.S. 2002. SREBPs: Activators of the complete program of cholesterol and fatty acid synthesis in the liver. *J Clin Invest*, 109: 1125–1131.

Ischiropoulos, H. 1998. Biological tyrosine nitration: A pathophysiological function of nitric oxide and reactive oxygen species. *Arch Biochem Biophys*, 356: 1–11.

Kakimoto, M., Inoguchi, T., Sonta, T., Yu, H.Y., Imamura, M., Etoh, T., Hashimoto, T., and Nawata, H. 2002. Accumulation of 8-hydroxy-2'-deoxyguanosine and mitochondrial DNA deletion in kidney of diabetic rats. *Diabetes*, 51: 1588–1595.

Kone, B.C. 1997. Nitric oxide in renal health and disease. *Am J Kidney Dis*, 30: 311–333.

Kopple, J.D., Gordon, S.I., Wang, M., and Swendseid, M.E. 1977. Factors affecting serum and urinary guanidinosuccinic acid levels in normal and uremic subjects. *J Lab Clin Med*, 90: 303–311.

Nakagawa, T., and Yokozawa, T. 2002. Direct scavenging of nitric oxide and superoxide by green tea. *Food Chem Toxicol*, 40: 1745–1750.

Nakazawa, H., Fukuyama, N., Takizawa, S., Tsuji, C., Yoshitake, M., and Ishida, H. 2000. Nitrotyrosine formation and its role in various pathological conditions. *Free Radic Res*, 33: 771–784.

Natelson, S., and Sherwin, J.E. 1979. Proposed mechanism for urea nitrogen re-utilization: Relationship between urea and proposed guanidine cycles. *Clin Chem*, 25: 1343–1344.

Natelson, S., Tseng, H.Y., and Sherwin, J.E. 1978. On the biosynthesis of guanidino-succinate. *Clin Chem*, 24: 2108–2114.

Orita, Y., Tsubakihara, Y., Ando, A., Nakata, K., Takamitsu, Y., Fukuhara, Y., and Abe, H. 1978. Effect of arginine or creatinine administration on the urinary excretion of methylguanidine. *Nephron*, 22: 328–336.

Paller, M.S., Hoidal, J.R., and Ferris, T.F. 1984. Oxygen free radicals in ischemic acute renal failure in the rat. *J Clin Invest*, 74: 1156–1164.

Radi, R., Beckman, J.S., Bush, K.M., and Freeman, B.A. 1991. Peroxynitrite-induced membrane lipid peroxidation: The cyto-toxic potential of superoxide and nitric oxide. *Arch Biochem Biophys*, 288: 481–487.

Scheen, A.J. 2003. Current management strategies for coexisting diabetes mellitus and obesity. *Drugs*, 63: 1165–1184.

Sun, L., Halaihel, N., Zhang, W., Rogers, T., and Levi, M. 2002. Role of sterol regulatory element-binding protein 1 in regulation of renal lipid metabolism and glomerulosclerosis in diabetes mellitus. *J Biol Chem*, 277: 18919–18927.

Vlassara, H., Fuh, H., Makita, Z., Krungkrai, S., Cerami, A., and Bucala, R. 1992. Exogenous advanced glycosylation end products induce complex vascular dysfunction in normal animals: A model for diabetic and aging complications. *Proc Natl Acad Sci USA*, 89: 12043–12047.

Vlassara, H., Striker, L.J., Teichberg, S., Fuh, H., Li, Y.M., and Steffes, M. 1994. Advanced glycation end products induce glomerular sclerosis and albuminuria in normal rats. *Proc Natl Acad Sci USA*, 91: 11704–11708.

Wills, M.R. 1985. Uremic toxins, and their effect on intermediary metabolism. *Clin Chem*, 31: 5–13.

Wright, Jr., E., Scism-Bacon, J.L., and Glass, L.C. 2006. Oxidative stress in type 2 diabetes: The role of fasting and postprandial glycaemia. *Int J Clin Pract*, 60: 308–314.

Yokozawa, T., Chen, C.P., Rhyu, D.Y., Tanaka, T., Park, J.C., and Kitani, K. 2002. Potential of sanguiin H-6 against oxidative damage in renal mitochondria and apoptosis mediated by peroxynitrite *in vivo*. *Nephron*, 92: 133–141.

Yokozawa, T., Cho, E.J., Nakagawa, T., Terasawa, K., and Takeuchi, S. 2000. Inhibitory effect of green tea tannin on free radical-induced injury to the renal epithelial cell line, LLC-PK$_1$. *Pharm Pharmacol Commun*, 6: 521–526.

Yokozawa, T., Chung, H.Y., He, L.Q., and Oura. H. 1996b. Effectiveness of green tea tannin on rats with chronic renal failure. *Biosci Biotechnol Biochem*, 60: 1000–1005.

Yokozawa, T., Dong, E., Nakagawa, T., Kashiwagi, H., Nakagawa, H., Takeuchi, S., and Chung, H.Y. 1998. *In vitro* and *in vivo* studies on the radical-scavenging activity of tea. *J Agric Food Chem*, 46: 2143–2150.

Yokozawa, T., Dong, E., and Oura, H. 1997. Proof that green tea tannin suppresses the increase in the blood methylguanidine level associated with renal failure. *Exp Toxic Pathol*, 49: 117–122.

Yokozawa, T., Fujitsuka, N., and Oura, H. 1991. Studies on the precursor of methylguanidine in rats with renal failure. *Nephron*, 58: 90–94.

Yokozawa, T., Oura, H., Shibata, T., Ishida, K., Kaneko, M., Hasegawa, M., Sakanaka, S., and Kim, M. 1996a. Effects of green tea tannin in dialysis patients. *J Trad Med*, 13: 124–131.

10

Green Tea Polyphenols Improve Bone and Muscle Quality

Olivier M. Dorchies and Urs T. Ruegg

Contents

10.1 Introduction

While there are numerous reports on the action of green tea extract and the polyphenols it contains on the inhibition of tumor development, inflammation, and on other beneficial effects, little has been reported regarding such effects on bone and diseased muscle. In this chapter, we will briefly review data from the literature on bone and describe our own results with green tea on skeletal muscle.

There is a distinctive difference between green tea extract (GTE), which contains caffeine as well as various minerals and vitamins, and green tea polyphenols (GTPs), which consist of a mixture of pure polyphenols. Some of the effects caused by GTE may not result from the polyphenols. For instance, the caffeine contained in GTE is a blocker of purinergic receptors in the central nervous system, and by their inhibition caffeine causes stimulation of peripheral sympathetic activity, leading, among other effects, to increased blood pressure.

10.2 Considerations about Molecular Targets for Green Tea Polyphenols

Green tea polyphenols are well known as radical scavengers and antioxidants, as described in other chapters in this book. However, in addition to these qualities, they have other effects resulting from interactions with more or less well-defined molecular targets. Most, if not all, compounds used in therapy of disease are not absolutely specific for the envisaged target, but interact with other targets as well. The affinities for these are reflected by the concentrations needed to obtain the desired or undesired therapeutic action. This requires pharmacological measurements to be quantitative. It is particularly important that experiments be done in a dose or concentration range that is in proximity to the concentration found in *in vivo* situations. Note that bioavailability of the polyphenols contained in green tea in humans is below 1% (Chow et al., 2001), and peak blood concentrations that are reached after oral administration of GTPs are about 0.3 μg/ml, corresponding to a concentration of about 1 μM (or 10^{-6} M) (Chow et al., 2003). Results should be interpreted with this in mind, and effects observed at much higher concentrations, especially in *in vitro* systems, should be interpreted with caution. The present chapter outlines this approach using the example of the action of GTPs on muscular disease.

10.3 Potential of Green Tea Polyphenols for Treating Bone Diseases

The skeleton of vertebrates consists of a particular form of calcium phosphate, hydroxyapatite, and various filamentous proteins; both are essential for the support of posture and to coordinate movements. In osteoporosis, the inorganic matter of bone is reduced and becomes more porous, leading to increased sensitivity to fractures. Osteoporosis is by far the most prevalent bone disease. A large number of disorders affect the protein component of bone, the tendons, and the cartilage. The most commonly known disorder is scurvy, caused by insufficient hydroxylation of collagen, due to the lack of the most well-known antioxidant, ascorbic acid (vitamin C), a cofactor for collagen hydroxylation.

There are few reports on green tea components on bone. Shen and colleagues have recently published a series of papers (reviewed in Shen et al., 2011b) that describe the beneficial effects of GTPs on bone health and structure. These authors showed that supplementation by GTE, together with exercise, increased bone formation and improved the biomarkers and bone turnover rate (Shen et al., 2011c). Oral GTE supplementation, exercise, and the combination of the two also improved muscle strength in postmenopausal women with osteopenia (Shen et al., 2011c). Similar improvements were observed in both ovariectomized and orchidectomized rats, suggesting that GTPs have the ability to protect bone structure irrespective of the sex (Shen et al., 2011a).

In rodents, GTPs have been shown to reduce the risk of developing a type of arthritis similar to human rheumatoid arthritis (Haqqi et al., 1999). Not only was the group of mice given polyphenols less likely to develop arthritis, but also, in those that did develop the condition, the disease occurred later and was milder than in the placebo group. Of the animals drinking polyphenols, less than half developed arthritis, compared with almost all in the control group. In their initial publication, these investigators (Haqqi et al., 1999) state: "Based on our data it is tempting to suggest that green tea in general, and the polyphenols present therein in particular, may prove to be a useful supplement/addition with other agents for the treatment of arthritis." Further work, summarized in Singh et al. (2010), shows that GTPs act on a variety of cell types involved in rheumatoid arthritis and on the inflammatory mediators of the affected joints. These mediators are kinases, transcription factors, and proteases (Singh et al., 2010); some of these will be discussed in the context of muscle disease below.

10.4 Green Tea Polyphenols and Skeletal Muscle

There is anecdotal evidence that green tea helps to decrease adipose tissue and increases muscle weight and force. With regard to its effect on adipose tissue, GTE is often coined a "fat burner." However, there is currently no reliable scientific information on this property, and we consider it wishful thinking. On the other hand, it has been shown that GTE improves endurance capacity and increases muscle lipid oxidation in mice (Murase et al., 2004). The human results are apparently encouraging as well, but it might be that these are mostly due to stimulation of the central nervous system by caffeine (Smith et al., 2010). What these results have in common is that exercise is required for GTE to have an effect. This has also been shown in aging mice (Murase et al., 2004). Finally, (–)-epigallocatechin-3-gallate (EGCG) has been reported to attenuate skeletal muscle atrophy caused by experimentally induced cancer cachexia in mice (Wang et al., 2011).

Regarding possible mechanisms of action, these appear to be due to multiple targets, all finally affecting gene expression. A very promising result, because of the low concentrations showing an effect and of the relevance of the target, is the report that green tea catechins are potent sensitizers of ryanodine receptors (Feng et al., 2010). These receptors are mediating contractions when the muscle is stimulated by the motor nerve. If sensitized, the same nervous input would result in a larger contraction, which could be of some benefit in a variety of conditions.

10.5 Green Tea Polyphenols Acting on Dystrophic Muscle

Oxidative stress appears to play a role in the progression of the pathology of some of the muscular disorders, of which there are over 200 (Kaplan, 2010), and most, if not all, are of genetic origin. This is also the case with the muscular dystrophies, in which there is a progressive loss of muscle function caused wasting of the affected muscles. Until about 5 years ago, it was unclear whether the involvement of oxidative stress was a link in the chain of events leading to muscle degeneration or merely an epiphenomenon, that is, an event taking place in parallel during the course of the disease but without involvement in causing the disease. For this reason, we have investigated the effects of antioxidants to slow down or halt disease progression.

We chose to use a decaffeinated fraction of purified GTPs, as it is known to contain some of the most potent antioxidants, is nontoxic, is readily available, and could be given to patients without lengthy and costly procedures, such as those required for a new drug application. Thus, we administered low doses of GTPs (0.01 and 0.05% of Sunphenon DCF in chow) to dystrophic mice, the so-called mdx^{5Cv} mice. These mice are the most commonly used animal models of Duchenne muscular dystrophy (DMD), the most frequently occurring and most devastating of the muscular dystrophies (Blake et al., 2002). We found an improvement in muscle quality (Buetler et al., 2002), better than the one obtained earlier with creatine (Passaquin et al., 2002; Ruegg et al., 2002). GTPs and their major component, EGCG, were further investigated with respect to their effects on muscle function (Dorchies et al., 2006). After 1 week of treatment, histology showed a delay in necrosis of the extensor digitorum longus muscle, and mechanical properties of triceps suræ muscles were increased to levels close to those of normal mice (Figure 10.1). After a 5-week exposure to these agents, the gain in function was 50 to 65% (Dorchies et al., 2006). These findings were confirmed and extended by others who found that not only the structure of the leg muscles was improved, but also that of the diaphragm (Nakae et al., 2008). Recently, we demonstrated that GTPs reduced the number of lipofuscine granules, a marker of cumulated oxidative stress, in various muscles of the mdx mouse (Nakae et al., 2012). Also, we established that the oral route was the most efficient for alleviating the symptoms of the mdx dystrophic mouse, which is of importance in the perspective of a clinical use on Duchenne boys (Nakae et al., 2012). After 15 months of treatment with GTPs or EGCG, a reduction of kyphosis was noted, suggesting a reduction of the dystrophic process at the level of the paraspinal and back muscles (Dorchies et al., unpublished). These results demonstrate that administration of GTPs or EGCG to dystrophic mice protects muscle against the first massive wave of necrosis, and stimulates muscle adaptation toward a stronger and more resistant phenotype. The results also underline the causal role of oxidative stress in the pathogenesis of muscular dystrophy, at least in our *in vivo* model.

On the other hand, little is known about a more specific mechanism of action of the GTPs besides antioxidant activity. There are numerous studies on GTE and EGCG effects that are then brought into connection with a mechanism. However, only few such effects have been described to occur in the low micromolar or nanomolar range where an *in vivo* action is likely to take place in view of the low blood concentrations (<1 μM) obtained when administering polyphenols or EGCG (Chow et al., 2003; Suzuki et al., 2004). The binding site with the highest affinity for EGCG is the 67LR, a 67 kDa laminin receptor (IC_{50} 40 nM), suggesting convincingly that a "nonantioxidant mechanism" exists. This receptor might be the target mediating the relevant antitumor effects of EGCG (Fujimura et al., 2007; Tachibana et al., 2004). Interactions between the 67LR and laminin are involved in metastatic forms of many cancers, and an increase in the expression of 67LR has been found in a variety of cancers (Nelson et al., 2008).

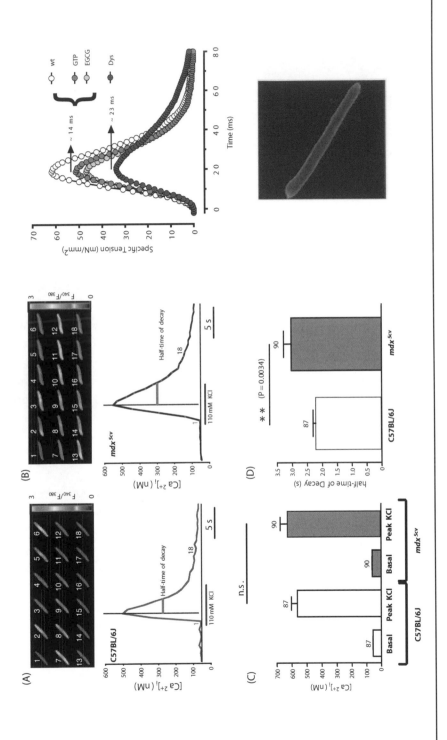

Figure 10.1

Calcium transients and effects of GTPs and EGCG on muscle twitch contraction. Left: KCl-induced cytosolic Ca^{2+} transients in isolated flexor digitorum brevis (FDB) fibers from control C57BL/6J and mdx^{5Cv} mice. (A, B) Calcium transients evoked by pressure ejection of a high KCl solution onto isolated C57BL/6J and mdx^{5Cv} fibers, respectively. Top panels in (A) and (B) show time series of F340/F380 ratio images of isolated fibers. Bottom panels represent global cytosolic Ca^{2+} transients triggered by high-KCl calculated from the whole fiber perimeter. Light gray circles on curves correspond to recording time of the 18 ratio images. (C) Plot of average [Ca^{2+}]c at baseline and peak of KCl-induced Ca^{2+} transients for both C57BL/6J and mdx^{5Cv} fibers. (D) Plot of average half-time of decay of KCl-induced Ca^{2+} transients in C57BL/6J and mdx^{5Cv} fibers analyzed in (C). Numbers of fibers tested (from 6 mice for C57BL/6J fibers and from 10 mice for mdx^{5Cv} fibers) are indicated on bar graphs. Right (above): Force and speed of contraction and relaxation of triceps suræ muscle in anaesthetized mice after 5-week treatment with GTPs (0.25%) or EGCG (0.10%) in chow. Electrically induced isometric contractions of the triceps suræ muscles were recorded. Peak twitch tension, time to peak, time for half-relaxation from the peak (RT½), and optimal tetanic tension were determined. These were normalized against the cross-sectional area of the muscle to express the specific force. For each test condition, the difference from untreated dystrophic mice is expressed as percent change. Dystrophic mdx^{5Cv} mice (Dys, black circles) showed a 50% reduction in force and about a 30% prolonged RT½ compared to normal mice (wt, white circles). Five-week treatment with GTPs (dark gray circles)/EGCG (light gray circles) attenuated these differences. Statistical significance: *, $p \le 0.01$. Right (below): Confocal section of an mdx^{5Cv} fiber (length about 0.15 mm) immunostained for iPLA$_2$β.

Perhaps of interest in the context of muscle, we found sevenfold higher levels of this receptor in cultures of mdx myotubes than in control myotubes (Dorchies et al., 2009).

Elevated levels of tumor necrosis factor (TNF) are involved in a number of pathologies, including muscular dystrophies. TNF production is controlled by nuclear factor (NF) κB, a transcription factor whose activity is stimulated by oxidative stress. Therefore, both NF-κB (Messina et al., 2006) inhibition and TNF scavenging by antibodies or soluble TNF receptor (Hodgetts et al., 2006; Radley et al., 2008) are valid approaches to treat these diseases. Indeed, it has been shown that GTE reduces TNF levels and necrosis in dystrophic mice (Evans et al., 2010).

An important pathway leading to degradation of intracellular proteins is the proteasome, whose inhibition can improve cell and tissue survival. The proof of concept in mdx mice has been made with a prototype proteasome inhibitor (Bonuccelli et al., 2003). Of interest, the chymotrypsin-like proteasome activity is, after the 67 kDa laminin receptor, the second-highest-affinity target of EGCG with an IC50 of 86 to 194 nM (Nam et al., 2001). Indeed, a report using tea polyphenols and a number of analogs as proteasome inhibitors (Smith et al., 2004) suggests that this pathway might be of some importance and bears therapeutic potential for EGCG.

10.5.1 A Vicious Cycle between Oxidative Stress and Ca^{2+} Influx

There is much evidence that an initial Ca^{2+} influx occurring into dystrophic muscle initiates a vicious cycle affecting local Ca^{2+} release, leading to downstream consequences culminating in necrosis and apoptosis (Allen and Whitehead, 2011; Basset et al., 2004; Dowling et al., 2004; Leijendekker et al., 1996). As outlined above, another early event is oxidative stress, occurring in the form of superoxide anion radical. But how are these two signaling components, Ca^{2+} influx and oxidative stress, related? It appears now that the increased production of superoxide anion radical comes from nicotinamide adenine dinucleotide phosphate (NADPH) oxidases (NOXes). Two of the five isoforms of this enzyme family, namely, type 2 (NOX2) and type 4 (NOX4), are expressed in muscle, and these are two- to threefold overexpressed in mdx muscle (Spurney et al., 2008), leading to increased output of superoxide anion radical. This aggressive but short-lived radical, in the presence of nitric oxide, can form peroxynitrite, which in turn may lead to damage via tyrosine nitrosylation of several proteins, including those involved in Ca2+ signaling (Hidalgo et al., 2006; Rando et al., 1998; Whitehead et al., 2006).

The pathway by which the Ca^{2+} influx occurs is not via classical L-type voltage-sensitive channels, which are sensitive to organic Ca^{2+} channel blockers, such as dihydropyridines (e.g., nifedipine), diltiazem, or verapamil. It is no surprise that two clinical trials with such channel blockers showed no improvement of the pathology characteristic for DMD (Brooke et al., 1984; Moxley et al., 1987). The Ca^{2+} influx takes place through nonspecific cation channels that are largely voltage insensitive, most likely through store-operated channels (SOCs) (Edwards et al., 2010; Iwata et al., 2009; Whitehead et al., 2006). Such SOCs are activated when the intracellular store, the sarco/endoplasmic reticulum, is emptied due to release of its content into the cytosol.

In an attempt to link oxidative stress and Ca^{2+} influx, Shirokova, Niggli, and coworkers have studied the formation and duration of very local $[Ca^{2+}]$ increases, so-called calcium sparks. They found that mechanically induced stress induced sparks that were sixfold larger in mdx than wild-type cells, and that inhibitors of NOXes blocked these sparks (Fanchaouy et al., 2009; Jung et al., 2008). It appears, therefore, quite likely that these local $[Ca^{2+}]$ increases occurring under the cell membrane (sarcolemma) are responsible for activating NOXes causing increased oxidative stress and the downstream deleterious consequences on the fate of the cell.

With respect to our current understanding of disease initiation and progression, important information became recently available. The group of H. Sies showed that several polyphenols, including those found in green tea, inhibit NOXes at low micromolar concentrations, EGCG being the most potent one, more so than the prototype NOX inhibitor apocynin (Steffen et al., 2008). Such an inhibition of NOX enzymes would block the downstream ill effects on the cell. As outlined above, superoxide anion radical generated by NOX affects cellular Ca^{2+} homeostasis. The targets include oxidation of ryanodine receptors (Oba et al., 2002; Xia et al., 2000), causing stimulation of Ca^{2+} release from the sarcoplasmic reticulum (Jung et al., 2008) and, according to our hypothesis, stimulation of the Ca^{2+}-insensitive isoform of phospholipase A_2 (iPLA$_2\beta$), an enzyme, which we suspect to contribute to the pathogenesis of dystrophy. Our hypothesis relating oxidative stress and Ca^{2+} homeostasis in dystrophic muscle is illustrated in Figure 10.2.

10.5.2 How Can Green Tea Polyphenols Counteract Muscle Disorders?

This possibility of a connection involving oxidative stress, Ca^{2+}, and iPLA$_2\beta$ has been strengthened by the fortuitous discovery regarding the action of

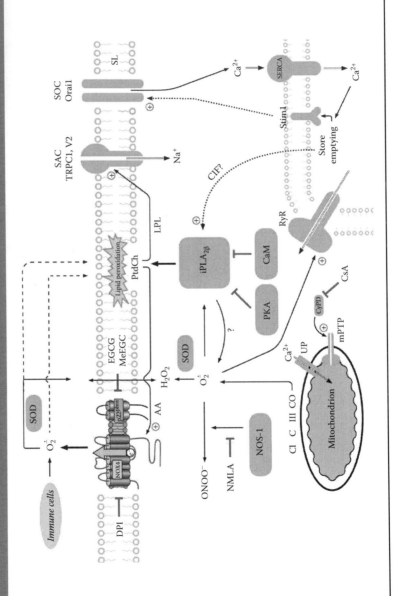

Figure 10.2
Oxidative stress and calcium handling in dystrophic muscle.

EGCG on excessive Ca^{2+} influx. Thus, when comparing recordings from Ca^{2+} imaging of isolated muscle fibers with those from *in vivo* investigations on muscle contraction-relaxation, we noted that the rates from peak to baseline were slowed in both dystrophic muscle fibers and in hind limb muscles, respectively (Figure 10.1). Of note, bromoenol lactone (BEL), an inhibitor of $iPLA_2$, normalized this phenomenon on Ca^{2+} influx, and green tea polyphenols did the same regarding muscle relaxation (Figure 10.2). Even though this is only circumstantial evidence, these observations point to a possible role of green tea polyphenols in inhibiting excessive Ca^{2+} influx occurring via store-operated channels (Figure 10.2).

As described above, oxidative stress appears to play a major role in the pathogenesis of DMD. In the past, mitochondria were thought to be the major producers of oxidative stress, mainly due to the release of the reactive superoxide anion radical from cytochrome oxidase, the terminal enzyme of the respiratory chain, but other mitochondrial producers, such as monoamine oxidase, have been suggested, although this evidence is based only on a single and not specific inhibitor (Menazza et al., 2010). Recently, it has been shown that—at least in muscle—NOXes are the main producers of superoxide anion radicals (Shkryl et al., 2009).

According to our hypothesis, these radicals attack a number of targets, including $iPLA_2$ (Figure 10.2). An inhibition of excessive reactive oxygen species (ROS) production, for example, by GTPs or EGCG, or of the deleterious ROS action, seems therefore an appropriate strategy to counteract the disease. Because of our encouraging results, we investigated the mechanism of action of GTPs or EGCG.

To achieve this, survival of primary cultures of muscle cells exposed to oxidative stress was investigated. The survival was increased when these were pretreated with GTPs or EGCG. On-line fluorimetric measurements using dichlorofluorescein (DCFH), a probe sensitive to oxidative stress, indicated that GTPs and EGCG protected the cells in a concentration-dependent manner (Dorchies et al., 2009). While after 48 h of exposure to GTPs or EGCG, activities of several antioxidant enzymes and expression of utrophin and dysferlin remained unaltered, and GTPs and EGCG increased the content of reduced glutathione. Interestingly, GTPs and EGCG decreased both the expression (–30%) and the activity of $iPLA_2$ (Figure 10.3). These results indicate that GTPs and EGCG protect dystrophic myotubes from oxidative stress by improving the glutathione balance and inhibiting $iPLA_2$ (Dorchies et al., 2009).

In our current working hypothesis, interplay exists between abnormal production of ROS and impairment of Ca^{2+} homeostasis. ROS producers are

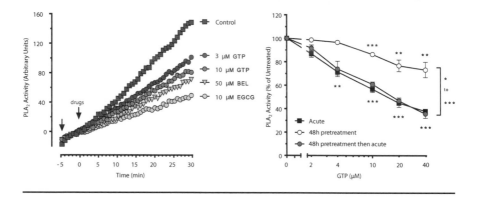

FIGURE 10.3
Inhibition of phospholipase A$_2$ by GTE, EGCG, and the suicide substrate inhibitor BEL. Measurements were made using primary cultures of mdx^{5Cv} myotubes and the fluorogenic PLA$_2$ substrate PED6. Left panel: GTE and EGCG at 10 µM inhibited PLA$_2$ activity almost as potently as 50 µM of the commonly used PLA$_2$ inhibitor BEL. Right panel: Acute exposure to GTP as well as long-term exposure to GTP for 48 hours prior to measurements inhibited PLA$_2$ activity in a concentration-dependent manner.

NOXes, enzymes located in the plasma membrane or in intracellular membranes, producing superoxide anion radical (O2$^-$•). NOX2 and NOX4 are expressed in the plasma membrane (PM) or in the sarcoplasmic reticulum (SR) membrane of muscle cells. Immune cells such as neutrophils can also produce superoxide or related ROS. Superoxide dismutase (SOD) converts O2$^-$• into hydroperoxide (H$_2$O$_2$), and both of them can damage membranes causing lipid peroxidation. Nitric oxide (NO•) produced by nitric oxide synthase-1 (NOS-1), which is lowered in DMD, can convert O2$^-$• into peroxynitrite (ONOO$^-$). Mitochondria produce oxidative stress via insufficiently operating complexes I and III (CI, CIII). These organelles contain Ca^{2+} uniporters (UPs) through which Ca^{2+} can enter into polarized mitochondria. The mitochondrial permeability transition pore (mPTP) can be activated by Ca^{2+} overload or ROS and can be inhibited via cyclophilin D by cyclosporin A (CsA) and the more potent CsA analog Debio 025 (Alisporivir). O2$^-$• stimulates the activity of the Ca^{2+}-independent phospholipase A2 (iPLA$_2$), possibly via lipid peroxidation. The lipid messengers produced by iPLA$_2$ acting on phosphatidyl choline (PtdCh) are arachidonic acid (AA) and lysophospholipids (LPLs).

AA can stimulate certain isoforms of NOXes, and some LPLs are known to stimulate store-operated Ca^{2+} entry (SOC), possibly by acting on Orai1.

It is becoming known that at the molecular level SOCs are composed of Orai1, the major protein forming the Ca^{2+} release activated Ca^{2+} (CRAC) channel, which is Ca^{2+} selective. Transient receptor potential (TRP) channels are nonselective channels allowing mostly sodium to enter the cells. Members of the canonical subfamily of TRPs (TRPC) channels, most likely TRPC1 but perhaps others as well, also play a role in Ca^{2+} handling. Store refilling via SOCs occurs when the Ca^{2+} stores are emptying (see right-hand lower corner). This causes Stim1 located in the SR membrane to activate SOC/Orai1. Alternatively or in parallel, release of the Ca^{2+} influx factor (CIF) sets in; this factor removes the inhibition by calmodulin (CaM) on $iPLA_2$, thus elevating LPLs and AA production.

Inhibitors are shown in red. These include diphenyliodonium (DPI), which inhibits all NADPH-dependent enzymes. As discussed in the text, some polyphenols, such as EGCG or methylated EGC (MeEGC), inhibit NOXes, and therefore the activation by $O2^{-\bullet}$ of $iPLA_2$. Cylic AMP-dependent protein kinase Λ (PKA) also blocks $iPLA_2$. The conversion of arginine into citrulline and NO• can be inhibited by arginine derivatives such as N-methyl-L-arginine (NMLA).

10.5.3 Which Proteins Are Affected by the Green Tea Components in Muscle?

Our investigations into the possible benefits of green tea components have shown that the muscles investigated are not durably protected from necrosis, occurring at different rates in the mdx mouse. However, the muscles attain a state of greater resistance, as we have shown (Dorchies et al., 2006). How exactly this increased resistance to multiple contractions in the form of tetani is achieved via oral administration of GTPs or EGCG is not fully understood. Investigations on the levels of isoforms of myosin heavy chains did not show any change in relative protein expression levels. Various other proteins, including desmin, were investigated, and no significant changes in quantity or isoforms were observed. The dystrophin homolog protein utrophin, known to be elevated in mdx muscle, was not further elevated in the gastrocnemius after feeding mice with GTPs or EGCG (Dorchies et al., 2006), whereas it was only slightly more abundant in the diaphragm (Nakae et al., 2012). Thus, we conclude that proteins involved in the contractile machinery or those contributing to membrane stability are not the major targets in the ameliorating effects of our compounds studied in mdx mice (Dorchies et al., 2009).

Therefore, we turned to proteins involved in signaling. As it is known that increased Ca^{2+} entry and cytosolic calcium overload occurring during individual contractions of muscle are an important link in the chain leading to muscle degeneration, we decided to study potential effects of GTPs and EGCG on the mechanisms involved in calcium entry. There are a number of Ca^{2+} conducting channels, both voltage sensitive and voltage insensitive. No effect of GTE or EGCG on L-type channel activity was found. This is not astonishing, as preclinical as well as clinical trials using the calcium antagonists nifedipine and diltiazem showed these compounds to be ineffective (Brooke et al., 1984; Moxley et al., 1987).

The large class of voltage-insensitive channels includes the group of TRP channels as well as the newly discovered Orai channels (Figure 10.2). Previous studies have shown that TRPC1 is attached to syntrophin (Vandebrouck et al., 2007), which itself links to dystrophin, and in the absence of the latter, syntrophin becomes released from the plasma membrane, also causing disturbances in the function of TRPC1. Of particular interest is the TRPV2 channel, whose activity has been attenuated by overexpressing dominant negative TRPV2 protein (Iwata et al., 2009). The dystrophic mice, having this inactive channel, show a very clear, although not full, recovery from the dystrophic phenotype (Iwata et al., 2009), thus making TRPV2 a very likely candidate for the excessive Ca^{2+} entry. Our investigations on the expression levels of these channels, however, gave only negative results: GTPs and EGCG left these unaltered. However, we found that GTPs and EGCG normalize the exaggerated calcium influx into various excised muscles, including the diaphragm, as well as isolated muscle fibers (unpublished data). Therefore, it is possible that GTPs and EGCG ameliorate calcium handling through a reduction of the activity of various cation channels instead of a downregulation of their expression.

Studies in DMD patients (Lindahl et al., 1995) as well as in mdx mice (Boittin et al., 2006) showed that phospholipase A2 is upregulated in dystrophic muscle, compared to control muscle, during the onset of the disease, ranging from 2 to 6 years in DMD boys (Lindahl et al., 1995) and 1 to 2 months in mdx mice, as our preliminary studies indicate. As mentioned above, we found a role of a particular isoform of this enzyme in activating SOCs. Also, an approximately 30% reduction in the expression level of PLA_2 was found in muscle cultures exposed to low concentrations of GTE or EGCG. These findings suggest that the green tea components diminished the deleterious Ca^{2+} influx via a reduction in the expression level of iPLA2, thus reducing the production of the second messenger, namely, lysophosphatidyl choline, which activates SOCs (Boittin et al., 2010).

10.6 Conclusions

The polyphenol components of green tea have a fairly large number of molecular targets, but only a few of them are of therapeutic relevance. This limitation is due to the fact that often in *in vitro* investigations high concentrations are used that cannot be reached even if very high doses of these compounds are administered to experimental animals or to humans. Another aspect is that correlations between the target and its involvement in a given pathology (target validation) are difficult; this can be done by inhibiting the target by specific pharmacological means or by deleting or inactivating the target protein. Finally, an interaction with a target has acute effects, but most of the beneficial properties are due to long-term downstream events on signaling causing alterations in gene expression. There is certainly a lot more work to be done to enhance our understanding on the mechanisms of action of the green tea components, a prerequisite to specific therapeutic applications in bone, muscle, and other diseases.

A clinical study with EGCG in DMD patients in collaboration with NeuroCure at Charité Hospital in Berlin was initiated in late 2010. Results should become available in 2013.

References

Allen, D.G., and Whitehead, N.P. 2011. Duchenne muscular dystrophy—What causes the increased membrane permeability in skeletal muscle? *Int J Biochem Cell Biol*, 43: 290–294.

Basset, O., Boittin, F.X., Dorchies, O.M., et al. 2004. Involvement of inositol 1,4,5-trisphosphate in nicotinic calcium responses in dystrophic myotubes assessed by near-plasma membrane calcium measurement. *J Biol Chem*, 279: 47092–47100.

Blake, D.J., Weir, A., Newey, S.E., and Davies, K.E. 2002. Function and genetics of dystrophin and dystrophin-related proteins in muscle. *Physiol Rev*, 82: 291–329.

Boittin, F.X., Petermann, O., Hirn, C., et al. 2006. Ca2+-independent phospholipase A2 enhances store-operated Ca2+ entry in dystrophic skeletal muscle fibers. *J Cell Sci*, 119: 3733–3742.

Boittin, F.-X., Shapovalov, G., Hirn, C., and Ruegg, U.T. 2010. Phospholipase A2-derived lysophosphatidylcholine triggers Ca2+ entry in dystrophic skeletal muscle fibers. *Biochem Biophys Res Commun*, 391: 401–406.

Bonuccelli, G., Sotgia, F., Schubert, W., et al. 2003. Proteasome inhibitor (MG-132) treatment of mdx mice rescues the expression and membrane localization of dystrophin and dystrophin-associated proteins. *Am J Pathol*, 163: 1663–1675.

Brooke, M.H., Fenichel, G.M., Griggs, R.C., et al. 1984. A trial of nifedipine in Duchenne muscular dystrophy. *Neurology*, 34: 290.

Buetler, T.M., Renard, M., Offord, E.A., Schneider, H., and Ruegg, U.T. 2002. Green tea extract decreases muscle necrosis in mdx mice and protects against reactive oxygen species. *Am J Clin Nutr*, 75: 749–753.

Chow, H.H., Cai, Y., Alberts, D.S., et al. 2001. Phase I pharmacokinetic study of tea polyphenols following single-dose administration of epigallocatechin gallate and Polyphenon E. *Cancer Epidemiol Biomarkers Prev*, 10: 53–58.

Chow, H.H., Cai, Y., Hakim, I.A., et al. 2003. Pharmacokinetics and safety of green tea polyphenols after multiple-dose administration of epigallocatechin gallate and Polyphenon E in healthy individuals. *Clin Cancer Res*, 9: 3312–3319.

Dorchies, O.M., Wagner, S., Buetler, T.M., and Ruegg, U.T. 2009. Protection of dystrophic muscle cells with polyphenols from green tea correlates with improved glutathione balance and increased expression of 67LR, a receptor for (–)-epigallocatechin gallate. *Biofactors*, 35: 279–294.

Dorchies, O.M., Wagner, S., Vuadens, O., et al. 2006. Green tea extract and its major polyphenol (–)-epigallocatechin gallate improve muscle function in a mouse model for Duchenne muscular dystrophy. *Am J Physiol Cell Physiol*, 290: C616–C625.

Dowling, P., Doran, P., and Ohlendieck, K. 2004. Drastic reduction of sarcalumenin in Dp427 (dystrophin of 427 kDa)-deficient fibres indicates that abnormal calcium handling plays a key role in muscular dystrophy. *Biochem J*, 379: 479–488.

Edwards, J.N., Friedrich, O., Cully, T.R., et al. 2010. Upregulation of store-operated Ca2+ entry in dystrophic mdx mouse muscle. *Am J Physiol Cell Physiol*, 299: C42–C50.

Evans, N.P., Call, J.A., Bassaganya-Riera, J., Robertson, J.L., and Grange, R.W. 2010. Green tea extract decreases muscle pathology and NF-kB immunostaining in regenerating muscle fibers of mdx mice. *Clin Nutr*, 29: 391–398.

Fanchaouy, M., Polakova, E., Jung, C., et al. 2009. Pathways of abnormal stress-induced Ca2+ influx into dystrophic mdx cardiomyocytes. *Cell Calcium*, 46: 114–121.

Feng, W., Cherednichenko, G., Ward, C.W., et al. 2010. Green tea catechins are potent sensitizers of ryanodine receptor type 1 (RyR1). *Biochem Pharmacol*, 80: 512–521.

Fujimura, Y., Umeda, D., Yano, S., et al. 2007. The 67kDa laminin receptor as a primary determinant of anti-allergic effects of O-methylated EGCG. *Biochem Biophys Res Commun*, 364: 79–85.

Haqqi, T.M., Anthony, D.D., Gupta, S., et al. 1999. Prevention of collagen-induced arthritis in mice by a polyphenolic fraction from green tea. *Proc Natl Acad Sci USA*, 96: 4524–4529.

Hidalgo, C., Sanchez, G., Barrientos, G., and Aracena-Parks, P. 2006. A transverse tubule NADPH oxidase activity stimulates calcium release from isolated triads via ryanodine receptor type 1 S-glutathionylation. *J Biol Chem*, 281: 26473–26482.

Hodgetts, S., Radley, H.G., Davies, M.J., and Grounds, M.D. 2006. Reduced necrosis of dystrophic muscle by depletion of host neutrophils, or blocking TNFa function with Etanercept in mdx mice. *Neuromuscul Disord*, 16: 591–602.

Holle, A.W., and Engler, A.J. 2011. More than a feeling: Discovering, understanding, and influencing mechanosensing pathways. *Curr Opin Biotechnol*, 22(5): 648–654.

Iwata, Y., Katanosaka, Y., Arai, Y., Shigekawa, M., and Wakabayashi, S. 2009. Dominant-negative inhibition of Ca2+ influx via TRPV2 ameliorates muscular dystrophy in animal models. *Hum Mol Genet*, 18: 824–834.

Jung, C., Martins, A.S., Niggli, E., and Shirokova, N. 2008. Dystrophic cardio-myopathy: Amplification of cellular damage by Ca2+ signalling and reactive oxygen species-generating pathways. *Cardiovascular Res*, 77: 766–773.

Kaplan, J.C. 2010. The 2011 version of the gene table of neuromuscular disorders. *Neuromuscul Disord*, 20: 852–873.

Khairallah, R.J., Shi, G., Sbrana, F., et al. 2012. Microtubules underlie dysfunction in Duchenne muscular dystrophy. *Sci Signal*, 5(236): ra56.

Leijendekker, W.J., Passaquin, A.C., Metzinger, L., and Ruegg, U.T. 1996. Regulation of cytosolic calcium in skeletal muscle cells of the mdx mouse under conditions of stress. *Br J Pharmacol*, 118: 611–616.

Lindahl, M., Backman, E., Henriksson, K.G., Gorospe, J.R., and Hoffman, E.P. 1995. Phospholipase A2 activity in dystrophinopathies. *Neuromuscul Disord*, 5: 193–199.

Menazza, S., Blaauw, B., Tiepolo, T., et al. 2010. Oxidative stress by monoamine oxidases is causally involved in myofiber damage in muscular dystrophy. *Hum Mol Genet*, 19: 4207–4215.

Messina, S., Bitto, A., Aguennouz, M., et al. 2006. Nuclear factor kappa-B blockade reduces skeletal muscle degeneration and enhances muscle function in Mdx mice. *Exp Neurol*, 198: 234–241.

Moxley, R.T., Brooke, M.H., Fenichel, G.M., et al. 1987. Clinical investigation in Duchenne dystrophy. VI. Double-blind controlled trial of nifedipine. *Muscle Nerve*, 10: 22–33.

Murase, T., Haramizu, S., Shimotoyodome, A., Nagasawa, A., and Tokimitsu, I. 2004. Green tea extract improves endurance capacity and increases muscle lipid oxidation in mice. *Am J Physiol Regul Integr Comp Physiol*, 288(3): R708 R715.

Nakae, Y., Dorchies, O.M., Stoward, P.J., et al. 2012. Quantitative evaluation of the beneficial effects in the mdx mouse of epigallocatechin gallate, an antioxidant polyphenol from green tea. *Histochem Cell Biol*, 137(6): 811–827.

Nakae, Y., Hirasaka, K., Goto, J., et al. 2008. Subcutaneous injection, from birth, of epigallocatechin-3-gallate, a component of green tea, limits the onset of muscular dystrophy in mdx mice: A quantitative histological, immunohistochemical and electrophysiological study. *Histochem Cell Biol*, 129: 489 501.

Nam, S., Smith, D.M., and Dou, Q.P. 2001. Ester bond-containing tea polyphenols potently inhibit proteasome activity *in vitro* and *in vivo*. *J Biol Chem*, 276: 13322–13330.

Nelson, J., McFerran, N.V., Pivato, G., et al. 2008. The 67 kDa laminin receptor: Structure, function and role in disease. *Biosci Rep*, 28: 33–48.

Oba, T., Kurono, C., Nakajima, R., et al. 2002. H2O2 activates ryanodine receptor but has little effect on recovery of releasable Ca2+ content after fatigue. *J Appl Physiol*, 93: 1999–2008.

Passaquin, A., Renard, M., Kay, L., et al. 2002. Creatine supplementation reduces skeletal muscle degeneration and enhances mitochondrial function in mdx mice. *Neuromuscul Disord*, 12: 174–182.

Radley, H.G., Davies, M.J., and Grounds, M.D. 2008. Reduced muscle necrosis and long-term benefits in dystrophic mdx mice after cV1q (blockade of TNF) treatment. *Neuromuscul Disord*, 18: 227–238.

Rando, T.A., Disatnik, M.-H., Yu, Y., and Franco, A. 1998. Muscle cells from mdx mice have an increased susceptibility to oxidative stress. *Neuromuscul Disord*, 8: 14–21.

Ruegg, U.T., Nicolas-Metral, V., Challet, C., et al. 2002. Pharmacological control of cellular calcium handling in dystrophic skeletal muscle. *Neuromuscul Disord*, 12: S155–S161.

Shen, C.-L., Cao, J., Dagda, R., et al. 2011a. Supplementation with green tea polyphenols improves bone microstructure and quality in aged, orchidectomized rats. *Calcif Tissue Int*, 88: 455–463.

Shen, C.L., Chyu, M.C., Yeh, J., et al. 2011c. Effect of green tea and Tai Chi on bone health in postmenopausal osteopenic women: A 6-month randomized placebo-controlled trial. *Osteoporos Int*, 1–12.

Shen, C.-L., Yeh, J.K., Cao, J.J., Chyu, M.-C., and Wang, J.-S. 2011b. Green tea and bone health: Evidence from laboratory studies. *Pharmacol Res*, 64: 155–161.

Shkryl, V., Martins, A., Ullrich, N., et al. 2009. Reciprocal amplification of ROS and Ca2+ signals in stressed mdx dystrophic skeletal muscle fibers. *Pflugers Arch*, 458: 915–928.

Singh, R., Akhtar, N., and Haqqi, T.M. 2010. Green tea polyphenol epigallocatechin-3-gallate: Inflammation and arthritis. *Life Sci*, 86: 907–918.

Smith, A.E., Lockwood, C.M., Moon, J.R., et al. 2010. Physiological effects of caffeine, epigallocatechin-3-gallate, and exercise in overweight and obese women. *Appl Physiol Nutr Metab*, 35: 607–616.

Smith, D., Daniel, K., Wang, Z., et al. 2004. Docking studies and model development of tea polyphenol proteasome inhibitors: Applications to rational drug design. *Proteins*, 54: 58–70.

Spurney, C.F., Knoblach, S., Pistilli, E.E., et al. 2008. Dystrophin-deficient cardiomyopathy in mouse: Expression of Nox4 and Lox are associated with fibrosis and altered functional parameters in the heart. *Neuromuscul Disord*, 18: 371–381.

Steffen, Y., Gruber, C., Schewe, T., and Sies, H. 2008. Mono-O-methylated flavanols and other flavonoids as inhibitors of endothelial NADPH oxidase. *Arch Biochem Biophys*, 469: 209–219.

Suzuki, M., Tabuchi, M., Ikeda, M., Umegaki, K., and Tomita, T. 2004. Protective effects of green tea catechins on cerebral ischemic damage. *Med Sci Monit*, 10: BR166–BR174.

Tachibana, H., Koga, K., Fujimura, Y., and Yamada, K. 2004. A receptor for green tea polyphenol EGCG. *Nat Struct Mol Biol*, 11: 380–381.

Vandebrouck, A., Sabourin, J., Rivet, J., et al. 2007. Regulation of capacitative calcium entries by a1-syntrophin: Association of TRPC1 with dystrophin complex and the PDZ domain of a1-syntrophin. *FASEB J*, 21: 608–617.

Wang, H., Lai, Y.-J., Chan, Y.-L., Li, T.-L., and Wu, C.-J. 2011. Epigallocatechin-3-gallate effectively attenuates skeletal muscle atrophy caused by cancer cachexia. *Cancer Lett*, 305: 40–49.

Whitehead, N.P., Yeung, E.W., and Allen, D.G. 2006. Muscle damage in mdx (dystrophic) mice: Role of calcium and reactive oxygen species. *Clin Exp Pharmacol Physiol*, 33: 657–662.

Xia, R., Stangler, T., and Abramson, J.J. 2000. Skeletal muscle ryanodine receptor is a redox sensor with a well defined redox potential that is sensitive to channel modulators. *J Biol Chem*, 275: 36556–36561.

11

Role of Green Tea Polyphenols in Strengthening the Immune System

Jack F. Bukowski

Contents

11.1 Introduction

Green tea polyphenols (GTPs) are composed of catechins and their derivatives, and are dominated quantitatively and in antioxidant capacity by epigallocatechin gallate (EGCg). The chemical structure and reactivity of catechins are described elsewhere in this volume. This chapter reviews the effect of catechins (GTPs) on strengthening immune function, beginning with brief mentions of earlier work on inhibitory function, and the role in autoimmunity for contrast and comparison, followed by a more in-depth review of function enhancement, and concluding with analysis of enhancement versus inhibition.

11.2 GTP-Mediated Inhibition of Inflammation

A large body of research, in fact, most GTP research, describes the anti-inflammatory effects of GTPs. Most are animal studies, or *in vitro* studies involving either human or animal cells. In most cases, the result is an amelioration of the damaging effects of inflammation resulting from a variety of inflammatory insults. Some insults are infections or reactions caused by bacteria, viruses, and parasites, or their products. Other insults are associated with toxic chemicals or radiation. The mechanism of this GTP-mediated protection often comes from inhibition of intracellular pathways leading to the production of potent Th-1 cytokines such as tumor necrosis factor (TNF) α (Yang et al., 1998), interleukin-1 (IL-1), and interferon gamma (IFN) γ (Wang et al., 2006), as well as IL-8 (Chen et al., 2002; Kim et al., 2006; Netsch et al., 2006), and neutrophil chemotaxis (Dona et al., 2003; Takano et al., 2004).

In many cases, the chief mechanism of Th1 cytokine downregulation appears to be inhibition of nuclear factor (NF) κB activity (Wheeler et al., 2004; Yang et al., 1998). Inhibition of NFκB may also be a mechanism of reactive oxygen species (ROS) and inducible nitric oxide synthase (iNOS) inhibition (Afaq et al., 2003; Aktas et al., 2004; Alvarez et al., 2002; Aneja et al., 2004; Cui et al., 2004; Lee, et al., 2003; Li et al., 2004; Lyu and Park, 2005; Murakami et al., 2003; Song et al., 2003; Surh et al., 2001; Wheeler et al., 2004). Several studies show inhibition of nitrite production by down-regulating NFκB activity (Chan et al., 1995, 1997; Lin and Lin, 1997).

EGCg can also bind to CD11b, inhibiting the adhesion of CD8$^+$ T cells to intercellular adhesion molecule 1 (ICAM-1) (Kawai et al., 2004). EGCg inhibits the glycosylation of Toll-like receptor 4 (TLR-4), thereby inhibiting *Heliobacter pylori*-induced apoptosis of gastric epithelial cells (Lee et al., 2004). The latter is an example of inhibiting microbe-induced damage without affecting the microbe itself, or the microbe-specific immune response.

11.3 GTPs Lessen Autoimmunity

GTPs may also decrease autoimmunity. EGCg decreases autoantibodies to SS-B/La and SS-A/Ro autoantigens associated with Sjogren's syndrome and systemic lupus erythemetosis (Hsu et al., 2005). In a nonobese diabetic mouse model of Sjogren's syndrome, EGCg reduced submandibular inflammation and autoantibody levels. It was not clear if nonautoimmune function was preserved (Gillespie et al., 2008). In a collagen-induced mouse model of arthritis, EGCg treatment decreased joint inflammation and histologic scores (Haqqi et al., 1999; Morinobu et al., 2008). EGCg can suppress TNF-mediated production of matrix metalloproteinases in synovial fibroblasts from rheumatoid arthritis patients (Yun et al., 2008). EGCg suppressed TNF-α production by T cells and inflammation in experimental autoimmune encephalitis, as well as neurotoxic ROS (Aktas et al., 2004).

11.4 GTP-Mediated Enhancement of Immune Response

Much less has been published on immune-enhancing effects. Several studies show that EGCg enhances the mitogenic activity of B lymphocytes, but not T lymphocytes, and that this mitogenic activity was dependent on the presence of red blood cells (Hu et al., 1992; Zenda et al., 1997). In a mouse macrophage cell line, pretreatment with EGCg enhances cyclooxygenase-2 and prostaglandin E2 production by enhancing the activity of extracellular signal-regulated kinase (ERK) and protein-tyrosine phosphatase signaling pathways (Lyu and Park, 2005; Park et al., 2001).

EGCg also blocks ultraviolet B (UV-B)-induced ROS in skin, including in humans (Katiyar et al., 1999a). In a mouse model of UV-B damage, however, IL-10 and IL-12 production are reduced and enhanced, respectively, by EGCg (Katiyar et al., 1999b). In addition, EGCg treatment at once decreased the number of CD11b infiltrating leukocytes, while preserving the number of major histocompatability complex (MHC) class II-positive antigen-presenting cells (APCs) (Katiyar, Mukhtar, et al., 2001). This combination of immune enhancement, with prevention of resident APC loss versus ROS suppression, may represent an important balance between maintaining the skin immune response to prevent infection, and containment of excessive inflammatory responses that may lead to skin tissue damage and carcinogenesis (Katiyar, 2003; Katiyar, Mukhtar, et al., 2001). In mice treated with EGCg, UV-B-induced skin tumors had more infiltrating CD8+ T cells than placebo-treated mice (Mantena et al., 2005). EGCg-dependent DNA repair in skin is absent in IL-12 knockout mice, indicating that in some cases immune enhancement can lead to a decrease rather than an increase in tissue damage (Meeran et al., 2006).

In a mouse model of *Legionella pneumophila* infection of macrophages, pretreatment with micromolar concentrations of EGCg results in enhanced IL-12, IFN-γ, and TNF-α responses, with a reduced IL-10 response. Bacterial growth was decreased, and direct antibacterial effect of EGCg was ruled out. The most likely explanation is the EGCg-mediated enhancement of macrophage antibacterial phagosome activity (Matsunaga et al., 2001). Nicotine and cigarette smoke condensate-mediated suppression of *L. pneumophila*-induced Th-1 cytokine production can be reversed by pretreatment with EGCg (Matsunaga et al., 2002a, 2002b; Yamamoto et al., 2004). The pattern was somewhat different when murine dendritic cells were stimulated with lipopolysacharride (LPS). Though TNF-α production was increased in a dose-dependent manner, IL-12 production was decreased (Rogers et al., 2005). Thus, the nature and dose of the stimulant, as well as the responding cell types, may determine the effect of EGCg treatment.

In a mouse tumor model whereby a human papilloma virus (HPV)-induced tumor is successfully treated with a DNA vaccine, ingestion of EGCg enhances antitumor efficacy while increasing the number of tumor-specific CD8+ cytotoxic T lymphocytes (CTLs) (Kang et al., 2007). The mechanism of action may be enhanced tumor cell apoptosis. This leads to enhanced uptake of tumor antigens by antigen-presenting cells, with more vigorous generation of CTL. Consistent with the overall immunoenhancement may be an EGCg-mediated increase in Th1 cytokine secretion, as demonstrated in other models. At high doses of EGCg, there were fewer CTLs than without EGCg. As previously discussed, there are many models demonstrating EGCg-mediated immunosuppression. Thus, depending on the experimental system and the immune parameters being measured, and the dose, EGCg may either enhance or suppress immunity.

A similar result was achieved using vaccinia virus as immunotherapy, in combination with EGCg. Tumor-specific CD8+ CTL numbers were enhanced, and therapy was more successful than without EGCg (Song et al., 2007). Thus, EGCg may be a nontoxic chemotherapeutic agent that also enhances tumor immunity. This experience is in contrast to most chemotherapy, which serves to immunosuppress.

Preventive treatment of healthy subjects with tea beverage or an admixture of L-theanine and EGCg prevents cold and flu symptoms, while enhancing an arm of innate immunity. Subjects have significantly shorter symptom duration than those given placebo (Kamath et al., 2003; Rowe et al., 2007). L-theanine is catabolized *in vivo* to yield ethylamine (Asatoor, 1966), which enhances the response of memory T lymphocytes with $\gamma\delta$ T cell receptors to microbial challenge (Bukowski et al., 1999). Since memory T cells preferentially respond to IL-12, it is likely that EGCg enhances the overall $\gamma\delta$ T cell response by increasing IL-12 production, most likely early in the infection. In those who became sick with a cold, it is likely that the role of EGCg was to limit inflammation later in the infection. Further study is needed to confirm the mechanism by which EGCg limits cold symptoms.

In senescent mice whose natural killer (NK) cell activity wanes with age, dietary GTPs maintain NK cell activity and reduce the number of lung metastases from melanoma cells, as compared to senescent mice given placebo (Shimizu et al., 2010). Rainbow trout fed EGCg demonstrate enhanced phagocytic activity by leukocytes (Thawonsuwan et al., 2010).

Healthy and prematurely aging mice fed cereal rich in polyphenols (including catechins) show enhancement in protective lymphocyte immune functions such as NK cell activity, lymphoproliferation, and IL-2 secretion. Macrophages exhibited increased chemotaxis, phagocytosis, and microbicidal activity (Alvarez et al., 2006a, 2006b; De la Fuente et al., 2011).

11.5 Conclusions

GTPs have a variety of effects on the immune system, which may depend on the dose, experimental system, immune parameters being studied, and time points at which study takes place. A substantial literature supports an anti-inflammatory role of GTPs. Inhibition of Th1 cytokine production, chemotaxis, and scavenging of ROS are the main modes of inhibition. Inhibition of NFκB is a prominent mechanism. GTPs may therefore be particularly useful as a treatment for chronic inflammatory diseases such as multiple sclerosis, rheumatoid arthritis, psoriasis, inflammatory bowel diseases, and asthma.

On the other hand, another group of studies suggests that in disease models that do not feature excessive inflammation, GTPs may enhance immune function, both adaptive and innate. Such diseases or conditions include metastatic cancer, sunburn, and certain viral and bacterial infections. Further study is needed to explore the possible benefits of GTPs as a preventative or as a vaccine enhancer.

References

Afaq, F., Adhami, V.M., Ahmad, N., and Mukhtar, H. 2003. Inhibition of ultraviolet B-mediated activation of nuclear factor kappaB in normal human epidermal keratinocytes by green tea constituent (–)-epigallocatechin-3-gallate. *Oncogene*, 22: 1035–1044.

Aktas, O., Prozorovski, T., Smorodchenko, A., Savaskan, N.E., Lauster, R., Kloetzel, P.M., Infante-Duarte, C., Brocke, S., and Zipp, F. 2004. Green tea epigallocatechin-3-gallate mediates T cellular NF-kappa B inhibition and exerts neuroprotection in autoimmune encephalomyelitis. *J Immunol*, 173: 5794–5800.

Alvarez, E., Leiro, J., and Orallo, F. 2002. Effect of (–)-epigallocatechin-3-gallate on respiratory burst of rat macrophages. *Int Immunopharmacol*, 2: 849–855.

Alvarez, P., Alvarado, C., Mathieu, F., Jimenez, L., and De la Fuente, M. 2006a. Diet supplementation for 5 weeks with polyphenol-rich cereals improves several functions and the redox state of mouse leucocytes. *Eur J Nutr*, 45: 428–438.

Alvarez, P., Alvarado, C., Puerto, M., Schlumberger, A., Jimenez, L., and De la Fuente, M. 2006b. Improvement of leukocyte functions in prematurely aging mice after five weeks of diet supplementation with polyphenol-rich cereals. *Nutrition*, 22: 913–921.

Aneja, R., Hake, P.W., Burroughs, T.J., Denenberg, A.G., Wong, H.R., and Zingarelli, B. 2004. Epigallocatechin, a green tea polyphenol, attenuates myocardial ischemia reperfusion injury in rats. *Mol Med*, 10: 55–62.

Asatoor, A.M. 1966. Tea as a source of urinary ethylamine. *Nature*, 210: 1358–1360.

Bukowski, J.F., Morita, C.T., and Brenner, M.B. 1999. Human gamma delta T cells recognize alkylamines derived from microbes, edible plants, and tea: Implications for innate immunity. *Immunity*, 11: 57–65.

Chan, M.M., Fong, D., Ho, C.T., and Huang, H.I. 1997. Inhibition of inducible nitric oxide synthase gene expression and enzyme activity by epigallocatechin gallate, a natural product from green tea. *Biochem Pharmacol*, 54: 1281–1286.

Chan, M.M., Ho, C.T., and Huang, H.I. 1995. Effects of three dietary phytochemicals from tea, rosemary and turmeric on inflammation-induced nitrite production. *Cancer Lett*, 96: 23–29.

Chen, P.C., Wheeler, D.S., Malhotra, V., Odoms, K., Denenberg, A.G., and Wong, H.R. 2002. A green tea-derived polyphenol, epigallocatechin-3-gallate, inhibits IkappaB kinase activation and IL-8 gene expression in respiratory epithelium. *Inflammation*, 26: 233–241.

Cui, Y., Kim, D.S., Park, S.H., Yoon, J.A., Kim, S.K. Kwon, S.B., and Park, K.C. 2004. Involvement of ERK and p38 MAP kinase in AAPH-induced COX-2 expression in HaCaT cells. *Chem Phys Lipids*, 129: 43–52.

De la Fuente, M., Medina, S., Baeza, I., and Jimenez, L. 2011. Improvement of leucocyte functions in mature and old mice after 15 and 30 weeks of diet supplementation with polyphenol-rich biscuits. *Eur J Nutr*, 50(7): 563–573.

Dona, M., Dell'Aica, I., Calabrese, F., Benelli, R., Morini, M., Albini, A., and Garbisa, S. 2003. Neutrophil restraint by green tea: Inhibition of inflammation, associated angiogenesis, and pulmonary fibrosis. *J Immunol*, 170: 4335–4341.

Gillespie, K., Kodani, I., Dickinson, D.P., Ogbureke, K.U., Camba, A.M., Wu, M., Looney, S., Chu, T.C., Qin, H., Bisch, F., Sharawy, M., Schuster, G.S., and Hsu, S.D. 2008. Effects of oral consumption of the green tea polyphenol EGCG in a murine model for human Sjogren's syndrome, an autoimmune disease. *Life Sci*, 83: 581–588.

Haqqi, T.M., Anthony, D.D., Gupta, S., Ahmad, N., Lee, M.S., Kumar, G.K., and Mukhtar, H. 1999. Prevention of collagen-induced arthritis in mice by a polyphenolic fraction from green tea. *Proc Natl Acad Sci USA*, 96: 4524–4529.

Hsu, S., Dickinson, D.P., Qin, H., Lapp, C., Lapp, D., Borke, J., Walsh, D.S., Bollag W.B., Stoppler, H., Yamamoto, T., Osaki, T., and Schuster, G. 2005. Inhibition of autoantigen expression by (–)-epigallocatechin-3-gallate (the major constituent of green tea) in normal human cells. *J Pharmacol Exp Ther*, 315: 805–811.

Hu, Z.Q., Toda, M., Okubo, S., Hara, Y., and Shimamura, T. 1992. Mitogenic activity of (–)epigallocatechin gallate on B-cells and investigation of its structure-function relationship. *Int J Immunopharmacol*, 14: 1399–1407.

Kamath, A.B., Wang, L., Das, H., Li, L., Reinhold, V.N., and Bukowski, J.F. 2003. Antigens in tea-beverage prime human Vgamma 2Vdelta 2 T cells *in vitro* and *in vivo* for memory and nonmemory antibacterial cytokine responses. *Proc Natl Acad Sci USA*, 100: 6009–6014.

Kang, T.H., Lee, J.H., Song, C.K., Han, H.D., Shin, B.C., Pai, S.I., Hung, C.F., Trimble, C., Lim, J.S., Kim, T.W., and Wu, T.C. 2007. Epigallocatechin-3-gallate enhances CD8+ T cell-mediated antitumor immunity induced by DNA vaccination. *Cancer Res*, 67: 802–811.

Katiyar, S.K. 2003. Skin photoprotection by green tea: Antioxidant and immuno-modulatory effects. *Curr Drug Targets Immune Endocr Metabol Disord*, 3: 234–242.

Katiyar, S.K., Bergamo, B.M., Vyalil, P.K., and Elmets, C.A. 2001. Green tea poly-phenols: DNA photodamage and photoimmunology. *J Photochem Photobiol B*, 65: 109–114.

Katiyar, S.K., Challa, A., McCormick, T.S., Cooper, K.D., and Mukhtar, H. 1999a. Prevention of UVB-induced immunosuppression in mice by the green tea polyphenol (–)-epigallocatechin-3-gallate may be associated with alterations in IL-10 and IL-12 production. *Carcinogenesis*, 20: 2117–2124.

Katiyar, S.K., Matsui, M.S., Elmets, C.A., and Mukhtar, H. 1999b. Polyphenolic antioxidant (–)-epigallocatechin-3-gallate from green tea reduces UVB-induced inflammatory responses and infiltration of leukocytes in human skin. *Photochem Photobiol*, 69: 148–153.

Katiyar, S.K., and Mukhtar, H. 2001. Green tea polyphenol (–)-epigallocatechin-3-gallate treatment to mouse skin prevents UVB-induced infiltration of leuko-cytes, depletion of antigen-presenting cells, and oxidative stress. *J Leukoc Biol*, 69: 719–726.

Kawai, K., Tsuno, N.H., Kitayama, J., Okaji, Y., Yazawa, K., Asakage, M., Hori, N., Watanabe, T., Takahashi, K., and Nagawa, H. 2004. Epigallocatechin gallate attenuates adhesion and migration of CD8+ T cells by binding to CD11b. *J Allergy Clin Immunol*, 113: 1211–1217.

Kim, I.B., Kim, D.Y., Lee, S.J., Sun, M.J., Lee, M.S., Li, H., Cho, J.J., and Park, C.S. 2006. Inhibition of IL-8 production by green tea polyphenols in human nasal fibroblasts and A549 epithelial cells. *Biol Pharm Bull*, 29: 1120–1125.

Lee, K.M., Yeo, M., Choue, J.S., Jin, J.H., Park, S.J., Cheong, J.Y., Lee, K.J., Kim, J.H., and Hahm, K.B. 2004. Protective mechanism of epigallocatechin-3-gallate against *Helicobacter pylori*-induced gastric epithelial cytotoxicity via the blockage of TLR-4 signaling. *Helicobacter*, 9: 632–642.

Lee, S.J., Lee, I.S., and Mar, W. 2003. Inhibition of inducible nitric oxide synthase and cyclooxygenase-2 activity by 1,2,3,4,6-penta-O-galloyl-beta-D-glucose in murine macrophage cells. *Arch Pharm Res*, 26: 832–839.

Li, R., Huang, Y.G., Fang, D., and Le, W.D. 2004. (–)-Epigallocatechin gallate inhibits lipopolysaccharide-induced microglial activation and protects against inflamma-tion-mediated dopaminergic neuronal injury. *J Neurosci Res*, 78: 723–731.

Lin, Y.L., and Lin, J.K. 1997. (–)-Epigallocatechin-3-gallate blocks the induction of nitric oxide synthase by down-regulating lipopolysaccharide-induced activity of transcription factor nuclear factor-kappa B. *Mol Pharmacol*, 52: 465–472.

Lyu, S.Y., and Park, W.B. 2005. Production of cytokine and NO by RAW 264.7 macrophages and PBMC *in vitro* incubation with flavonoids. *Arch Pharm Res*, 28: 573–581.

Mantena, S.K., Roy, A.M., and Katiyar, S.K. 2005. Epigallocatechin-3-gallate inhibits photocarcinogenesis through inhibition of angiogenic factors and activation of CD8+ T cells in tumors. *Photochem Photobiol*, 81:1174–1179.

Matsunaga, K., Klein, T.W., Friedman, H., and Yamamoto, Y. 2001. *Legionella pneumophila* replication in macrophages inhibited by selective immunomodulatory effects on cytokine formation by epigallocatechin gallate, a major form of tea catechins. *Infect Immun*, 69: 3947–3953.

Matsunaga, K., Klein, T.W., Friedman, H., and Yamamoto, Y. 2002a. Epigallocatechin gallate, a potential immunomodulatory agent of tea components, diminishes cigarette smoke condensate-induced suppression of anti-*Legionella pneumophila* activity and cytokine responses of alveolar macrophages. *Clin Diagn Lab Immunol*, 9: 864–871.

Matsunaga, K., Klein, T.W., Friedman, H., and Yamamoto, Y. 2002b. *In vitro* therapeutic effect of epigallocatechin gallate on nicotine-induced impairment of resistance to *Legionella pneumophila* infection of established MH-S alveolar macrophages. *J Infect Dis*, 185: 229–236.

Meeran, S.M., Mantena, S.K., and Katiyar, S.K. 2006. Prevention of ultraviolet radiation-induced immunosuppression by (–)-epigallocatechin-3-gallate in mice is mediated through interleukin 12-dependent DNA repair. *Clin Cancer Res*, 12: 2272–2280.

Morinobu, A., Biao, W., Tanaka, S., Horiuchi, M., Jun, L., Tsuji, G., Sakai, Y., Kurosaka, M., and Kumagai, S. 2008. (–)-Epigallocatechin-3-gallate suppresses osteoclast differentiation and ameliorates experimental arthritis in mice. *Arthritis Rheum*, 58: 2012–2018.

Murakami, A., Takahashi, D., Koshimizu, K., and Ohigashi, H. 2003. Synergistic suppression of superoxide and nitric oxide generation from inflammatory cells by combined food factors. *Mutat Res*, 523–524: 151–161.

Netsch, M.I., Gutmann, H. Aydogan, C., and Drewe, J. 2006. Green tea extract induces interleukin-8 (IL-8) mRNA and protein expression but specifically inhibits IL-8 secretion in caco-2 cells. *Planta Med*, 72: 697–702.

Park, J.W., Choi, Y.J., Suh, S.I., and Kwon, T.K. 2001. Involvement of ERK and protein tyrosine phosphatase signaling pathways in EGCG-induced cyclooxygenase-2 expression in Raw 264.7 cells. *Biochem Biophys Res Commun*, 286: 721–725.

Rogers, J., Perkins, I., van Olphen, A., Burdash, N., Klein, T.W., and Friedman, H. 2005. Epigallocatechin gallate modulates cytokine production by bone marrow-derived dendritic cells stimulated with lipopolysaccharide or muramyldipeptide, or infected with *Legionella pneumophila*. *Exp Biol Med (Maywood)*, 230: 645–651.

Rowe, C.A., Nantz, M.P., Bukowski, J.F., and Percival, S.S. 2007. Specific formulation of *Camellia sinensis* prevents cold and flu symptoms and enhances gamma,delta T cell function: A randomized, double-blind, placebo-controlled study. *J Am Coll Nutr*, 26: 445–452.

Shimizu, K., Shimizu, K.N., Hakamata, W., Unno, K., Asai, T., and Oku, N. 2010. Preventive effect of green tea catechins on experimental tumor metastasis in senescence-accelerated mice. *Biol Pharm Bull*, 33: 117–121.

Song, C.K., Han, H.D., Noh, K.H., Kang, T.H., Park, Y.S., Kim, J.H., Park, E.S., Shin, B.C., and Kim, T.W. 2007. Chemotherapy enhances CD8(+) T cell-mediated antitumor immunity induced by vaccination with vaccinia virus. *Mol Ther*, 15: 1558–1563.

Song, E.K., Hur, H., and Han, M.K. 2003. Epigallocatechin gallate prevents auto-immune diabetes induced by multiple low doses of streptozotocin in mice. *Arch Pharm Res*, 26: 559–563.

Surh, Y.J., Chun, K.S., Cha, H.H., Han, S.S., Keum, Y.S., Park, K.K., and Lee, S.S. 2001. Molecular mechanisms underlying chemopreventive activities of anti-inflammatory phytochemicals: Down-regulation of COX-2 and iNOS through suppression of NF-kappa B activation. *Mutat Res*, 480–481: 243–268.

Takano, K., Nakaima, K., Nitta, M., Shibata, F., and Nakagawa, H. 2004. Inhibitory effect of (–)-epigallocatechin 3-gallate, a polyphenol of green tea, on neutrophil chemotaxis *in vitro* and *in vivo*. *J Agric Food Chem*, 52: 4571–4576.

Thawonsuwan, J., Kiron, V., Satoh, S., Panigrahi, A., and Verlhac, V. 2010. Epigallocatechin-3-gallate (EGCG) affects the antioxidant and immune defense of the rainbow trout, *Oncorhynchus mykiss*. *Fish Physiol Biochem*, 36: 687–697.

Wang, Y., Mei, Y., Feng, D., and Xu, L. 2006. (–)-Epigallocatechin-3-gallate protects mice from concanavalin A-induced hepatitis through suppressing immune-mediated liver injury. *Clin Exp Immunol*, 145: 485–492.

Wheeler, D.S., Catravas, J.D. Odoms, K., Denenberg, A., Malhotra, V., and Wong, H.R. 2004. Epigallocatechin-3-gallate, a green tea-derived polyphenol, inhibits IL-1 beta-dependent proinflammatory signal transduction in cultured respiratory epithelial cells. *J Nutr*, 134: 1039–1044.

Yamamoto, Y., Matsunaga, K., and Friedman, H. 2004. Protective effects of green tea catechins on alveolar macrophages against bacterial infections. *Biofactors*, 21: 119–121.

Yang, F., de Villiers, W.J., McClain, C.J., and Varilek, G.W. 1998. Green tea poly-phenols block endotoxin-induced tumor necrosis factor-production and lethality in a murine model. *J Nutr*, 128: 2334–2340.

Yun, H.J., Yoo, W.H., Han, M.K., Lee, Y.R., Kim, J.S., and Lee, S.I. 2008. Epigallocatechin-3-gallate suppresses TNF-alpha-induced production of MMP-1 and -3 in rheumatoid arthritis synovial fibroblasts. *Rheumatol Int*, 29: 23–29.

Zenda, N., Okubo, S., Hu, Z.Q., Hara, Y., and Shimamura, T. 1997. Erythrocyte-dependent mitogenic activity of epigallocatechin gallate on mouse splenic B cells. *Int J Immunopharmacol*, 19: 399–403.

12

Green Tea Polyphenols in Allergic Remedies

Hirofumi Tachibana

Contents

12.1 Introduction

Allergies are disorders of the immune system. In healthy individuals, the immune system is poised between maximizing protection from unwelcome invasion or internal imbalance, and minimizing harmful overreaction to external or internal events. In allergic diseases, this delicate balance is disturbed with consequent adverse effects on morbidity and mortality. Most of us are familiar with the consequences of allergic diseases. Allergic diseases, such as asthma, rhinitis, eczema, and food allergies, are reaching epidemic proportions in both the developed and developing world (Holgate, 1999). The costs to public health and its reflection on economy are massive and growing, and a major research effort has been underway for many years to understand these diseases. Seasonal allergic rhinitis (SAR) is a very common disease in developed countries, and its occurrence has been increasing in recent years. The prevalence of Japanese cedar pollinosis, the most common SAR in Japan, is estimated to be approximately 16.2% of the population (Okuda, 2003).

In allergic diseases, polarization of T lymphocyte responses and enhanced secretion of cytokines are involved in regulation of immunoglobulin (Ig) E, mast cells, basophils, and eosinophils, ultimately leading to inflammation and disease. The development of allergic diseases depends on a number of factors, from genetic to environmental, that make it difficult to decide where to focus to find proper cures and prevention.

Many drugs are now in development for the treatment of atopic diseases, including asthma, allergic rhinitis, and atopic dermatitis (Barnes, 1999). These treatments are based on improvements in existing therapies or on a better understanding of the cellular and molecular mechanisms involved in allergic diseases. Since medical costs for treating the diseases are large and possible adverse effects of available medication are not negligible, there is a greater demand for physiologically functional food factors for allergy prevention.

Natural products and their derivatives have historically proven to be an invaluable source of therapeutic agents (Koehn and Carter, 2005). New therapies for allergic diseases have been developed by discovering new types of therapeutic chemical substances derived from animals, plants, and microbes. Advances in understanding the molecular mechanisms of allergies have also led to identification of the natural products, including food factors, for therapeutic treatment of allergic diseases. Although there are many possible therapeutic targets, we have selected (1) IgE production, (2) degranulation, and (3) high-affinity IgE receptor as the targets for the identification of possible therapeutic green tea constituents (Figure 12.1). This chapter is limited to and summarizes the green tea polyphenols with antiallergic activities.

12.2 IgE Production Inhibitor in Green Tea

IgE is believed to be one of the major mediators of immediate hypersensitivity reactions that underlie atopic conditions such as urticaria and eczema, seasonal allergy, food allergy, asthma, and anaphylaxis. In many individuals, total serum IgE level correlates roughly with the severity of the disease (Corry and Kheradmand, 1999). The Ig heavy chain gene segments encode constant regions, which rearrange by class switch recombination, resulting

FIGURE 12.1
IgE-mediated immune response and its suppressive targets.

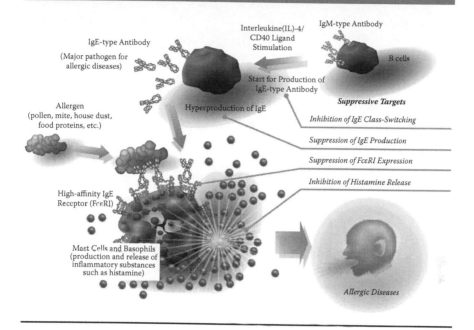

in the production of one of other isotypes (Kataoka et al., 1980). Induction of isotype switching to a particular heavy chain constant region correlates with the transcriptional activation of that particular gene in germline configuration (Stavezer-Nordgren and Sirlin, 1986). Induction of germline transcripts is necessary to target a switch region for recombination and switching. In the case of IgE class switching, interleukin (IL)-4 activates the ε germline promoter, which leads to the expression of ε germline transcription (εGT) through the tyrosine phosphorylation pathway to activate the signal transducers and activators of transcription (STAT) 6, as shown in Figure 12.2 (Quelle et al., 1995). These findings strongly suggest that the inhibitors of εGT expression can suppress the initiation of IgE synthesis by inhibiting IgE class switch recombination (Kaplan et al., 1996).

We focused on tea leaves as a promising source for effective antiallergic agents. To search for tea constituents that are able to suppress IgE synthesis, we examined various substances purified from tea for their effect on IL-4-mediated εGT expression. For screening, we used the human B cell line DND39, which has been shown to express εGT upon IL-4 stimulation (Ichiki et al., 1992). DND39 cells were treated with IL-4 or tea samples, and then

FIGURE 12.2
Molecular basis for suppression of IgE production.

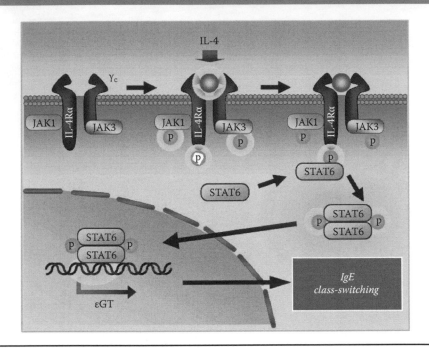

assessed for levels of εGT. Various tea substances were tested, and at least two substances were found to be able to reduce the IL-4-induced εGT expression. One of the substances was determined to be strictinin (1-*O*-galloyl-4,6-(–)-hexahydroxydiphenoyl-β-D-glucose) (Nonaka et al., 1984), and the structure of strictinin is shown in Figure 12.3 (Tachibana et al., 2001).

Strictinin dose-dependently inhibited IL-4-induced εGT expression in DND39 cells. Based on this observation, we tested the effect of strictinin

FIGURE 12.3
The chemical structure of strictinin.

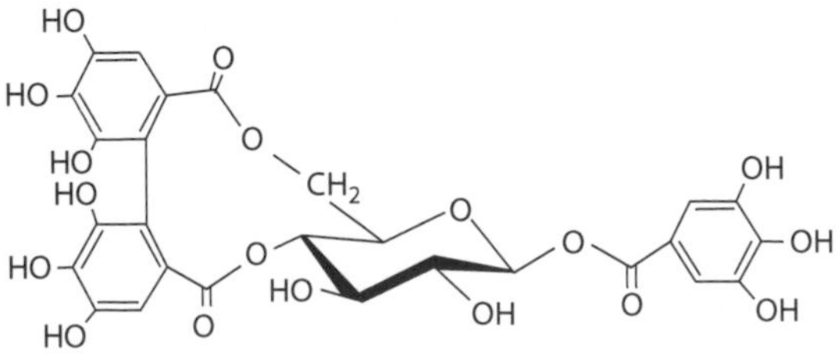

on the expression of εGT in human peripheral blood mononuclear cells (PBMCs) obtained from either healthy or atopic donors. We found that strictinin inhibited IL-4-induced εGT expression in the cells from healthy donors. Interestingly, cells isolated from atopic donors were able to express εGT without additional IL-4 stimulation, and strictinin was also able to inhibit εGT expression in their cells.

The *in vitro* findings were extended to the *in vivo* mouse model. Mice were first treated with ovalbumin to induce an IgE response and were then administered water containing or not containing 0.5 mg strictinin every 2 days for 8 days. The amount of ovalbumin-specific IgE was shown to be reduced for the strictinin administered mice compared with the water-only control mice. Moreover, the amount of other allergen-specific Ig isotypes (IgG and IgM) was not significantly affected. These results suggest that *in vivo* strictinin selectively downregulates the production of antigen-specific IgE antibody, which is the isotype regulated by IL-4.

To examine the mechanism responsible for the suppression of IL-4-induced εGT expression by strictinin, the effect of strictinin on the STAT6 phosphorylation status was examined. No tyrosine-phosphorylated STAT6 was observed in the unstimulated control, and phosphorylation occurred when exogenous IL-4 was added. However, STAT6 tyrosine phosphorylation was shown to be reduced in strictinin-treated cells in a dose-dependent manner. This observation suggests that strictinin inhibits STAT6 tyrosine phosphorylation, which in turn blocks εGT expression and the subsequent IgE class switching necessary for driving IgE production.

12.3 Degranulation Inhibitors in Green Tea

Green tea contains polyphenols, which include flavanols, flavonol glycosides, and phenolic acids. Most of the green tea polyphenols are flavanols, commonly known as catechins. (+)-Catechin (C), (–)-epicatechin (EC), (–)-epigallocatechin (EGC), (–)-epicatechin-3-O-gallate (ECG), and (–)-epigallocatechin-3-O-gallate (EGCG) are the major green tea catechins that can act as degranulation inhibitors.

12.3.1 EGCG and ECG

Tea catechins have been reported to inhibit degranulation from rat mast cells and rat basophilic leukemia RBL-2H3 cells (Matsuo et al., 1996). Some sensitivity differences between human and rat mast cells to stimulate degranulation have been reported. Therefore, we investigated tea constituents that have an inhibitory effect on the degranulation of human basophils. We demonstrated that IL-4 induces differentiation in KU812 cells to morphologically and functionally mature into human basophilic cells (Hara et al., 1998); thus, these IL-4-treated KU812 cells were used as human mature basophils in this study. Degranulation of KU812 cells was monitored by measuring the released histamine.

To examine whether histamine release from human basophilic KU812 cells was affected by tea catechins, cells were pretreated with each catechin, and then the cells were stimulated with the calcium ionophore A23187. Significant inhibition of histamine release was observed in cells pretreated with catechin containing the galloyl group (ECG and EGCG). The inhibitory effect of the two galloyl-type catechins on the histamine release was observed when more than 1 μM was used, and the inhibitory effect of EGCG was higher than that of ECG. These results suggested that the galloyl group may be essential for catechins to exert an inhibitory effect on histamine release.

To elucidate the intracellular mechanism of how tea catechins inhibit the release of histamine, the effect of ECG and EGCG on the intracellular Ca^{2+} concentration in KU812 cells was examined. Both catechins could not inhibit the increase of the intracellular Ca^{2+} level in KU812 cells after stimulation with A23187. This result suggested that the effect of catechins on histamine release occurs after the elevation of the intracellular Ca^{2+} concentration.

The increase of intracellular Ca^{2+} activates degranulation-related cytoskeleton, and the rearrangement of cytoskeleton requires the interaction between myosin and actin filaments (Choi et al., 1994). Phosphorylation of the myosin regulatory light chain (MRLC) at Thr18/Ser19 has been shown to regulate the association between myosin II and F-actin (Goeckeler and Wysolmerski, 1995). Thr18/Ser19 phosphorylation of MRLC has been reported to be temporally correlated with degranulation, and the inhibition of MRLC phosphorylation has been shown to impair the degranulation (Ludowyke et al., 1989). To investigate the possibility of whether MRLC is involved in the inhibitory effect of catechins on histamine release induced by A23187, we examined the effect of catechins on MRLC phosphorylation. EGC showed no inhibitory effect on MRLC phosphorylation, but both ECG and EGCG clearly reduced the level of phosphorylated MRLC. Stronger inhibition of phosphorylation was observed in the EGCG-treated cells. The level of phosphorylated MRLC was reduced upon treatment with EGCG at a concentration of more than 1 μM. Taken together, these results indicated that the ability of tea catechins to inhibit histamine release correlated well with the suppressive effect for MRLC phosphorylation. The ability of catechins to inhibit the phosphorylation of MRLC may be directly involved in the intracellular mechanism for inhibition of degranulation.

To explore a potential role of lipid rafts on plasma membrane in the suppressive effect of EGCG on degranulation, a study of raft integrity was performed using MβCD, a cholesterol-removing agent that disturbs raft function (Kilsdonk et al., 1995). After treatment of KU812 cells with MβCD, the binding of EGCG was significantly reduced in MβCD-pretreated cells. We further examined whether the reduced binding of EGCG caused by the defect of raft integrity could affect the inhibition of histamine release. EGCG's ability to inhibit degranulation was also reduced by the MβCD treatment. These results indicated that the ability of MβCD to abrogate the inhibitory effect of EGCG on histamine release may be due to the lowering of cell surface binding.

12.3.2 Methylated Catechins

Since the polyphenol fraction of the cultivar "Benihomare" oolong (semi-fermented) tea has been shown to include some antiallergic factors (Maeda-Yamamoto et al., 1998), the polyphenol fraction of "Benihomare" (*Camellia sinensis* L. cv. Benihomare) oolong was used as a source of anti-degranulation agents. KU812 cells were incubated with the polyphenol fractions, and the amount of histamine in the supernatant was measured (Tachibana et al., 2000). A constituent diminished the histamine release from the cells, and ^{1}H-nuclear magnetic resonance (^{1}H-NMR) spectrum of the active constituent was in agreement with the data of epigallocatechin -3-*O*-(3-*O*-methyl)-gallate (EGCG3″Me) (Figure 12.4), which was isolated from tea leaves such as Tung-ting oolong tea or cultivars. "Benifuuki" has

FIGURE 12.4
Chemical structures of EGCG and methylated EGCG.

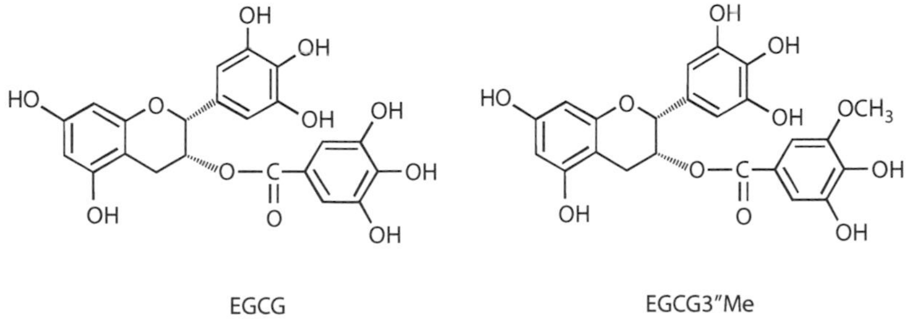

EGCG   EGCG3″Me

been shown to inhibit allergic reactions *in vitro* (Sano et al., 1999). The inhibitory effects of *O*-methylated EGCG on mouse type I and IV allergies *in vivo* are more potent than those of EGCG (Sano et al., 1999; Suzuki et al., 2000). Consistent with the *in vivo* results, the methylated catechin inhibited IgE/Ag-induced activation of mouse mast cells: histamine release, leukotriene release, and cytokine production and secretion after FcεRI crosslinking were all inhibited (Maeda-Yamamoto et al., 2004). As a molecular basis for the catechin-mediated inhibition of mast cell activation, Lyn, Syk, and Bruton's tyrosine kinase, known to be critical for early activation events, is shown to be inhibited by the methylated catechin. *In vitro* kinase assays using purified proteins show that the methylated catechin can directly inhibit the above protein tyrosine kinases (Maeda-Yamamoto et al., 2004). EGCG3″Me is absorbed efficiently and is more stable than EGCG in animal and human plasma, suggesting the reason for the methylated derivative of EGCG as having potent inhibitory activities to allergies *in vivo*.

12.3.3 A Cell Surface EGCG Receptor

It should be noted that most of the effects of EGCG in cell culture systems and cell-free systems have been obtained with considerably high concentrations compared to those observed in the plasma or tissues of animals or in human plasma after administration of green tea or EGCG (Yang et al., 2006). The pharmacokinetic studies in humans indicate that the peak plasma concentration after a single dose of EGCG is <1.0 μM. Furthermore, the intracellular levels of EGCG are much lower than the concentrations observed in the extracellular levels. Searching for high-affinity proteins that bind to EGCG is the first step to understanding the molecular and biochemical mechanisms of the anticancer effects of tea polyphenols. We found that all-trans-retinoic acid (ATRA) enhances the binding of EGCG to the cell surface of cancer cells. To identify candidates through which EGCG inhibits cell growth, we employed a subtraction cloning strategy using complementary DNA libraries constructed from cells treated with or without ATRA. We were able to isolate a single target receptor that allows EGCG to bind to the cell surface (Tachibana et al., 2004). An analysis of the DNA sequence identified this unknown cell surface candidate as 67 kDa laminin receptor (67LR). The 67LR is expressed on a variety of tumor cells, and the expression level of this protein strongly correlates with the risk of tumor invasion and metastasis (Menard et al., 1997). The predicted K_d value for the binding of EGCG to the 67LR protein is 39.9 nM. Flow cytometric analysis using the 67LR antibody showed that KU812 cells expressed the 67LR on the cell surface. Most of the 67LR protein was found to exist in the raft fraction rather than the nonraft fraction on KU812 cells (Fujimura et al., 2005), and this distribution pattern correlated well with the plasma membrane-associated EGCG level after treating the cells with EGCG (Fujimura et al., 2004). To investigate whether the 67LR can confer a sensitivity to EGCG at physiologically relevant concentration, we treated the 67LR-transfected cells with two concentrations of EGCG (0.1 and 1.0 µM), which is similar to the amount of EGCG found in human plasma after drinking more than just two or three cups of tea (Yang, 1997). The growth of the transfected cells was inhibited at both these concentrations. In addition, this growth suppressive effect was completely eliminated upon treatment with the anti-67LR antibody before the addition of EGCG. To compare the ability of 67LR to mediate a response for other tea constituents, caffeine and other tea polyphenols were examined. All these other compounds were shown to be unable to affect the growth of 67LR-expressing cells, and also could not bind to the cell surface. EGCG is the only gallate (gallic acid ester) tested here, suggesting that the gallate moiety may be critical for 67LR binding and subsequent activity. Next, we investigated the effect of oral administration of EGCG on subcutaneous tumor growth in C57BL/6N mice challenged with 67LR-ablated B16 cells (Umeda et al., 2008). We confirmed both silencing of 67LR by stable RNAi in B16 cells and attenuation of the inhibitory effect of 1 µM EGCG on cell growth in 67LR-ablated B16 cells *in vitro*. Tumor growth was significantly retarded in EGCG-administered mice implanted with the B16 cells harboring a control shRNA, whereas tumor growth was not affected by EGCG in the mice implanted with 67LR-ablated B16 cells, suggesting that 67LR functions as an EGCG receptor not only *in vitro* but also *in vivo*. Together, these observations demonstrate that the cell surface 67LR is the receptor for antitumor action of EGCG at the physiologically relevant concentration. The discovery of EGCG receptor as 67LR has solved some of the discrepancies of the cancer-preventing activity of EGCG between *in vitro* data and *in vivo* data.

12.3.4 Antidegranulation Action of Catechins through the EGCG Receptor

As mentioned above, the cell surface EGCG receptors 67LR and EGCG have been shown to be distributed predominantly in lipid rafts on KU812 cells, and the raft-associated 67LR was demonstrated to be the cell surface target molecule for EGCG to mediate the suppressive effect on the FcεRI expression. We hypothesized that the inhibitory effect of EGCG on degranulation is also mediated by cell surface binding to the 67LR (Fujimura et al., 2006). After treatment of KU812 cells with the anti-67LR antibody, cells were incubated with EGCG, and further challenged with A23187. The reductive effect of EGCG on the histamine release was almost completely inhibited in cells treated with the anti-67LR antibody. Experiments using the 67LR-downregulated cells revealed a significant abrogation of the inhibitory effect of EGCG on degranulation. Furthermore, the lowering effect of EGCG on the phosphorylation of MRLC was also inhibited by either treatment with the anti-67LR antibody or 67LR knockdown. These findings indicate that the inhibitory effect of EGCG on degranulation was caused by a modification of myosin cytoskeleton through the binding of EGCG to 67LR on the cell surface.

Activation of mast cells and basophils results in extensive changes in cell morphology, the formation of F-actin, and extensive redistribution of myosin and actin filaments within the cells (Pfeiffer et al., 1985). Generally, degranulation-inducing stimuli have been known to induce a dramatic change of actin cytoskeleton, such as membrane ruffling, a wavy form of membrane (Edgar and Bennett, 1997). When the cells were stimulated with A23187 in the presence of EGCG, membrane ruffling was inhibited and a biased F-actin accumulation was observed. Furthermore, this EGCG-induced actin remodeling was abolished in both anti-67LR antibody-treated cells and 67LR-knocked down cells. These findings suggest that the inhibitory effect of EGCG on the A23187-induced degranulation may be caused by a modification of actin cytoskeleton, and the effect is mediated by the binding of EGCG to the 67LR on the cell surface.

We also investigated the structure-activity relationship of major green tea catechins and their corresponding epimers on cell surface binding and inhibitory effect on histamine release (Fujimura et al., 2008). Galloylated catechins, EGCG, (−)-gallocatechin-3-O-gallate (GCG), ECG, and (−)-catechin-3-O-gallate (CG), showed the cell surface binding to the human basophilic KU812 cells, but their nongalloylated forms did not. Binding activities of pyrogallol-type catechins (EGCG and GCG) were higher than those of catechol-type catechins (ECG and CG). These patterns were also observed in their inhibitory effects on histamine release. Downregulation of 67LR expression caused a reduction of both activities of galloylated catechins. These results suggest that both the galloyl moiety and the B-ring hydroxylation pattern contribute to the exertion of biological activities of tea catechins and their 67LR dependencies.

12.4 Suppressors for the High-Affinity IgE Receptor FcɛRI Expression

The high-affinity IgE receptor, FcɛRI, plays a key role in a series of acute and chronic human allergic reactions such as atopic dermatitis, bronchial asthma, allergic rhinitis, and food allergy (Blank et al., 1991). In humans and rodents, this receptor is found at high levels on basophils and mast cells where activation by cross-linking of the allergen-specific IgE bound to FcɛRI with multivalent allergens produces several mediators, including histamine, proteases, chemotactic factors, and arachidonic acid metabolites responsible for FcɛRI-dependent allergic reactions. Studies on the FcɛRI α chain knockout mice demonstrated that IgE cannot bind to the cell surface of mast cells; thereby degranulation through IgE binding was not induced (Dombrowicz et al., 1993). Thus, it is expected that the downregulation of FcɛRI expression in mast cells and basophils leads to the attenuation of the IgE-mediated allergic symptoms. However, an evaluation of antiallergic activity of natural products based on the suppression of FcɛRI expression has not been reported yet.

In order to examine whether the expression of FcɛRI on the cell surface of KU812 cells was affected by the major tea catechins (C, EC, EGC, ECG, and EGCG), the cells were treated with each catechin. Among the tea catechins tested, a decrease of the cell surface expression of FcɛRI was observed only in EGCG-treated cells, while there was no decrease in the level of FcɛRI expression upon treatment with the other catechins (Fujimura et al., 2001).

To examine whether suppression of the cell surface expression of the FcɛRI α chain by EGCG is due to the decrease in the amount of total cellular FcɛRI α chain level, we performed an immunoblot analysis. The amount of total cellular FcɛRI α chain level decreased upon treatment with EGCG, suggesting that the EGCG-induced decrease in the amount of cellular FcɛRI α chain level may be associated with the suppression of the cell surface expression. The FcɛRI α and γ mRNA levels in the EGCG-stimulated cells were shown to be significantly reduced. These results suggested that the suppressive effect of EGCG on the cell surface expression of FcɛRI was at least related to the downregulation of the expression of the FcɛRI α and γ mRNA.

We previously showed that the O-methylated derivative of EGCG, (–)-epigallocatechin-3-O-(3-O-methyl)-gallate, has potent antiallergic activity. We found that (–)-epigallocatechin-3-O-(3-O-methyl)-gallate was also able to decrease the cell surface expression of FcεRI (Fujimura et al., 2002). The methylated EGCG treatment inhibited the FcεRI cross-linking-induced histamine release. These results suggested that the methylated EGCG can negatively regulate basophil activation through the suppression of FcεRI expression. EGCG has been reported to undergo methylation, and (–)-4′-O-methyl-epigallocatechin-3-O-(4-O-methyl) gallate (EGCG4′4″diMe) has been shown to be a major metabolite of EGCG in plasma (Lambert et al., 2003), and EGCG4′4″diMe did not demonstrate a suppressive effect in KU812 cells (Yano et al., 2007).

We investigated the possibility that 67LR is involved in mediating the suppressive effect of EGCG on the expression of FcεRI (Fujimura et al., 2005). To elucidate whether 67LR is involved in the suppressive effect of EGCG on the expression of FcεRI, we constructed 67LR-overexpressed cells by stable transfection of a 67LR gene expression vector into KU812 cells. Flow cytometric analysis showed that EGCG suppressed the FcεRI expression compared to the nontreated cells, and this reductive effect was enhanced in the 67LR-overexpressed cells. These results indicated that the level of 67LR expression is able to affect the suppressive effect of EGCG on FcεRI cell surface expression. Previously, we described that the inhibition of ERK1/2 phosphorylation was responsible for the suppressive effect elicited by EGCG on the FcεRI expression (Fujimura et al., 2004). To clarify whether the reductive action of ERK1/2 phosphorylation by EGCG is transduced through the binding with 67LR, KU812 cells transfected with either the 67LR vector or the empty vector were treated with EGCG to examine the level of phosphorylated ERK1/2. The suppressive effect of EGCG on the phosphorylated ERK1/2 level in 67LR-overexpressed cells was higher than that of the empty vector-transfected cells.

We also constructed 67LR-downregulated cells using RNA interference (RNAi)-mediated gene silencing. In the empty vector-transfected cells, the relative FcεRI expression in the cells treated with EGCG was decreased compared with the nontreated cells. However, the 67LR-downregulated cells showed a slight reduction of FcεRI expression. These results suggest that the suppressive effect of EGCG was inhibited by the knockdown of 67LR. Furthermore, the ability of EGCG to decrease the phosphorylation of ERK1/2 was reduced in the 67LR-knocked down cells. These results indicate that the effect of EGCG on ERK1/2 phosphorylation correlates with the expression of 67LR, which implies that 67LR is the molecule responsible for transducing the EGCG's downregulatory signaling of the FcεRI.

12.5 Antiallergic Effect of Green Tea in Human Studies

A double-blinded clinical study on subjects with Japanese cedar pollinosis was carried out to evaluate the effects and safety of "Benifuuki" green tea, containing EGCG3"Me, and a combination of "Benifuuki" green tea and ginger extract, compared with "Yabukita" green tea as a placebo (Yasue et al., 2005a). The "Benifuuki" and "Yabukita" green teas contained 44.7 and 0 mg of EGCG3"Me, 176.1 and 202.8 mg of EGCG, 71.4 and 84.6 mg of caffeine, and 432 and 425 mg of total catechin per 3 g, respectively. As the cedar pollen increased, the symptoms of pollinosis became exacerbated in the placebo group compared with the "Benifuuki" group. During the most severe cedar pollen period, 11 weeks after starting the treatment, symptoms such as nose blowing and itchy eyes showed significant improvement in the "Benifuuki" group compared with the control group ($p < 0.05$). Among the groups, there were no clinically relevant changes in the hematological examination, general biochemical examination, total IgG antibody titer, serum iron content, or interview results throughout the intake period. It has been also concluded that drinking "Benifuuki" green tea for over 1.5 months was useful in improving some Japanese cedar pollinosis symptoms while causing no effects in normal immune responses in subjects with seasonal rhinitis. A consecutive administration test with an excess amount of the "Benifuuki" green tea was also performed on 35 patients with mild allergic rhinitis (Yasue et al., 2005b). The patients were daily given 1500 ml of "Benifuuki" green tea for 4 weeks. Nasal and ocular scores tended to improve and no adverse effect was noted. To clarify an initiation time of the drinking in order to optimize the beneficial effects of the tea, the efficacy of "Benifuuki" green tea in Japanese cedar pollinosis has been examined during the most prevalent season for allergic rhinitis in Japan, focusing on the effects of the drinking period on symptom improvement (Maeda-Yamamoto et al., 2009). An open-label, single-dose, randomized, parallel-group study was performed on 38 subjects with Japanese cedar pollinosis. The subjects were randomly assigned to long-term (December 27, 2006 to April 8, 2007, 1.5 months before pollen exposure) or short-term (February 15, 2007 (after cedar pollen dispersal) to April 8, 2007) drinking of a "Benifuuki" tea drink containing 34 mg O-methylated catechin per day. Each subject recorded his or her daily symptom scores in a diary. The primary efficacy variable was the mean weekly nasal symptom medication score during the study period. The nasal symptom medication score in the long-term intake group was significantly lower than that of the short-term intake group at the peak of pollen dispersal. The symptom scores for throat pain, nose blowing, tears, and hindrance to activities of daily living were significantly better in the long-term group than the short-term group. In particular, the differences in the symptom scores for throat pain and nose blowing between the two groups were marked. These results suggested that drinking "Benifuuki" tea for 1.5 months prior to the cedar pollen season is effective in reducing symptom scores for Japanese cedar pollinosis.

12.6 Conclusions

We identified several antiallergic natural products derived from plant polyphenols, and possible mechanisms by which tea catechin EGCG exerts degranulation inhibition and FcεRI suppression could be proposed. Figure 12.5 denotes our proposed scheme of the antiallergic action of EGCG through the cell surface receptor 67LR, which concerns the reduction of ERK1/2 and MRLC phosphorylation levels in basophils by EGCG. A plastic-bottled beverage made by "Benifuuki" leaf extract has been commercialized. The

FIGURE 12.5
A scheme of the antiallergic action of EGCG through the cell surface receptor 67LR.

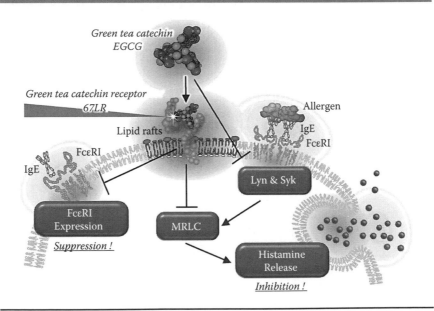

findings mentioned above highlight the therapeutic potential of these natural products and their analogues as potential pharmaceutical agents that may be useful to inhibit the progression of IgE-mediated allergic responses. Recently, we also found that 67LR mediates anti-inflammatory action of EGCG on Toll-like receptor (TLR) 2 and 4 ligands activation of downstream signaling pathways and target gene expressions in macrophages (Byun et al., 2010, 2011). More definitive information on the relationship between EGCG sensing pathways and beneficial effects of EGCG on allergy and inflammation will emerge from human intervention trials.

References

Barnes, P.J. 1999. Therapeutic strategies for allergic diseases. *Nature*, 402: B31–B38.

Blank, U., Ra, C., and Kinet J.-P. 1991. Characterization of the truncated α chain products from human, rat and mouse high affinity receptor for immunoglobulin E. *J Biol Chem*, 266: 2639–2646.

Byun, E.-H., Fujimura, Y., Yamada, K., and Tachibana, H. 2010. TLR 4 signaling inhibitory pathway induced by green tea polyphenol epigallocatechin-3-gallate through 67-kDa laminin receptor. *J Immunol*, 185: 33–45.

Byun, E.-H., Omura, T., Yamada, K., and Tachibana, H. 2011. Green tea polyphenol epigallocatechin-3-gallate inhibits TLR2 signaling induced by peptidoglycan through the polyphenol sensing molecule 67-kDa laminin receptor. *FEBS Lett*, 585: 814–820.

Choi, O.H., Adelstein, R.S., and Beaven, M.A. 1994. Secretion from rat basophilic RBL-2H3 cells is associated with diphosphorylation of myosin light chains by myosin light chain kinase as well as phosphorylation by protein kinase C. *J Biol Chem*, 269: 536–541.

Corry, D.B., and Kheradmand, F. 1999. Induction and regulation of the IgE response. *Nature*, 402: B18–B23.

Dombrowicz, D., Flamand, V., Brigman, K., Koller, B.H., and Kinet, J.-P. 1993. Abolition of anaphylaxis by targeted disruption of the high affinity immunoglobulin E receptor alpha chain gene. *Cell*, 75: 969–976.

Edgar, A.J., and Bennett, J.P. 1997. Circular ruffle formation in rat basophilic leukemia cells in response to antigen stimulation. *Eur J Cell Biol*, 73: 132–140.

Fujimura, Y., Tachibana, H., Maeda-Yamamoto, M., Miyase, T., Sano, M., and Yamada, K. 2002. Antiallergic tea catechin, (–)-epigallocatechin-3-*O*-(3-*O*-methyl)gallate, suppresses FcεRI expression in human basophilic KU812 cells. *J Agric Food Chem*, 50: 5729–5734.

Fujimura, Y., Tachibana, H., and Yamada, K. 2001. A tea catechin suppresses the expression of the high affinity IgE receptor FcεRI in the human basophilic KU812 cells. *J Agric Food Chem*, 49: 2527–2531.

Fujimura, Y., Tachibana, H., and Yamada, K. 2004. Lipid raft-associated catechin suppresses the FcεRI expression by inhibiting phosphorylation of the extracellular signal-regulated kinase1/2. *FEBS Lett*, 556: 204–210.

Fujimura, Y., Tachibana, H., and Yamada, K. 2005. A lipid raft-associated 67 kDa laminin receptor mediates suppressive effect of epigallocatechin-3-O-gallate on FcεRI expression. *Biochem Biophys Res Commun*, 336: 674–681.

Fujimura, Y., Umeda, D., Kiyohara, Y., Sunada, Y., Yamada, K., and Tachibana, H. 2006. The involvement of the 67 kDa laminin receptor-mediated modulation of cytoskeleton in the degranulation inhibition induced by epigallocatechin-3-O-gallate. *Biochem Biophys Res Commun*, 348: 524–531.

Fujimura, Y., Umeda, D., Yamada, K., and Tachibana, H. 2008. The impact of the 67 kDa laminin receptor on both cell-surface binding and anti-allergic action of tea catechins. *Arch Biochem Biophys*, 476: 133–138.

Goeckeler, Z.M., and Wysolmerski, R.B. 1995. Myosin light chain kinase-regulated endothelial cell contraction: The relationship between isometric tension, actin polymerization, and myosin phosphorylation. *J Cell Biol*, 130: 613–627.

Hara, T., Yamada, K., and Tachibana, H. 1998. Basophilic differentiation of the human leukemia cell line KU812 upon treatment with interleukin-4. *Biochem Biophys Res Commun*, 247: 542–548.

Holgate, S.T. 1999. The epidemic of allergy and asthma. *Nature*, 402: B2–B4.

Ichiki, T., Takahashi, W., and Watanabe, T. 1992. The effect of cytokines and mitogens on the induction of Cε germline transcripts in a human Burkitt lymphoma B cell line. *Int Immunol*, 4: 747–754.

Kaplan, M.H., Schindler, U., Smiley, S.T., and Grusby, M.J. 1996. Stat6 is required for mediating responses to IL-4 and for development of Th2 cells. *Immunity*, 4: 313–319.

Kataoka, T., Kawakami, T., Takahashi, N., and Honjo, T. 1980. Rearrangement of immunoglobulin γ1-chain gene and mechanism for heavy-chain class switch. *Proc Natl Acad Sci USA*, 77: 919–923.

Kilsdonk, E.P., Yancey, P.G., Stoudt, G.W., Bangerter, F.W., Johnson, W.J., Phillips, M.C., and Rothblat, G.H. 1995. Cellular cholesterol efflux mediated by cyclodextrins. *J Biol Chem*, 270: 17250–17256.

Koehn, F.E., and Carter, G.T. 2005. The evolving role of natural products in drug discovery. *Nature Rev Drug Discov*, 4: 206–220.

Lambert, J.D., Lee, M.J., Lu, H., Meng, X., Hong, J.J., Seril, D.N., Sturgill, M.G., and Yang, C.S. 2003. Epigallocatechin-3-gallate is absorbed but extensively glucuronidated following oral administration to mice. *J Nutr*, 133: 4172–4177.

Ludowyke, R.I., Peleg, I., Beaven, M.A., and Adelstein, R.S. 1989. Antigen-induced secretion of histamine and the phosphorylation of myosin by protein kinase C in rat basophilic leukemia cells. *J Biol Chem*, 264: 12492–12501.

Maeda-Yamamoto, M., Ena, K., Monobe, M., Shibuichi, I., Shinoda, Y., Yamamoto, T., and Fujisawa, T. 2009. The efficacy of early treatment of seasonal allergic rhinitis with Benifuuki green tea containing O-methylated catechin before pollen exposure: An open randomized study. *Allergol Int*, 58: 437–444.

Maeda-Yamamoto, M., Inagaki, N., Kitaura, J., Chikumoto, T., Kawahara, H., Kawakami, Y., Sano, M., Miyase, T., Tachibana, H., Nagai, H., and Kawakami, T. 2004. O-methylated catechins from tea leaves inhibit multiple protein kinases in mast cells. *J Immunol*, 172: 4486–4492.

Maeda-Yamamoto, M., Kawahara, H., Matsuda, N., Nesumi, K., Sano, M., Tsuji, K., Kawakami, Y., and Kawakami, T. 1998. Effects of tea infusions of various or different manufacturing types on inhibition of mouse mast cell activation. *Biosci Biotechnol Biochem*, 62: 2277–2279.

Matsuo, N., Yamada, K., Shoji, K., Mori, M., and Sugano, M. 1996. Effect of tea polyphenols on histamine release from rat basophilic leukemia (RBL-2H3) cells: The structure-inhibitory activity relationship. *Allergy*, 52: 58–64.

Menard, S., Castronovo, V., Tagliabue, E., and Sobel, M.E. 1997. New insights into the metastasis-associated 67 kD laminin receptor. *J Cell Biochem*, 67: 155–165.

Nonaka, G.-I., Sakai, R., and Nishioka, I. 1984. Hydrolysable tannins and proanthocyanidines from green tea. *Phytochemistry*, 23: 1753–1755.

Pfeiffer, J.R., Seagrave, J.C., Davis, B.H., Deanin, G.G., and Oliver, J.M. 1985. Membrane and cytoskeletal changes associated with IgE-mediated serotonin release from rat basophilic leukemia cells. *J Cell Biol*, 101: 2145–2155.

Okuda, M. 2003. Epidemiology of Japanese cedar pollinosis throughout Japan. *Ann Allergy Asthma Immunol*, 91: 288–296.

Quelle, F.W., Shimoda, K., and Thierfelder, W. 1995. Cloning of murine Stat6 and human Stat6, Stat proteins that are tyrosine phosphorylated in responses to IL-4 and IL-3 but are not required for mitogenesis. *Mol Cell Biol*, 15: 3336–3343.

Sano, M., Suzuki, M., Miyase, T., Yoshino, K., and Maeda-Yamamoto, M. 1999. Novel antiallergic catechin derivatives isolated from oolong tea. *J Agric Food Chem*, 47: 1906–1910.

Stavezer-Nordgren, J., and Sirlin, S. 1986. Specificity of immunoglobulin heavy chain switch correlates with activity of germline heavy chain genes prior to switching. *EMBO J*, 5: 95–102.

Suzuki, M., Yoshino, K., Maeda-Yamamoto, M., Miyase, T., and Sano, M. 2000. Inhibitory effects of tea catechins and *O*-methylated derivatives of (–)-epigallocatechin-3-*O*-gallate on mouse type IV allergy. *J Agric Food Chem*, 48: 5649–5653.

Tachibana, H., Koga, K., Fujimura, Y., and Yamada, K. 2004. A receptor for green tea polyphenol EGCG. *Nat Struct Mol Biol*, 11: 380–381.

Tachibana, H., Kubo, T., Miyase, T., Tanino, S., Yoshimono, M., Sano, M., Maeda-Yamamoto, M., and Yamada, K. 2001. Identification of an inhibitor for interleukin 4-induced germline transcription and antigen-specific IgE production *in vivo*. *Biochem Biophys Res Commun*, 280: 53–60.

Tachibana, H., Sunada, Y., Miyase, T., Sano, M., Yamamoto-Maeda, M., and Yamada, K. 2000. Identification of a methylated tea catechin as inhibitor of degranulation in human basophilic KU812 cells. *Biosci Biotechnol Biochem*, 64: 452–454.

Umeda, D., Yano, S., Yamada, K., and Tachibana, H. 2008. Green tea polyphenol epigallocatechin-3-gallate signaling pathway through 67-kDa laminin receptor. *J Biol Chem*, 283: 3050–3058.

Yang, C.S. 1997. Inhibition of carcinogenesis by tea. *Nature*, 389: 134–135.

Yang, C.S., Sang, S., Lambert, J.D., Hou, Z., Ju, J., and Lu, G. 2006. Possible mechanisms of the cancer-preventive activities of green. *Mol Nutr Food Res*, 50: 170–175.

Yano, S., Fujimura, Y., Umeda, D., Miyase, T., Yamada, K., and Tachibana, H. 2007. Relationship between the biological activities of methylated derivatives of (–)-epigallocatechin-3-*O*-gallate (EGCG) and their cell surface binding activities. *J Agric Food Chem*, 55: 7144–7148.

Yasue, M., Ikeda, M., Nagai, H., Sato, K., Mitsuda, H., Maeda-Yamamoto, M., Sakamoto, A., Yabune, M., Kajimoto, Y., Kajimoto, O., and Tamura, M. 2005a. The efficacy and safety of "Benifuuki" green tea containing *O*-methylated catechin: Clinical study in patients with mild perennial allergic rhinitis (in Japanese). *Nippon Shokuhin Shinsozai Kenkyu Kaishi*, 8: 65–80.

Yasue, M., Ohtake, Y., Nagai, H., Sato, K., Mitsuda, H., Maeda-Yamamoto, M., Yabune, M., Nakagawa, S., Kajimoto, Y., and Kajimoto, O. 2005b. The clinical effects and the safety of the intakes of "Benifuuki" green tea in patients with perennial allergic rhinitis (in Japanese). *J Jpn Soc Clin Nutr*, 27: 33–51.

13

Green Tea Polyphenols and Gut Health

Theertham P. Rao, Tsutomu Okubo, Mahendra P. Kapoor, and Lekh R. Juneja

Contents

13.1 Introduction

Next to the brain, the gut is often considered to be the second most important organ of the body, being the center for whole body nutrition and well-being. However, unlike the brain, which is not exposed directly to environmental factors, the gastrointestinal system is the primary site of interaction between the host immune system and numerous external factors, including both beneficial and pathogenic microorganisms, poisonous substances, pollutants, and carcinogens. The ancient physician Hippocrates (460–370 b.c.) said, "All diseases begin in the gut," and Phillipe Pinel (1745–1828) said, "The primary seat of insanity lies generally in the region of the stomach and intestines." Such statements emphasize the importance of gut and body health from very early times. The gut is considered to be the largest immune organ needed to maintain one's health (Kraehenbuhl and Neutra, 1992). In the modern world, factors including stress, poor diet, illness, pharmaceuticals, and even the natural aging process can have a detrimental effect on the gut, which may lead to discomforts such as constipation, diarrhea, bloating, gas, and infections. Anything having an effect on gut health in turn has a major influence on the immunity system, bowel management, weight management, and the overall health of the whole body. The old saying of "You are what you eat," which is derived from "*Dis-moi ce que tu manges, je te diraj ce que tu es*" (Anthelem Brillat Savarin's *Physiologie du Gout ou Meditations de Gastronomie Transcendante*, 1826), or "Tell me what you eat and I will tell you what you are," emphasizes the importance of the gut and the relation between food and one's behavior.

Research on overall gut health has traditionally focused on probiotics and intestinal microflora. In such a complex system, however, the interaction of many food substances greatly affects the health of the gut. In the last few decades extensive research on plant phenolics, in particular, green tea polyphenols, has revealed a number of health benefits. In this chapter we have extensively reviewed the function and benefits of green tea polyphenols on gut health.

13.2 Green Tea Catechins and Gut Microflora

The human gut harbors a number of bacterial cells that outnumber host cells by a factor of 10^{14}, with more than 1000 bacterial species that have a direct effect on the digestion and absorption of nutrients, gut integrity, immunity, and more than 20 different types of hormones. Previous studies have well established the link between intestinal microflora and immunity (Kosiewicz et al., 2011 and references therein), inflammation (Hakansson and Molin, 2011), allergy (Kalliomaki, 2009), metabolic disorders (Sanz et al., 2010), and gastrointestinal disorders (Lutgendorff et al., 2008). The interaction of intestinal microflora with the immune system plays a major role in both the health of the gut and the health of the host. The microflora in the gut is largely divided into harmful bacteria (*Clostridium difficile*, *Salmonella*, *Bacteroides*, *Escherichia coli*, and *Enterococcus* species) of pathogenic nature and beneficial bacteria (*Bifidobacterium* and *Lactobacillus* species), which help in the development and maintenance of healthy gut functions. An imbalance between beneficial and harmful bacterial species in gut microflora has been the subject of gut health for many years. Recently, increasing the number of beneficial bacterial species via direct supplementation of probiotics or prebiotics has been considered important modulation of gut microflora and gut health.

Probiotic bacteria such as *Lactobacillus* species (*L. acidophilus*, *L. casei* GG, *L. bulgaricus*), *Bifidobacterium* species (*B. bifidum*, *B. breve*, *B. infants*, *B. lactis*), *Streptococcus thermophilus*, and *Saccharomyces boulardii* have exerted different beneficial effects (Fuller, 1989; Girardin and Seidman, 2011; Lebba et al., 2011). Each strain is believed to have specific immunomodulatory properties (Deshpande et al., 2007). *L. fermentum* was recently reported to boost the immune health of long-distance runners by protecting them from respiratory illnesses. This *Lactobacillus* strain was associated with an enhancement in the activity of T cells, the key players in the immune system (Cox et al., 2008). In other clinical studies *Lactobacillus* species have been found to modulate the immune response against grass pollen in subjects suffering from hay fever (Snel et al., 2011; Wassenberg et al., 2011). The combination of two particular strains, GR-1 (*L. rhamnosus*) and RC-14 (*L. reuteri*), was found to provide the greatest benefit for the relief and prevention of bacterial vaginosis, a common problem affecting about 30% of women between the ages of 14 and 49 (MacPhee et al., 2010). In gut health, probiotic strains have been reported to play an important role in reducing abdominal pain linked to stress (Diop et al., 2008), constipation (Bekkali et al., 2007), diarrhea (Franks 2011; Guandalini, 2011; Weihua and Jing, 2011), necrotizing enterocolitis (Deshpande et al., 2007; Ganguli and Walker, 2011), intestinal inflammation (Vanderhoof and Mitmesser, 2010), irritable bowel syndrome (Lee and Bak, 2011), and other gastrointestinal disorders (Chen and Walker, 2011). Though various studies confirm the beneficial effects of probiotic strains, scientists often caution on the generalization of their category due to multiple complex interactions that can greatly alter their efficacy (Deshpande et al., 2007).

On the other hand, the harmful bacterial species of *Bacteroides* and *Clostridium* genera in the gut and *Helicobacter pylori*, *E. coli*, and *Salmonella typhimurium* of external origin exert strong pathogenic effects and great consequence to the human body. Antibacterial drugs are effective to inhibit most of these pathogens, but they also have a detrimental effect on the beneficial bacterial species. Supplementation of probiotics is often recommended along with antibiotics to maintain gut health. Inhibition of pathogenic bacterial species and maintenance of beneficial bacterial species are equally important to maintain gut health. The supplementation of probiotics together with prebiotics, including dietary fiber, may help to increase the number of beneficial bacterial species; however, they have seldom had any inhibitory effect on the harmful bacterial species. A mechanism for not only inhibiting the harmful bacterial species but also increasing the number of beneficial bacterial species in the gut is a more logical strategic approach to achieve a sustained benefit for gut health. Scientists have recently come to the belief that a combination of prebiotics, antioxidants, and antimicrobial compounds is potentially more relevant for gut health than probiotics alone. In the recent past, various *in vitro* and animal studies have suggested that plant phenolics can play a great role in inhibiting the activities of various harmful bacterial species (Lee et al., 2006; Tzounis et al., 2008).

Several reports elucidated the antimicrobial properties of green tea polyphenols against various pathogenic bacterial species, including *Streptococcus mutans* (Sakanaka et al., 1989), *Streptococci* (Sakanaka et al., 1990), *Porphyromonas gingivalis* (Katoh 1995; Sakanaka et al., 1996), *Bacillus stearothermophilus* and *Clostridium thermoaceticum* (Sakanaka et al., 2000), *C. perfringens* (Ahn et al., 1991), *Salmonella* (Hara-Kudo et al., 2001; Lee et al., 2006), *Clostridium* and *Bacillus* spores (Hara-Kudo et al., 2005), *E.coli* O157:H7 (Isogai et al., 1998; Hara, 1997; Hara-Kudo et al., 2001; Lee et al., 2006; Sugita-Konishi et al., 1999), and *H. pylori* (Takabayashi et al., 2004). It is important to note that green tea polyphenols further had a selective inhibitory activity against pathogenic bacterial species of intestinal microflora. In 1991, Ahn et al. reported the selective inhibitory nature of green tea polyphenols against the growth of the pathogenic Clostridia species, including *C. difficile*, *C. paraputrificum*, and *C. perfringens*, while at the same time having no inhibition of the beneficial bifidobacteria, by examining their growth under an *in vitro* medium. In another study, Kakuda et al. (1991) reported concentration-dependent growth of *Bifidobacterium adolescentis* without any growth promotion of *C. perfringens*, *Bacteroides fragilis*, and *Eubacterium lentum* with water extracts of green tea. Later, Okubo et al. (1992) confirmed these findings through an *in vivo* system by examining the human fecal microflora after intake of green tea polyphenols (0.4 g/person, three times/day) over a 4-week period. They indicated that the intake of green tea polyphenols significantly reduced the counts of *Clostridium* species, while increasing the counts of *Bifidobacterium* species in the feces. In an animal study, Ishihara et al. (2001) observed a similar effect of high fecal counts of *Bifidobacterium* and *Lactobacillus* species and a decrease of *C. perfringens* in Holstein calves following the intake of green tea extracts. In 2006, Lee et al. examined the growth of pathogenic, commensal (normal), and probiotics found in the intestine under the influence of 31 phenolic compounds derived from tea. They found that the phenolic compounds in general have an inhibitory effect on all intestinal bacteria, but the degree of inhibition varied, depending on the bacterial species and chemical structure of the compound. They further found that the growth of the pathogenic *E. coli*, *S. typhimurium*, and *Bacteroides* and *Clostridium* genera was strongly inhibited, while the growth of probiotic *Bifidobacterium* and *Lactobacillus* strains was less affected by the tea compounds. The above studies suggest that green tea polyphenols help to improve the microflora balance by the preferential growth effect on beneficial probiotic bacteria while inhibiting the pathogenic bacterial species of the intestine.

In most studies, ester-linked galloyl moieties such as epigallocatechin gallate (EGCg), epicatechin gallate (ECg), and gallocatechin gallate (GCg) of green tea polyphenols showed strong antimicrobial properties against the pathogenic bacteria, thus suggesting a linkage between the structural and functional influences of green tea polyphenols (Ishihara et al., 2001; Katoh, 1995; Sakanaka et al., 1996; Shimamura et al., 2007). Ikigai et al. (1993) suggested that green tea polyphenols damage the lipid bilayers of pathogen bacteria, thus hindering their growth. In the case of *E. coli*, the inhibition of their growth and the release of vero-toxins (Sugita-Konishi et al., 1999) were thought to be due to damage of cell membranes by green tea polyphenols (Hara-Kudo et al., 2001). Later the same group (Hara-Kudo et al., 2005) observed the damage of vegetative membrane, including membrane of spore-forming bacterial pathogens by green tea polyphenols. Some studies postulated that the modulation of cytokine (Lee et al., 2005) and hydrogen peroxide production (Arakawa et al., 2004) might lead to hyperacidification (Cueva et al., 2010) as a mechanistic reason for the antimicrobial activities of the polyphenols and their metabolites. In a recent review of the bacterial inhibitory mechanism of green tea polyphenols, especially that of EGCg, it was concluded that such activity was due to their ability to bind with the peptide structure of bacterial species (Shimamura et al., 2007). According to various studies, the minimum inhibitory concentrations (MICs) of green tea extracts and EGCg against various genera of pathogenic species are within the range of 100 to 1000 µg/ml. As phenolic compounds would invariably transform into smaller metabolites in the intestinal system, the potential of antibacterial activities of catechins could be attributed not only to the specific molecules but also to their metabolites, such as phenolic acids (Cueva et al., 2010; Selma et al., 2009).

13.3 Green Tea Catechins and Gastric Inflammation

Intestinal epithelium plays an important role as a barrier to antigens and microorganisms (Nagler-Anderson, 2001). Disruption of intestinal epithelium, which sometimes is referred to as leaky gut, facilitates the paracellular and transcellular permeability of micro- or macromolecules that leads to several gut disorders, such as irritable bowel syndrome, ulcerative colitis, and Crohn's disease (Meddings, 1997), causing severe inflammation. The pathogenic bacterial species such as *H. pylori* and *E. coli* are well-known factors of serious stomach inflammation that may eventually lead to ulcers and gastric cancer. The deleterious effects of *H. pylori* are involved in various pathways such as stimulation of glycosylation of Toll-like receptors (TLR-4), which lead to gastric inflammation. Pathogenic *E. coli* infection or increased release of pro-inflammatory cytokines (IFN-γ, TNF-α, IL-4, IL-13) by various reasons results in the disruption of the epithelial barrier, resulting in increased permeability of micro- and macromolecules (McKay and Baird, 1999; Philpott et al., 1996). Upon reviewing various factors, Romier et al. (2009) concluded that intestinal inflammation is a complex network of immune and nonimmune cells, including epithelial, endothelial, mesenchymal, and nerve cells and their interaction, resulting in the upregulation of proinflammatory mediators such as cytokines and growth factors. The intestinal epithelium produces antimicrobial agents that act as chemical barriers against various pathogens (Ouellette and Bevins, 2001). The gut-associated lymphoid tissue (GALT) in the epithelium can redirect the antigens from lumen to subepthelium tissues for destruction (Kerneis et al., 1997). Also, intestinal epithelium integrity depends on the presence of TLRs and cytosolic Nod-like receptors (NLRs), which are capable of isolating pathogenic bacteria from normal bacteria (Magalhaes et al., 2007). Cox enzymes, a ubiquitous enzyme constitutively present in various cells such as endothelium and the gastrointestinal tract, are upregulated during the inflammation process by other inflammatory mediators (Claria and Romano, 2005). Additionally, several intracellular signal transduction cascades, particularly the signal pathways, such as nuclear factor kappa B (NF-κB) and mitogen-activated protein kinase (MAPK), play a crucial role in regulating inflammatory gene expression and intestinal cell defense programs, respectively (Barnes, 1997; Waetzig and Schreiber, 2003). Earlier, several studies and reviews emphasized the anti-inflammatory action of plant phenolics (Calixto et al., 2003, 2004; Dryden et al., 2006; Han et al., 2007; Nam, 2006; Rahman et al., 2006; Santangelo et al., 2007; Shapiro et al., 2007; Yoon and Baek, 2005). In this chapter, the anti-inflammatory properties of green tea extracts and their compounds were reviewed based on the studies conducted over the last decade.

Green tea polyphenols prevent the migration of *H. pylori* into the gastric epithelium, thus preventing inflammation. EGCg has been shown to inhibit glycosylation (Lee et al., 2004) and DNA damage (Ruggiero et al., 2007) caused by *H. pylori*. In various experimentally induced inflammation in *in vitro* systems, on different kinds of cells lines, including human colon epithelial cell lines, EGCg has effectively inhibited proinflammatory mediators such as cytokines IL-6 (Paradkar et al., 2004) and IL-8 (Porath et al., 2005; Romier et al., 2008), proteins MIP-3 α and TNF-α (Porath et al., 2005), and growth-related oncogenes like GRO-α (Porath et al., 2005). The intracellular anti-inflammatory effects of EGCg markedly recognized by its interference in signal pathways such as NF-κB and MAPKs cascade activities through control of the IκB phosphorylation (Navarro-Peran et al., 2008; Yang et al., 2001) and inhibition of intracellular proinflammatory mediator expressions such as COX2 expression and secretion of its end product PGE2 (Porath et al., 2005). In another study, Watson et al. (2004) indicated that EGCg prevented IFN-γ-induced increased epithelial permeability by blocking STAT-dependent events like phosphorylation in the gut epithelia. The above mechanistic anti-inflammatory actions of pure EGCg were also observed with standard green tea extracts in simulated *in vitro* inflammatory cell models (Netsch et al., 2006; Teschedi et al., 2004). In *in vivo* studies conducted mostly with stimulated colitis model animals, both EGCg (Mochizuki and Hasegawa, 2005; Sato et al., 1998) and green tea extracts (Mazzon et al., 2005; Oz et al., 2005; Varilek et al., 2001) were found to exert anti-inflammatory activity by suppressing the severity of colitis and related inflammation.

13.4 Green Tea Catechins and Gastritis

The intestinal mucosa that maintains the integrity of the colon is often subjected to oxidative stress generated by various inflammatory factors or damaged by pathogenic infections leading to severe gastritis. It was well noted that the infestation of *H. pylori* chronically infects gastric mucosa, causing gastritis in more than 50% of the population. One of the major factors is vacuolation of cytotoxin VacA from *H. pylori* causing this distress (Bernard et al., 2004; Telford et al., 1994). Ruggiero et al. (2007) reported the decrease of gastric epithelium damage induced by *H. pylori* by green tea polyphenols in mice. This effect was postulated to be due to the inhibition of cytotoxin VacA release. Matsubara et al. (2003) examined *H. pylori*-induced gastritis in Mongolian gerbils, and found a dose-dependent decrease of urease enzyme activity with the consumption of green tea, which is essential for *H. pylori* colonization.

At least three epidemiological studies, having a large number of populations infected with *H. pylori*, indicated a reduced risk of gastritis with increased daily consumption of green tea. Shibata et al. (2000) did a cross-sectional study with 636 Japanese people and identified that the risk of chronic atrophic gastritis was 3.73 times greater with *H. pylori* infection. They noticed a 37% reduced risk of gastritis with regular consumption of 10 cups of green tea per day. Another study examined 295 men and 443 women in a Japanese village, and found that *H. pylori* infection increased the risk of gastritis almost five times, but such risk, especially in the infected individuals, decreased with daily consumption of green tea (Iwahashi et al., 2002). In another epidemiological study conducted in China with 133 stomach cancer, 166 chronic gastritis, and 433 healthy control subjects, an inverse relationship was observed between green tea consumption and chronic gastritis (Setiawan et al., 2001). Epidemiological studies of green tea consumption in relation to gastrointestinal cancer or preneoplastic lesions were identified through on-line literature searches, and a protective effect of green tea on adenomatous polyps and chronic atrophic gastritis formations (Borrelli et al., 2004) was identified. *H. pylori* and other anaerobic bacterial proliferations in endothelium may reduce nitrate to nitrite, which can then react with other nitrogen-containing compounds to produce N-nitroso carcinogens causing gastritis and gastic cancers. So it was postulated that green tea catechins having stong antioxidant activitites might inhibit the nitrosation process for the improvement of gastritis.

Likewise, oxidative damage resulting from various factors of inflammation was associated with mucosal atrophy and gastritis (Weijl et al., 1997). The generation of free radicals that are responsible for oxidative stress is somewhat balanced by the action of endogenous antioxidant enzymes, such as superoxide dismutase (SOD), catalase, and glutathione peroxidase present in the gut wall or mucosa (Halliwell, 1996). However, it is often noticed that chronic infections, leading to the decrease of these enzymes concomitantly with reactive oxygen species (ROS), increase their potential to promote the proliferation of inflammation and mucosal atrophy (Dinu et al., 2009). Oxidative damage resulting in the decrease of glutathione and an increase of its reduced form of glutathione disulfide (GSSG) and gastric mucosal myeloperoxidase (MPO) activity in mice was suppressed with the administration of green tea polyphenols (Wessner et al., 2007). In another study, the ingestion of green tea reduced the fasting-induced mucosal atrophy in mice (Asfar et al., 2003). In rats, the long-term administration of green tea polyphenols increased the quantity of gastrointestinal mucosa (Ito et al., 2008); however, the underlying mechanism for this increase has not been identified yet.

13.5 Green Tea Catechins and Gastric Cancer

Gastric cancer is often recognized as being the second largest cause of cancer-associated deaths in the world. As described above, unabated gastric inflammation, gastritis, and inflammatory disorders such as inflammatory bowel disease (IBD), including chronic ulcerative colitis (UC) and Crohn's disease, predispose subjects to the development of colon and gastric cancer (Itzkowitz and Yio, 2004). On the other hand, exposure to mutagens/carcinogens such as nitrosamines, which are formed from secondary amines and nitrite under gastric pH conditions, are well-known etiological factors in the development of gastric cancer (Aiub et al., 2003; Groves et al., 2002). Numerous *in vitro* and *in vivo* studies have examined the effect of green tea catechins on various kinds of cancers, and the details of those studies are reviewed elsewhere in this book. In the case of gastric cancer, conflicting ameliorating effects of green tea polyphenols have been reported. In *in vitro* studies, nitrosation and formation of nitrosamines were inhibited by green tea polyphenols (Nakamura and Kawabata, 1981; Tanaka et al., 1998). Later Matsuda et al. (2006) noticed the inhibition of nitrosation with high concentrations of catechins, but not at low concentrations in both animals and humans. In another study, a *H. pylori*-infected gastric cancer mouse model found that regular oral administration of 0.5% polyphenols in drinking water for 2 weeks suppressed epithelial cell proliferation and apoptosis (Akai et al., 2007). A number of small population epidemiological case studies reported a reduced risk of gastric cancer with the consumption of green tea (Hoshiyama et al., 2004; Mu et al., 2005 and references therein). Kono et al. (1988) observed that people who drank 10 or more cups of green tea per day had a 40% lower risk of developing stomach cancer than those who drank 0 to 4 cups per day. Later, an inverse relationship between gastric cancer and green tea consumption was observed only in women but not in men (Inoue et al., 2009; Sasazuki et al., 2004). However, over the years several epidemiological studies and their pooled meta-analysis performed by various groups denied the relation between green tea consumption and gastric cancer (Borrelli et al., 2004; Koizumi et al., 2003; Myung et al., 2009; Tsubono et al., 2001; Zhou et al., 2008). The above studies indicate that green tea polyphenols could be effective in reducing the risk of gastric cancer by inhibiting the premodulating effects, such as inflammation and gastritis, that lead to gastric cancer rather than lowering the gastric cancer itself.

13.6 Metabolism and Bioavailability in the Gut

All food substances that we eat invariably transform into small molecules before their absorption into the body, and green tea polyphenols are no exception to this process. All of the above biological effects of green tea catechins and their components largely depend on their availability either in their true form or by their metabolites at the target site. Scalbert et al. (2002) disclosed a large influence of the structures of polyphenols on their bioavailability. Essentially, small molecules are readily absorbed through the gut barrier, whereas large molecules are poorly absorbed. Once absorbed, polyphenols are conjugated to glucuronide, sulfate, and mostly to methyl groups in the gut mucosa and inner tissues (Chow et al., 2001; Kohri et al., 2001; Lee et al., 2000; Yang et al., 1999). On the other hand, the remaining polyphenols are intensively metabolized by colon microflora into further smaller molecules or metabolites. Since green tea polyphenols have preferential prebiotic effects on gut microflora, the interaction of the polyphenols with microflora and their metabolites is also important to consider for their biological activities (Del Rio et al., 2010; Scalbert and Williamson, 2000; Selma et al., 2009). Numerous studies have evaluated the absorption and bioavailability of green tea polyphenols and their metabolites in the serum after oral administration in animals (Catterall et al., 2003; Kim et al., 2000; Okushio et al., 1996) and humans (Chow et al., 2001; Higdon and Frei, 2003; Ullmann et al., 2004; Yang et al., 1998). The detailed metabolism on the absorption and bioavailability of green tea polyphenols and their metabolites in plasma and various organs is discussed elsewhere in this book. A few studies have examined the distribution of polyphenols in the gut of animals, but to date none have been conducted on humans. Kim et al. (2000) observed substantial amounts of EGCg, EGC, and EC in the large intestine when rats were given 0.6% green tea polyphenols in their drinking water over a period of 28 days. Lambert et al. (2003) reported high concentrations of EGCg in the small intestine (46.2 ± 13.5 nmol/g) and colon (7.9 ± 2.4 nmol/g) in mice after oral administration of EGCg at 163.8 μmol/kg. A number of studies have reported a low oral bioavailability of green tea polyphenols through serum (Lu et al., 2003), and found that the majority of polyphenols interact and metabolize within gut wall and intestinal microflora, thus exerting their biological effects on gut health (Kutschera et al., 2011; Lee et al., 2006; Silberberg et al., 2006; van Duynhoven et al., 2011). The presystemic gut wall metabolism of green tea polyphenols was attributed to UDP-glucuronosyltransferase and phenolsulfotransferase located in the intestinal mucosa (Cai et al., 2002).

13.7 Conclusion

Green tea polyphenols are known for their strong antioxidant and antimicrobial properties, prebiotic, and other benefits for the improvement and maintenance of gut health. Green tea polyphenols exhibit preferable prebiotic effects through preferential inhibitory action against pathogenic bacterial species by destroying their cell lipid layers. Gastric inflammation triggered by various factors like bacterial infections and other gastrointestinal disorders is suppressed by green tea polyphenols through modulating proinflammatory mediators such as cytokines, protein expressions, and growth factors and interfering with the signal pathways. Gastritis, a default inflammatory discomfort arising from the damage of gastric mucosa and disruption of the gut wall, was ameliorated with the intake of green tea polyphenols largely through its antioxidant activity. According to the literature reviewed, green tea polyphenols show conflicting effects on gastric cancer, where few studies observed improvement, but the pooled analysis of several epidemiological studies uphold this benefit. Overall, the review of numerous studies recommends the daily intake of green tea polyphenols to maintain overall gut health.

References

Ahn, Y.J., Kawamura, T., Kim, M., Yamamoto, T., and Mitsuoka, T. 1991. Tea polyphenols: Selective growth inhibitors of *Clostridium* spp. *Agric Biol Biochem*, 55: 1425–1426.

Aiub, C.A., Pinto, L.F., and Felzenszwalb, I. 2003. N-Nitrosodiethylamine mutagensity at concentrations. *Toxicol Lett*, 145: 36–45.

Akai, Y., Nakajima, N., Ito, Y., Matsui, T., Iwasaki, A., and Arakawa, Y. 2007. Green tea polyphenols reduce gastric epithelial cell proliferation and apoptosis stimulated by *Helicobacter pylori* infection. *J Clin Biochem Nutr*, 40: 108–115.

Arakawa, H., Maeda, M., Okubo, S., and Shimamura, T. 2004. Role of hydrogen peroxide in bactericidal action of catechin. *Boil Pharm Bull*, 27: 277–281.

Asfar, S., Abdeenb, S., Dashtia, H., Khoursheeda, M., Al-Sayera, H., Mathewc, T., and Al-Baderb, A. 2003. Effect of green tea in the prevention and reversal of fasting-induced intestinal mucosal damage. *Nutrition*, 19: 536–540.

Barnes, P.J. 1997. Nuclear factor-kappa B. *Int J Biochem Cell Biol*, 29: 867–870.

Bekkali, N.L., Bongers, M.F., Van den Berg, M.M., Liem, O., and Benninga, M.A. 2007. The role of a probiotics mixture in the treatment of childhood constipation: A pilot study. *Nutr J*, 6: 17.

Bernard, D., Cappon, M.A., Del Giudice, G., Rappuoli, R., and Montecucco, C. 2004. The multiple cellular activities of the VacA cytotoxin of *Helicobacter pylori*. *Int J Med Microbiol*, 293: 589–597.

Borrelli, F., Capasso, R., Russo, A., and Ernst, E. 2004. Green tea and gastrointestinal cancer risk. *Alimentary Pharmacol Ther*, 19: 497–510.

Cai, Y., Anavy, N.D., and Chow, H.H.S. 2002. Contribution of presystemic hepatic extraction to the low oral bioavailability of green tea catechins in rats. *Drug Met Disposition*, 30: 1246–1249.

Calixto, J.B., Campos, M.M., Otuki, M.F., and Santos, A.R. 2004. Anti-inflammatory compounds of plant origin. Part II. Modulation of pro-inflammatory cytokines, chemokines and adhesion molecules. *Planta Med*, 70: 93–103.

Calixto, J.B., Otuki, M.F., and Santos, A.R. 2003. Anti-inflammatory compounds of plant origin. Part I. Action on arachidonic acid pathway, nitric oxide and nuclear factor kappa B (NF-kB). *Planta Med*, 69: 973–983.

Catterall, F., King, L.J., Clifford, M.N., and Loannides, C. 2003. Bioavailability of dietary doses of 3H-labelled tea antioxidants (+)-catechin and (–)-epicatechin in rat. *Xenobiotica*, 33: 743–753.

Chen, C.C., and Walker, W.A. 2011. Clinical applications of probiotics in gastro-intestinal disorders in children. *Nat Med J India*, 24: 153–160.

Chow, H.H., Cai, Y., Alberts, D.S., Hakim, I., Dorr, R., Shahi, F., Crowell, J.A., Ynag, C.S., and Hara, Y. 2001. Phase I pharmacokinetic study of tea poly-phenols following single-dose administration of epigallocatechin gallate and Polyphenon E. *Cancer Epidemiol Biomarkers Prev*, 10: 53–58.

Claria, J., and Romano, M. 2005. Pharmacological intervention of cycloxygenase-2 and 5-lipoxygenase pathways. Impact on inflammation and cancer. *Curr Pharm Des*, 11: 3431–3437.

Cox, A.J., Pyne, D.B., Saunders, P.U., and Fricker, P.A. 2008. Oral administration of probiotic *Lactobacillus fermentus* VRI-003 and mucosal immunity in endur-ance athletes. *Br J Sports Med*, 44: 222–226.

Cueva, C., Moreno-Arribas, M.V., Martin-Alvarez, P.J., Bills, G., Vicente, M.F., Basilio, A., Rivas, C.L., Requena, T., Rodriguez, J.M., and Bartolome, B. 2010. Antimicrobial activity of phenolic acids against commensal, probiotic and pathogenic bacteria. *Res Microbiol*, 161: 372–382.

Del Rio, D., Calani, L., Cordero, C., Salvatore, S., Pellegrini, N., and Brighenti, F. 2010. Bioavailability and catabolism of green tea flavan-3-ols in humans. *Nutrition*, 26: 1110–1116.

Deshpande, G., Rao, S., and Patole, S. 2007. Probiotics for prevention of necrotising enterocolotis in pretern neonates with very low birth weight: A systematic review of randomized controlled trials. *Lancet*, 369: 1614–1620.

Dinu, C., Diaconescu, C., and Avram, N. 2009. Gastric mucosa injury associated with oxidative stress. *Lucrari Stiintifice Medicina Veterinara*, 2: 369–374.

Diop, L., Guillou, S., and Durand, H. 2008. Probiotic food supplement reduces stress-induced gastrointestinal symptoms in volunteers: A double-blind, placebo-controlled, randomized trial. *Nutr Res*, 28: 1–5.

Dryden, G.W., Song, M., and McClain, C. 2006. Polyphenols and gastrointestinal diseases. *Curr Opin Gastroenterol*, 22: 165–170.

Franks, I. 2011. Probiotics: Probiotics and diarrhea in children. *Nat Rev Gastroenterol Hepatol*, 8: 602.

Fuller, R. 1989. Probiotics in man and animals. *J Appl Bacteriol*, 66: 365–378.

Ganguli, K., and Walker, W.A. 2011. Probiotics in the prevention of necrotizing enterocolitis. *J Clin Gastoenterol Suppl*, S133–S138.

Girardin, M., and Seidman, E.G. 2011. Indications for the use of probiotics in gastrointestinal diseases. *Dig Dis*, 29: 574–587.

Groves, F.D., Issaq, H., Fox, S., Jeffrey, A.M., Whysner, J., Zhang, L., You, W.C., and Fraumeni Jr., J.F. 2002. N-nitroso compounds and mutagens in Chinese fermented (sour) corn pancakes. *J AOAC Int*, 85: 1052–1056.

Guandalini, S. 2011. Probiotics for prevention and treatment of diarrhea. *J Clin Gastroenterol*, 45: S149–S153.

Hakansson, A., and Molin, G. 2011. Gut microbiota and inflammation. *Nutrients*, 3: 637–682.

Halliwell, B. 1996. Antioxidants in health and disease. *Ann Rev Nutr*, 16: 33–50.

Han, X., Shen, T., and Lou, H. 2007. Dietary polyphenols and their biological significance. *Int J Mol Sci*, 8: 950–988.

Hara, Y. 1997. Influence of tea catechins on digestive tract. *J Cell Biochem Suppl*, 27: 52–58.

Hara-Kudo, Y., Okubo, T., Tanaka, S., Chu, D.C., Juneja, L.R., Saito, N., and Sugita-Konishi, Y. 2001. Bacterial action of green tea extract and damage to the membrane of *Escherichia coli* 157:H7. *Biocontrol Sci*, 6: 57–61.

Hara-Kudo, Y., Yamasaki, A., Sasaki, M., Okubo, T., Minai, Y., Haga, M., Kondo, K., and Sugita-Konishi, Y. 2005. Antibacterial action of pathogenic bacterial spore by green tea catechins. *J Sci Food Agric*, 85: 2354–2361.

Higdon, J.V., and Frei, B. 2003. Tea catechins and polyphenols: heAlth effects, metabolism, and antioxidant functions. *Crit Rev Food Sci Nutr*, 43: 80–143.

Hoshiyama, Y., Kawaguchi, T., Miura, Y., Mizoue, T., Tokui, N., Yatsuya, H., Sakata, K., Kondo, T., Kikuchi, S., Toyoshima, H., Hayakawa, N., Tamakoshi, A., Ohno, Y., and Yoshimura, T. 2004. Japan collaborative cohort study group. A nested case-control study of stomach cancer in relation to green tea consumption in Japan. *Br J Cancer*, 90: 135–138.

Ikigai, H., Nakae, T., Hara, Y., and Shimamura, I. 1993. Bactericidal catechins damage the lipid bilayer. *Biochem Biophys Acta*, 1147: 132–136.

Inoue, M., Sasazuki, S., Wakai, K., Suzuki, T., Matsuo, K., Shimazu, T., Tsuji, I., Tanaka, K., Mizoue, T., Nagata, C., Tamakoshi, A., Sawada, N., and Tsugane, S. 2009. Green tea consumption and gastric cancer in Japanese: A pooled analysis of six cohort studies. *Gut*, 58: 1323–1332.

Ishihara, N., Chu, D.C., Akachi, S., and Juneja, L.R. 2001. Improvement of intestinal microflora balance and prevention of digestive and respiratory organ diseases in calves by green tea extracts. *Livestock Prod Sci*, 68: 217–229.

Isogai, E., Isogai, H., Takeshi, K., and Nishikawa, T. 1998. Protective effect of Japanese green tea extract on gnotobiotic mice infected with an *Escherichia coli* O157:H7 strain. *Microbial Immunol*, 42: 125–128.

Ito, Y., Ichikawa, T., Iwai, T., Saegusa, Y., Ikezawa, T., Goso, Y., and Ishihara, K. 2008. Effects of tea catechins on the gastrointestinal mucosa in rats. *J Agric Food Chem*, 56: 12122–12126.

Itzkowitz, S.H., and Yio, X. 2004. Inflammation and cancer IV. Colorectal cancer in inflammatory bowel disease: The role of inflammation. *Am J Physiol Gastrointest Liver Physiol*, 287: G7–G17.

Iwahashi, M., Momose, Y., Miyazaki, M., Shibata, K., and Une, H. 2002. Effects of cytotoxin-associated gene A (CagA) antibodies with *Helicobacter pylori* infection and lifestyle on chronic atrophic gastritis. *Nippon Koshu Eisei Zasshi*, 49: 1152–1158.

Kakuda, T., Matsuura, T., Mortelmans, K., and Parkhurst, R. 1991. Biological activity of tea extracts on *Bifidobacterium* proliferation. In *Proceedings of the International Symposium on Tea Science*, 357. Shizuoka: Kurofune Printing Co. Ltd.

Kalliomaki, M. 2009. The role of microbiota in allergy. *Ann Nestle*, 67: 19–25.

Katoh, H. 1995. Prevention of mouse experimental periodontal disease by tea catechins (in Japanese). *Nihon Univ J Oral Sci*, 21: 1.

Kerneis, S., Bogdanova, A., Kraehenbuhl, J.P., and Pringault, E. 1997. Conversion by Peyers patch lymphocytes of human enterocytes into M cells that transport bacteria. *Science*, 277: 949–952.

Kim, S., Lee, M.J., and Hong, J. 2000. Plasma and tissue levels of tea catechins in rats and mice during chronic consumption of green tea polyphenols. *Nutr Cancer*, 37: 41–48.

Kohri, T., Matsumoto, N., Yamakawa, M., Suzuki, M., Nanjo, F., Hara, Y., and Oku, N. 2001. Metabolic fate of (–)-[4-(3)H]epigallocatechin gallate in rats after oral administration. *J Agric Food Chem*, 49: 1673–1681.

Koizumi, Y., Tsubono, Y., Nakaya, N., Nishino, Y., Shibuya, D., Matsuoka, H., and Tsuji, I. 2003. No association between green tea and the risk of gastric cancer: Pooled analysis of two prospective studies in Japan. *Cancer Epidemiol Biomarkers Prev*, 12: 472–473.

Kono, S., Ikeda, M., Tokumdome, S., and Kuratsune, M. 1988. A case-control study of gastric cancer and diet in northern Kyushu, Japan. *Jpn J Cancer Res*, 79: 1067–1074.

Kosiewicz, M.M., Zirnheld, A.L., and Alard, P. 2011. Gut microflora, immunity and disease: A complex relationship. *Front Microbiol*, 2: 1–11.

Kraehenbuhl, J.P., and Neutra, M.R. 1992. Molecular and cellular basis of immune protection of mucosal surfaces. *Physiol Rev*, 72: 853–879.

Kutschera, M., Engst, W., Blaut, M., and Braune, A. 2011. Isolation of catechins-converting human intestinal bacteria. *J Appl Microbiol*, 111: 165–175.

Lambert, J.D., Lee, M.J., Lu, H., Meng, X., Hong, J.J., Seril, D.N., Sturgill, M.G., and Yang, C.S. 2003. Epigallocatechin-3-gallate is absorbed but extensively glucuronidated following oral administration to mice. *J Nutr*, 133: 4172–4177.

Lebba, V., Aloi, M., Cicitelli, F., and Cucchiara, S. 2011. Gut microflora and pediatric disease. *Dig Dis*, 29: 531–539.

Lee, B.J., and Bak, Y.T. 2011. Irritable bowel syndrome, gut microbiota and probiotics. *J Neurogastroenterol Motil*, 17: 252–266.

Lee, H.C., Jenner, A.M., Low, C.S., and Lee, Y.K. 2006. Effect of tea phenolics and their aromatic fecal bacterial metabolites on intestinal microbiota. *Res Microbiol*, 157: 876–884.

Lee, K.M., Yeo, M., Choue, J.S., Jin, J.H., Park, S.J., Cheong, J.Y., Lee, K.J., Kim, J.H., and Hahm, K.B. 2004. Protective mechanism of epigallocatechin-3-gallate against *Helicobacter pylori*-induced gastric epithelial cytotoxicity via blockage of TLR-4 signaling. *Helicobacter*, 9: 632–642.

Lee, M.J., Prabhu, S., Meng, X., Li, C., and Yang, C.S. 2000. An improved method for the determination of green and black tea polyphenols in biomatrices by high-performance liquid chromatography with coulometric array detection. *Anal Biochem*, 279: 164–169.

Lee, Y.S., Han, C.H., Kang, S.H., Lee, S.J., Kim, S.W., Shin, O.R., Sim, Y.C., Lee, S.J., and Cho, Y.H. 2005. Synergistic effect between catechin and ciprofloxacin on chronic bacterial prostrates rat model. *Int J Urol*, 12: 383–389.

Lu, H., Meng, X.F., Lee, M.J., Li, C., Maliakal, P., Yang, C.S. 2003. Bioavailability and biological activity of tea polyphenols. *Food Factors Health Promotion Dis Prev Symposium Ser*, 851: 9–15.

Lutgendorff, F., Akkermans, K.M., and Soderholm, J.D. 2008. The role of micro-biota and probiotics in stress-induced gastro-intestinal damage. *Curr Mol Med*, 8: 282–298.

MacPhee, R.A., Hummelen, R., Bisanz, J.E., Miller, W.L., and Reid, G. 2010. Probiotic strategies for the treatment and prevention of bacterial vaginosis. *Expert Opin Pharmacother*, 11: 2985–2995.

Magalhaes, J.G., Tattoli, I., and Girardin, S.E. 2007. The intestinal epithelial barrier: How to distinguish between the microbial flora and pathogens. *Semin Immunol*, 19: 106–115.

Matsubara, S., Shibata, H., Ishikawa, F., Yokokura, T., Takahashi, M., Sugimura, T., and Wakabayashi, K. 2003. Suppression of *Helicobactor pylori*-induced gastritis by green tea extract in Mongolian gerbils. *Biochem Biophys Res Commun*, 310: 715–719.

Matsuda, S., Uchida, S., Terashima, Y., Kuramoto, H., Serizawa, M., Deguchi, Y., Yanai, K., Sugiyama, C., Oguni, I., and Kinae, N. 2006. Effect of green tea on the formation of nitrosamines and cancer mortality. *J Health Sci*, 52: 211–220.

Mazzon, E., Muià, C., Paola, R.D., Genovese, T., Menegazzi, M., DeSarro, A., Suzuki, H., and Cuzzocrea, S. 2005. Green tea polyphenol extract attenuates colon injury induced by experimental colitis. *Free Radic Res*, 39: 1017–1025.

McKay, D.M., and Baird, A.W. 1999. Cytokine regulation of epithelial permeability and ion transport. *Gut*, 44: 283–289.

Meddings, J.B. 1997. Intestinal permeability in Crohn's disease. *Aliment Pharmacol Ther Suppl*, 3: 47–53.

Mochizuki, M., and Hasegawa, N. 2005. Protective effect of (–)-epigallocatechin gallate on acute experimental colitis. *J Health Sci*, 51: 362–364.

Mu, L.N., Lu, Q.Y., Yu, S.Z., Jiang, Q.W., Cao, W., You, N.C., Sctiawan, V.W., Zhou, X.F., Ding, B.G., Wang, R.H., Zhao, J., Cai, L., Rao, J.Y., Herber, D., and Zhang, Z.F. 2005. Green tea drinking and multigentic index on the risk of stomach cancer in Chinese population. *Int J Cancer*, 116: 972–983.

Myung, S.K., Bae, W.K., Oh, S.M., Kim, Y., Ju, W., Sung, J., Lee, Y.J., Ko, J.A., Song, J.I., and Choi, H.J. 2009. Green tea consumption and risk of stomach cancer: A meta-analysis of epidemiologic studies. *Int J Cancer*, 124: 670–677.

Nagler-Anderson, C. 2001. Man the barrier! Strategic defenses in the intestinal mucosa. *Nat Rev Immunol*, 1: 59–67.

Nakamura, M., and Kawabata, T. 1981. Effect of Japanese green tea on nitrosamine formation *in vitro*. *J Food Sci*, 46: 306–307.

Nam, N.H. 2006. Naturally occurring NF-kappa B inhibitors. *Mini Rev Med Chem*, 6: 945–951.

Navarro-Peran, E., Cabezas-Herrera, J., Sanchez-Del-Campo, L., Garcia-Canovas, F., and Rodriguez-Lopez, J.N. 2008. The anti-inflammatory and anti-cancer properties of epigallocatehin-3-gallate are mediated by folate cycle disruption, adenosine release and NF-kappaB suppression. *Inflamm Res*, 57: 472–478.

Netsch, M.I., Gutmann, H., Aydogan, C., and Drewe, J. 2006. Green tea extract induces interleukin-8 (IL-8) mRNA and protein expression but specifically inhibits IL-8 secretion in Caco-2 cells. *Planta Med*, 72: 697–702.

Okubo, T., Ishihara, N., Oura, A., Serit, M., Kim, M., Yamamoto, T., and Mitsuoka, T. 1992. *In vivo* effects of tea polyphenols intake on human intestinal microflora and metabolism. *Biosci Biotechnol Biochem*, 56: 588–591.

Okushio, K., Matsumoto, N., Kohri, T., Suzuki, M., Nanjo, F., and Hara, Y. 1996. Absorption of tea catechins into rat portal vein. *Biol Pharm Bull*, 19: 326–329.

Ouellette, A.J., and Bevins, C.L. 2001. Paneth cell defenses and innate immunity of the small bowel. *Inflamm Bowel Dis*, 7: 43–50.

Oz, H.S., Chen, T.S., McClain, C.J., and de Villiers, W.J. 2005. Antioxidants as novel therapy in a murine model of colitis. *J Nutr Biochem*, 16: 297–304.

Paradkar, P.N., Blum, P.S., Berhow, M.A., Baumann, H., and Kuo, S.M. 2004. Dietary isoflavons suppress endotoxin-induced inflammatory reaction in liver and intestine. *Cancer Lett*, 215: 21–28.

Philpott, D.J., McKay, D.M., Sherman, P.M., and Perdue, M.H. 1996. Infection of T84 cells with enteropathogenic *Escherichia coli* alters barrier and transport functions. *Am J Physiol Gastrointest Liver Physiol*, 270: G634–G645.

Porath, D., Riegger, C., Drewe, J., and Schwager, J. 2005. Epigallocatechin-3-gallate impairs chemokine production of human colon epithelial cell lines. *J Pharmacol Exp Ther*, 315: 1172–1180.

Rahman, I., Biswas, S.K., and Kirkham, P.A. 2006. Regulation of inflammation and redox signaling by dietary polyphenols. *Biochem Pharmacol*, 72: 1439–1452.

Romier, B., Schneider, Y.J., Larondelle, Y., and During, A. 2009. Dietary polyphenols can modulate the intestinal inflammatory response. *Nutr Rev*, 67: 363–378.

Romier, B., Van de Walle, J., During, A., Larondelle, Y., and Schneider, Y.J. 2008. Modulation of signaling nuclear factor-kB activation pathway by olyphenols in human intestinal caco-2 cells. *Br J Nutr*, 100: 542–551.

Ruggiero, P., Rossi, G., Tombola, F., Pancotto, L., Lauretti, L., del Giudice, G., and Zoratti, M. 2007. Red wine and green tea reduce *H. pylori*- or Vac-A-induced gastritis in a mouse model. *World J Gastroenterol*, 13: 349–354.

Sakanaka, S., Aizawa, M., Kim, M., and Yamamoto, T. 1996. Inhibitory effects of green tea polyphenols on growth and cellular adherence of an oral bacterium *Porphyromonas gingivalis*. *Biosci Biotechnol Biochem*, 60: 745–749.

Sakanaka, S., Juneja, L.R., and Taniguchi, M. 2000. Antimicrobial effects of green tea polyphenols on thermophilic spore-forming bacteria. *J Biosci Bioeng*, 90: 81–85.

Sakanaka, S., Kim, M., Taniguchi, M., and Yamamoto, T. 1989. Antimicrobial substances in Japanese green tea extract against *Streptococcus mutans*, a carcinogenic bacterium. *Agric Biol Biochem*, 53: 2307–2311.

Sakanaka, S., Sato, T., Kim, M., and Yamamoto, T. 1990. Inhibitory effects of green tea polyphenols on glucan synthesis and cellular adherence of carcinogenic streptococci. *Agric Biol Chem*, 54: 2925–2929.

Santangelo, C., Vari, R., Scazzocchio, B., Di Benedetto, R., Filesi, C., and Masella, R. 2007. Polyphenols, intercellular signaling and inflammation. *Ann Ist Super Sanita*, 43: 394–405.

Sanz, Y., Santacruz, A., and Gauffin, P. 2010. Gut microbiota in obesity and metabolic disorders. *Proc Nutr Soc*, 69: 434–441.

Sasazuki, S., Inoue, M., Hanaoka, T., Yamamoto, S., Sobue, T., and Tsugane, S. 2004. Green tea consumption and subsequent risk of gastric cancer by subsite: The JPHC study. *Cancer Causes Control*, 15: 483–491.

Sato, K., Kanazawa, A., Ota, N., Nakamura, T., and Fujimoto, K. 1998. Dietary supplementation of catechins and alpha-tocopherol accelerates the healing of trinitrobenzene sulfonic acid-induced ulcerative colitis in rats. *J Nutr Sci Vitaminol*, 44: 769–778.

Scalbert, A., Morand, C., Manach, C., and Rémésy, C. 2002. Absorption and metabolism of polyphenols in the gut and impact on health. *Biomed Pharmacother*, 56: 276–282.

Scalbert, A., and Williamson, G. 2000. Dietary intake and bioavailability of polyphenols. *J Nutr*, 130: 2073S–2085S.

Selma, M.V., Espin, J.C., Francisco, A., and Barber, T. 2009. Interaction between phenolics and gut microbiota: Role in human health. *J Agric Food Chem*, 57: 6485–6501.

Setiawan, V.W., Zhang, Z.F., Yu, G.P., Lu, Q.Y., Li, Y.L., Lu, M.L., Wang, M.R., Guo, C.H., Yu, S.Z., Kurtz, R.C., and Hsieh, C.C. 2001. Protective effect of green tea on the risks of chronic gastritis and stomach cancer. *Int J Cancer*, 92: 600–604.

Shapiro, H., Singer, P., Halpern, Z., and Bruck, R. 2007. Polyphenols in the treatment of inflammatory bowel disease and acute pancreatitis. *Gut*, 56: 426–435.

Shibata, K., Moriyama, M., Fukushima, T., Kaetsu, A., Miyazaki, M., and Une, M. 2000. Green tea consumption and chronic atrophic gastritis: A cross-sectional study in a green tea production village. *J Epidemiol*, 10: 310–316.

Shimamura, T., Shao, W.H., and Hu, Z.Q. 2007. Mechanism of action and potential for use of tea catechin as an anti-infective agent. *Anti-infective Agents Medical Chem*, 6: 57–62.

Silberberg, M., Morand, C., Mathevon, T., Besson, C., Manach, C., Scalbert, A., and Remsey, C. 2006. The bioavailability of polyphenols is highly governed by the capacity of the intestine and of the liver to secrete conjugated metabolites. *Eur J Nutr*, 45: 88–96.

Snel, J., Vissers, Y.M., Smit, B.A., Jongen, J.M., van der Meulen, E.T., Zwijsen, R., Ruinemans-Koerts, J., Jansen, A.P., Kleerebezem, M., and Savelkoul, H.F. 2011. Strain-specific immunomodulatory effects of *Lactobacillus plantarum* strains on birch-pollen-allergic subjects out of season. *Clin Exp Allergy*, 41: 232–242.

Sugita-Konishi, Y., Hara-Kudo, T., Amanao, G., Okubo, T., Aoi, N., Iwaki, M., and Kumagai, S. 1999. Epigallocatechin gallate and gallocatechin gallate in green tea catechins exhibit extracelluar release of vero toxins release from enterohemorrhagic *Escherichia coli* O157:H7. *Biochim Biophys Acta*, 1472: 42–50.

Takabayashi, F., Harada, N., Yamada, M., Murohisa, B., and Oguni, I. 2004. Inhibitory effect of green tea catechins in combination with sucralfate on *Helicobacter pyroli* infection in Mongolian gerbils. *J Gastroenterol*, 39: 61–63.

Tanaka, K., Hayatsu, T., Negishi, T., and Hayatsu, H. 1998. Inhibition of N-nitrosation of secondary amines *in vitro* by tea extracts and catechins. *Mut Res*, 412: 91–98.

Telford, J.L., Covacci, A., Ghiara, P., Montecucco, C., and Rappuoli, R. 1994. Gene structure of the *Helicobacter pylori* cytotoxin and evidence of its key role in gastric disease. *J Exp Med*, 179:1653–1658.

Teschedi, E., Menegazzi, M., Yao, Y., Suzuki, H., Forstermann, U., and Kleinert, H. 2004. Green tea inhibits human inducible nitric oxide synthase expression by down-regulating signal transducer and activator of transcription-1 alpha activation. *Mol Pharmacol*, 65: 111–120.

Tsubono, Y., Nishino, Y., Komatsu, S., Hsieh, C.C., Kanemura, S., Tsuji, I., Nakatsuka, H., Fukao, A., Satoh, H., and Hisamichi, S. 2001. Green tea and the risk of gastric cancer in Japan. *N Engl J Med*, 344: 632–636.

Tzounis, X., Vulevic, J., Kuhnle, G.C.G., George, T., Leonczak, J., Gibson, G.R., Kwick-Uribe, C., and Spencer, J.P.E. 2008. Flavonol monomer-induced changes to the human fecal microflora. *Br J Nutr*, 99: 782–792.

Ullmann, U., Haller, J., Decourt, J.D., Girault, J., Spitzer, V., and Weber, P. 2004. Plasma-kinetic characteristics of purified and isolated green tea catechins epigallocatechin gallate (EGCG) after 10 days repeated dosing in healthy volunteers. *Int J Vitam Nutr Res*, 74: 269–278.

Vanderhoof, J.A., and Mitmesser, S.H. 2010. Probiotics in the management of children with allergy and other disorders of intestinal inflammation. *Benef Microbes*, 1: 351–356.

Van Duynhoven, J., Vaughan, E.E., Jacobs, D.M., Kemperman, R.A., van Velzen, E.J.J., Gross, G., Roger, L.C., Possemiers, S., Smidle, A.K., Dore, J., Weterhus, J.A., and Van de Wiele, T. 2011. Metabolic fate of polyphenols in the human superorganism. *Proc Natl Acad Sci USA*, 108: 4531–4538.

Varilek, G.W., Yang, F., Lee, E.Y., deVilliers, W.J., Zhong, J., Oz, H.S., Westberry, K.F., and McClain, C.J. 2001. Green tea polyphenol extract attenuates inflammation in interleukin-2-deficient mice, a model of autoimmunity. *J Nutr*, 131: 2034–2039.

Waetzig, G.H., and Schreiber, S. 2003. Mitogen-activated protein kinases in chronic intestinal inflammation—Targeting ancient pathways to treat modern diseases. *Aliment Pharmocol Ther*, 18: 17–32.

Wassenberg, J., Nutten, S., Audran, R., Barbier, N., Aubert, V., Moulin, J., Mercenier, A., and Spertini, F. 2011. Effect of Lactobacillus paracasei ST11 on a nasal provocation test with grass pollen in allergic rhinitis. *Clin Exp Allergy*, 41: 565–573.

Watson, J.L., Ansari, S., Cameron, H., Wang, A., Akhtar, M., and McKay, D.M. 2004. Green tea polyphenol (–)-epigallocatechin gallate blocks epithelial barrier dysfunction provoked by IFN-γ but not by IL-4. *Am J Physiol Gastrointest Liver Physiol*, 287: G954–G961.

Weihua, L., and Jing, L. 2011. Probiotics for treating persistent diarrhea in children. *Am J Nurs*, 111: 61.

Weijl, N.I., Cleton, F.J., and Osanto, S. 1997. Free radicals and antioxidants in chemotherapy-induced toxicity. *Cancer Treat Rev*, 23: 209–240.

Wessner, B., Strasser, E.M., Koitz, N., Schmuckenschlager, C., Manhart, N.U., and Roth, E. 2007. Green tea polyphenol administration partly ameliorates chemotherapy-induced side effects in the small intestine of mice. *J Nutr*, 137: 3634–3640.

Yang, C.S., Chen, L., Lee, M.J., Balentine, D., Kuo, M., and Schantz, S.P. 1998. Blood and urine levels of tea catechins after ingestion of different amounts of green tea by human volunteers. *Cancer Epidemiol Biomarkers Prev*, 7: 351–354.

Yang, C.S., Lee, M.J., and Chen, L. 1999. Human salivary tea catechin levels and catechin esterase activities: Implication in human cancer prevention studies. *Cancer Epidemiol Biomarkers Prev*, 8: 83–89.

Yang, F., Oz, H.S., Barve, S., de Villiers, W.J., McClain, C.J., and Varilek, G.W. 2001. The green ea polyphenols epigallocatechin-3-gallate blocks nuclear factor-kappa B activation by inhibiting Ikappa B kinase activity in the intestinal epithelial cell line IEC-6. *Mol Pharmacol*, 60: 528–533.

Yoon, J.H., and Baek, S.J. 2005. Molecular targets of dietary polyphenols with anti-inflammatory properties. *Yonsei Med J*, 46: 585–596.

Zhou, Y., Li, N., Zhuang, W., Liu, G., Wu, T., Yao, X., Du, L., Wei, M., and Wu, X. 2008. Green tea and gastric cancer risk: Meta-analysis of epidemiological studies. *Asia Pacific J Clin Nutr*, 17: 159–165.

14

Green Tea Polyphenols in Oral Care

Kazuko Takada and Masatomo Hirasawa

Contents

14.1 Introduction

Dental caries, periodontal disease, and mycoses are the main infectious diseases of the human oral cavity. The important bacteria associated with these diseases are *Streptococcus mutants* and *Streptococcus sobrinus* for dental caries (Hamada and Slade, 1980; Loesche, 1986), *Porphyromonas gingivalis* and *Prevotella intermedia* for chronic periodontitis (Slots et al., 1986; Zambon et al., 1981), and *Candida albicans* for oral mycoses (Lynch, 1994; McCullough et al., 1996).

In this chapter, preventive effects of green tea polyphenols on dental caries, periodontal disease, and mycoses evidenced in both *in vitro* and *in vivo* tests are described.

14.2 Green Tea Polyphenols against Dental Caries

Many approaches, such as the elimination of cariogenic bacteria, increasing the resistance of the teeth, and modifying the diet, have been adopted to prevent dental caries (Hamada and Slade, 1980; Loesche, 1986). Some of these have been put to practical uses, while others remain at the level of research.

Green tea catechins are well known to inhibit the bacterial growth, glucan synthesis, and artificial plaque formation of *Streptococcus mutans* and *S. sobrinus in vitro*. Also, tea polyphenols reduce to dental caries in the rat model and decrease dental plaque formation in humans (Sakanaka, 1997).

14.2.1 Additive Antidental Caries Synergic Effects of Tea Polyphenols and Fluoride as Antidental Caries in Rats

The aim of this study is to examine whether tea catechin and fluoride show additive antidental caries effect. The incidence of dental caries in rats infected with *S. mutans* was reduced by the fed diet supplemented with fluoride (0.5 ppm). Further, when green tea extract (Sunphenon) and fluoride were fed in the diet, synergetic effects were observed (Figure 14.1). Caries were reduced approximately 80% in the presence of both Sunphenon and fluoride compared with the absence of both materials (Sato, 1995). The inhibition of caries formation in rats in this study could be explained by the inhibition of GTase activity as well as by the bactericidal activity of Sunphenon, and thus increasing resistance of the teeth by fluoride.

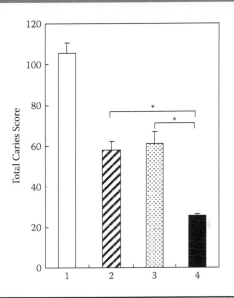

FIGURE 14.1
Total means caries score of rats infected with
S. mutans and fed with diet 2000 with or without
Sunphenon and fluoride. (1) Without Sunphenon and
fluoride; (2) with 0.1% Sunphenon; (3) with 0.5 ppm
fluoride; (4) with 0.1% Sunphenon and 0.5 ppm
fluoride. *, significance of difference, $p < 0.01$.

14.2.2 Inhibitory Effects of Lactic Acid Beverage Added with Tea Polyphenols on Dental Caries in Rats

Our previous studies show that feeding of chocolate, candy, caramel, and biscuits that Sunphenon is added to resulted in the reduction of caries development in rats. Antidental caries effects of Sunphenon added to a lactic acid beverage have also been examined (Kuroki et al., 1999). The incidence of dental caries in specific pathogen-free or gnotobiotic rats infected with S. mutans and fed with a lactic acid beverage containing Sunphenon was significantly lower than in similarly infected rats given a lactic acid beverage alone (Figure 14.2).

14.2.3 Synergic Effects of Tea Polyphenols and Hen Egg Yolk Antibody (HEY) for Anti-Streptococcus mutans in Rats

Mutual interaction of anti-S. mutans HEY and Sunphenon against rat dental caries was investigated (Tanaka, 1992). The incidence of dental caries in rats

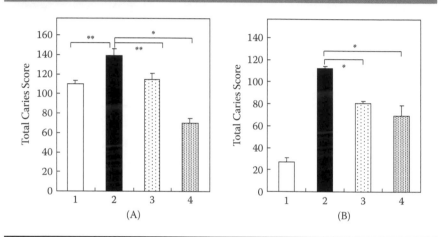

FIGURE 14.2

Total means caries scores of SPF (A) and gnotobiotic (B) rats infected with *S. mutans*, fed with a modified diet 2000 and with lactic acid beverage. (1) Fed with a plain drinking water-infected bacteria (A) or noninfected (B); (2) fed with lactic acid beverage (control); (3) containing 0.026% Sunphenon; (4) containing 0.052% Sunphenon. *, significance of difference, $p < 0.01$; **, significance of difference, $p < 0.05$.

infected with *S. mutans* fed with a diet supplemented with anti-*S. mutans* HEY and receiving drinking water that contained Sunphenon was significantly lower than in similarly infected rats given plain water only without Sunphenon (Figure 14.3).

14.2.4 Influence of Tea Polyphenols on the Adherence of Genus Streptococcus to Oral Cavity

Streptococcus species are the known predominant bacteria in the oral microflora. The effect of green tea catechins (Sunphenon) on adherence of the genus *Streptococcus* (*S. mutans*, *S. sobrinus*, *S. sanguinis*, *S. mitis*, and *S. salivarius*) to the oral cavity has been studied *in vitro* and *in vivo* (Kuribayashi et al., 1998; Wada, 1996). The number of adhered cells to S-HA and teeth surfaces or saliva-coated buccal epithelial cells and buccal mucosal surfaces of each streptococcal species was clearly reduced by the presence of Sunphenon. However, when each streptococcus species on saliva-coated hydroxyapatite (S-HA) and teeth surfaces or saliva-coated buccal epithelial cells and buccal

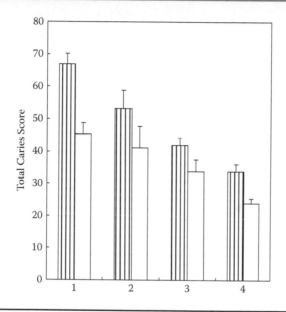

FIGURE 14.3
Total mean caries scores of SPF rats infected with
S. mutans fed with diet M2036 and drank with or without
0.05% Sunphenon water. Diet M2036 was replaced with
36% sucrose and 20% egg yolk powder instead of 56%
sucrose in diet 2000. The ratio of immunized yolk powder
was 0% (1), 10% (2), 33% (3), and 100% (4) ▥ water
without Sunphenon; ▪ water with Sunphenon.

mucosal surfaces was treated with Sunphenon, the significant removal or antibacterial effect of Sunphenon was not observed (Table 14.1).

14.2.5 Effects of Tea Polyphenols on Acid Production

Surprisingly, the positive effects of the acid production on bacterial growth and tooth decay were noticed depending on the diverse parameters of analyses (Hirasawa et al., 2006).

14.2.5.1 Influence of Tea Polyphenols on Inhibition of Bacterial Growth The effects of epigallocatechin gallate (EGCg) on the growth of *S. mutans* and *S. sobrinus* cultured in medium containing sucrose are illustrated in Figure 14.4. Although EGCg could also have a bactericidal effect on *S. mutans* and *S. sobrinus* cultured in medium without sucrose (data not shown), the numbers

TABLE 14.1

Influence of Green Tea Polyphenols on Oral Streptococcus Adhered to Each Hard and Soft Surface in *In Vitro* and *In Vivo* Experiments

| | Percent Survival, % | | | |
| | *In Vitro* | | *In Vivo* | |
Species	*HA*	*Buccal Epithelial Cells*	*Tooth*	*Cheek*
S. mutans	103.3	nd	83.5	nd
S. sobrinus	95.1	nd	98.8	nd
S. sanguinis	79.9	nd	97.6	nd
S. mitis	nd	73.8	nd	97.9
S. salivarius	nd	94.1	nd	78.8

nd, not done.

of *S. mutans* and *S. sobrinus* cultured in medium containing sucrose were preferably higher under the EGCg treatment (Figure 14.4B and D).

14.2.5.2 Inhibition of Acid Production from Streptococcus mutans EGCg effectively inhibited the pH reduction. The phenomenon was noticed when the packed cells of *S. mutans* and *S. sobrinus* were incubated with sucrose (Figure 14.4A and C). The duration to reach pH 4 was more than double with the incubation performed without the EGCg.

14.2.5.3 Inhibition of Acid Production in Dental Plaque The effect of EGCg solution on dental plaque pH was investigated in 15 volunteers, as described in Figure 14.5. The plaque pH time course follows a 2 min mouth rinse with 10% sugar solution and a 5 min mouth rinse with 2 mg/ml concentration of EGCg after 30 min. Obviously, the minimum mean pH values were significantly higher for the EGCg rinse than for the water control rinse. The minimum pH was 6.2 at 3 min after the EGCg rinse, while the minimum pH was 5.2 at 3 min after rinse without EGCg.

14.2.5.4 Inhibition of Lactate Dehydrogenase (LDH) Activity Lactate dehydrogenase (LDH) activity was measured using a LDH-UV test Wako kit, 500 μl of NADH substrate buffer, and 10 μl of LDH, and 90 μl of catechins of different molecular structures was added. A reduction in optical density at 340 nm was monitored for 5 min. Inhibition of LDH activities was observed

FIGURE 14.4

pH (A and C) and survival cells (B and D) curve by BHI containing 5% sucrose culture *S. mutans* (A and B) and *S. sobrinus* (C and D) cells with saliva treatment. ■, without EGCg; ◆, with EGCg.

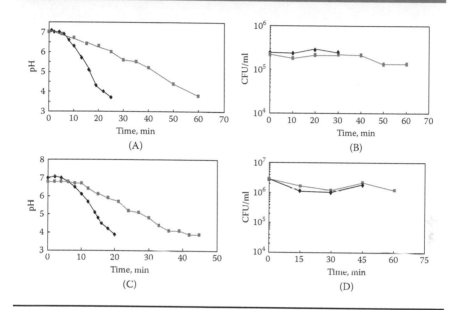

with the catechins complex as well as with galloyl radical-containing compounds; however, no LDH activity was noticed with compounds without any galloyl radicals. The LDH activity was found to be increased when the catechins were preincubated for at least 15 min prior to the reaction with the substrates. The percent inhibitions of LDH activity observed with EGCg, ECg, and catechins complex were 99, 92, and 80%, respectively (Figure 14.6).

14.3 Periodontal Disease Prevention

Black-pigmented, Gram-negative, obligate anaerobic rods, *Porphyromonas gingivalis* and *Prevotella* spp., have been implicated as a periodontal pathogen and are isolated in high numbers from periodontal lesions, and their proportion within microflora seems to be increased in heavily inflamed subgingival lesions. Tea polyphenols are well known to inhibit bacterial growth, collagenase activity, and adherence of bacteria to human buccal epithelial cells (Sakanaka et al., 1997).

FIGURE 14.5
Mean plaque pH time course in 15 volunteers following a 2 min mouth rinse with 10% sugar 30 min after a 5 min mouth rinse ■ with 2 mg/ml EGCg; ▥ without EGCg, with EGCg. *, significantly different at $p < 0.01$; **, significantly different at $p < 0.05$.

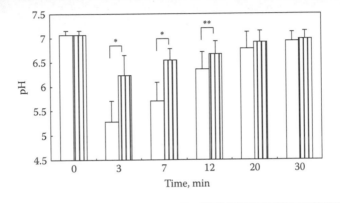

FIGURE 14.6
Percentage inhibition of lactate dehydrogenase activity by various catechins.

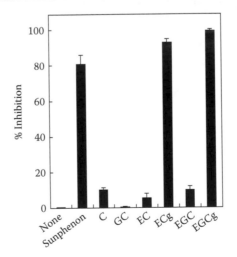

14.3.1 Effect of Slow-Release Local Delivery System of Tea Polyphenols in Periodontal Pockets

Hydroxypropylcellulose (HPC) powder with 5% Sunphenon was dissolved in ethanol and lyophilized to prepare the sheet. A dry sheet was cut into small strips (2 × 4 × 0.3 mm) in size and thickness. The strips containing green tea polyphenols were applied in the periodontal pockets of patients once a week for 8 weeks, as described elsewhere (Hirasawa et al., 2002).

14.3.1.1 Evaluation of Bacterial Number, Enzymatic Activity, and Pocket Depth The pocket depth and proportion of black-pigmented rods were markedly decreased in the green tea polyphenols group at week 8 compared to baseline (Figures 14.7 and 14.8). While the peptidase activities in the gingival fluid were maintained at lower levels during the experimental period (Table 14.2), the improvements of their clinical parameters indicated that green tea polyphenols continued to be released from HPC strips into the sub-gingival pocket throughout the experimental period. Green tea polyphenols also showed a bactericidal effect against black pigmented rods (BPRs). Overall, the application of tea polyphenols using a slow-release local delivery system was effective in improving periodontal status.

FIGURE 14.7
Changes of pocket depth by the application of HPC strips with or without tea polyphenols at baseline (Placebo) and week 8 (Test). ■, w/o Sunphenon; ▥, w/Sunphenon; *, significantly different at $p < 0.05$.

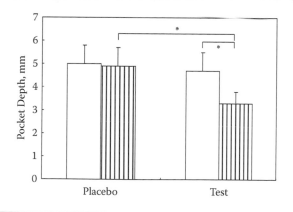

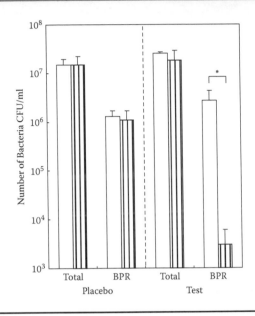

FIGURE 14.8
Colony counts of total cultivable bacteria and BPR from periodontal pockets at baseline (Placebo) and week 8 (Test). ■, w/o Sunphenon; ▥, w/Sunphenon; *, significantly different at $p < 0.01$.

TABLE 14.2
Time Course of the Changes in Peptidase Activities (U/ml)

Group	Experimental Period, Week				
	0	2	4	6	8
Test sites of nonscaled	0.52	0.53	0.54	0.55	0.56
Test sites of scaled	0.50	0.08	0.08	0.085*	0.087*
Placebo sites of scaled	0.47	0.08	0.085	0.13	0.22

*, significant difference at $p < 0.01$.

14.3.2 Antibacterial and Anticytotoxic Effect of Tea Polyphenols against A. actinomycetemcomitans

To examine the antibacterial and anticytotoxic effects of tea polyphenols on *Aggregatibacter actinomycetemcomitans*, bacteria cells (1 × 10⁶ CFU/ml) and green tea polyphenols (1 mg/ml) were incubated for 60 min at 37°C

FIGURE 14.9
Effect of tea polyphenols on cytotoxic activity of partial purified leukotoxin.

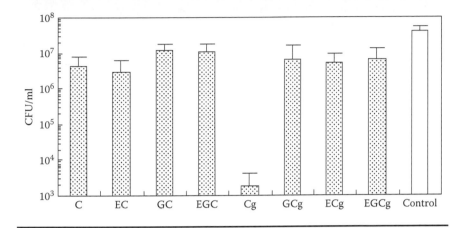

and plating. Only catechin gallate (Cg) among tea polyphenols showed the bactericidal effect against *A. actinomycetemcomitans* at such a concentration (Figure 14.9). Similarly, *Escherichia coli* has shown similar results as *A. actinomycetemcomitans*, wherein leukotoxin was purified from culture supernatant of *A. actinomycetemcomitans* by gel filtration using a Sephadex G-100 column. HL-60 (1 × 10⁵ cells), leukotoxin (0.33 µg protein), and green tea polyphenol (25 µg) were incubated for 30 min at 37°C and survival cells were counted. Leukotoxin activity was markedly inhibited through killing of HL-60 cells by green tea polyphenols containing galloyl residue (Figure 14.10). However, it was noticed that pretreated HL-60 cells with green tea polyphenols could not clearly inhibit the leukotoxin activity. In addition, the Cg solution remained stable at ambient temperature and did not change to brown in color even for more than 1 month.

14.4 Mycoses Prevention

C. albicans is part of normal microbial flora in humans and could be found in the oral cavity. *C. albicans* has been shown to play an important role in oral candidosis, denture stomatitis, and severe periodontitis (Lynch, 1994; Sakanaka and Yamamoto, 1997). The susceptibility of *C. albicans* to catechins under varying pH conditions and the synergism of the combination of catechins and antimycotics were evaluated (Hirasawa and Takada, 2004).

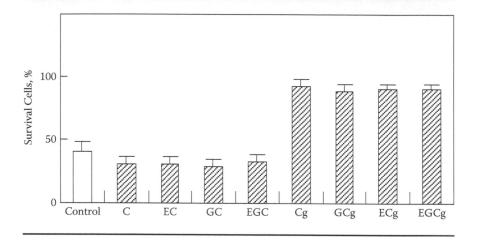

FIGURE 14.10
Bactericidal effects of tea polyphenols against
***A. actinomycetemcomitans* resting cells.**

14.4.1 *Inhibitory Effects of Tea Polyphenols on* Candida albicans

The antifungal effect of EGCg was pH dependent (Table 14.3). The anti-fungal action of EGCg was weakened under acidic conditions and activity reduced several fold at pH ranges varying from 6.0 to 7.0. Among catechins, pyrogallol catechins (EGCg, EGC, GC, and GCg) were more effective than catechol catechins (EC, ECg, C, and Cg) against *C. albicans* (Figure 14.11), whereby the actions of EGCg, EGC, and GC were fungicidal in nature.

14.4.2 *Synergic Antifungal Activity of Tea Polyphenols with Combination of Antimycotics*

C. albicans expresses multidrug efflux transporter (MET), which mediates the efflux of a broad range of compounds, including fluconazole. MET inhibitor, cyclosporine, and fluconazole showed a potent synergistic effect against *C. albicans*. A combined use of EGCg along with fluconazole was effective even against fluconazole-resistant *C. albicans*. The effective dose of fluconazole was decreased to number of folds compared to EGCg doses and even compared to the presence of EGCg alone (Table 14.4). Similar synergic antifungal activities of a combination of EGCg and amphotericin B have also been observed.

TABLE 14.3
MIC$_{90}$ of EGCg against
C. albicans at Varied pH

| pH | MIC$_{90}$ (mg/L) | |
	ATCC 90029	ATCC 96901[a]
6.0	2000	2000
6.5	1000–500	500
7.0	125–31.2	15.6

[a] Fluconazole-resistant strain.

FIGURE 14.11
Antifungal effect of various catechins against
C. albicans ATCC 90029 using resting cells with NaPB,
pH 7.0. ●, control; ■, C; ▲, ECg; ○, EGC; ◆, EGCg.

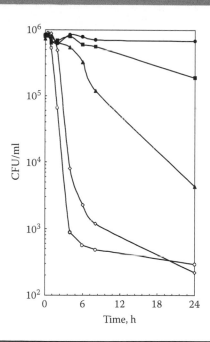

TABLE 14.4
The Effect of the Combination of Fluconazole (FLC) with EGCg on the Growth of *C. albicans* ATCC90029 and ATCC96901

		% Killed	
Strain	*FLC µg/ml*	*EGCg µg/ml*	
		0	25
90029	0	0	0
	0.125	0	92.3*
	0.25	11.8	95.6*
		0	12.5
96901	0	0	0
	10	8.7	98.5*
	50	39.5	99.9*

*, Values differ significantly ($p < 0.01$) from values without EGCg.

References

Hamada, S., and Slade, H.D. 1980. Biology, immunology and cariogenicity of *Streptococcus mutans*. *Microbiol Rev*, 44: 331–384.

Hirasawa, M., and Takada, K. 2004. Multiple effects of green tea catechin on the antifungal activity of antimycotics against *Candida albicans*. *J Antimicrob Chemother*, 53: 225–229.

Hirasawa, M., Takada, K., Makimura, M., and Otake, S. 2002. Improvement of periodontal status by green tea catechin using a local delivery system: A clinical pilot study. *J Periodont Res*, 37: 433–438.

Hirasawa, M., Takada, K., and Otake, S. 2006. Inhibition of acid production in dental plaque bacteria by green tea catechins. *Caries Res*, 40: 265–270.

Kuribayashi, S., Kono, Y., Fukatsu, A., Niki, Y., Uehara, Y., Matsumoto, H., Makimura, M., and Otake, S. 1998. Inhibitory effects of tea catechins on the adherence of genus *Streptococcus* to oral cavity (in Japanese). *Nihon Univ J Oral Sci*, 24: 231–235.

Kuroki, T., Shimpuku, Y., Saito, M., Niki, Y., Uehara, Y., Matsumoto, H., Takahashi, M., and Makimura, M. 1999. Inhibitory effects of lactic acid beverage added tea catechins on dental caries in rats (in Japanese). *Nihon Univ J Oral Sci*, 25: 476–482.

Loesche, W.J. 1986. Role of *Streptococcus mutans* in human dental decay. *Microbiol Rev*, 50: 353–380.

Lynch, D.P. 1994. Oral candidiasis: History, classification, and clinical presentation. *Oral Surg Oral Med Oral Pathol*, 78: 189–193.

McCullough, M.J., Ross, B.C., and Reade, P.C. 1996. *Candida albicans*: A review of its history, taxonomy, epidemiology, virulence attributes, and methods of strain differentiation. *Int J Oral Maxill Surg*, 25: 136–144.

Sakanaka, S. 1997. Green tea polyphenols for prevention of dental caries. In *Chemistry and applications of green tea*, ed. T. Yamamoto, et al., 81–101. Boca Raton, FL: CRC Press.

Sakanaka, S., and Yamamoto, T. 1997. Inhibitory effects of green tea polyphenols on growth and cellular adherence of a periodontal disease bacterium *Porphyromonas gigivalis*. In *Chemistry and applications of green tea*, ed. T. Yamamoto, et al., 103–108. Boca Raton, FL: CRC Press.

Sato, S. 1995. Inhibitory effects of fluoride and tea catechins on dental caries in rats (in Japanese). *Nihon Univ J Oral Sci*, 21: 341–349.

Slots, J., Bragd, L., Wikström, M., and Dahlén, G. 1986. The occurrence of *Actinobacillus actinomycetemcomitans*, *Bacteroides gingivalis* and *Bacteroides intermedius* in destructive periodontal disease in adults. *J Clin Periodontol*, 13: 570–577.

Tanaka, H. 1992. Inhibitory effects of tea catechins and anti-Streptococcus mutans hen egg yolk antibody on dental caries in rats (in Japanese). *Nihon Univ J Oral Sci*, 18: 562–570.

Wada, Y. 1996. Inhibitory effects of tea catechins on the adherence of genus *Streptococcus* to oral cavity (in Japanese). *Nihon Univ J Oral Sci*, 22: 361–369.

Zambon, J.J., Reynolds, H.S., and Slots, J. 1981. Black-pigmented *Bacteroides* spp. in the human oral cavity. *Infect Immun*, 32: 198–203.

15

Nutrigenomics and Proteomics of Tea Polyphenols

Molay K. Roy and Yoshinori Mine

Contents

15.1 Introduction

Food is the primary and fundamental requirement for the survival of all living organisms. Food provides energy, supplies building blocks, and maintains growth of living subjects, including humans. Beyond basic nutrition, food also provides health benefits and protects cellular machinery against deleterious effects of endogenous and exogenous factors that compromise human wellness by exploiting basic mechanisms of cell function. Green tea polyphenols, derived from the fresh leaves of the evergreen shrub *Camellia sinensis*, are a group of flavonoids commonly known as catechins that have been very useful in maintaining good health since the early history of human beings. However, it has only been recently that modern science has started to explore the benefits of tea polyphenols in protecting against the pathogenesis of oxidative stress, obesity, diabetes, cardiovascular diseases, cancer, inflammation, osteoporosis, aging, and so on. Through epidemiological studies, clinical trials, and laboratory tests, advanced food science has linked the benefit of tea consumption with modulated gene function or protein activity. Nutri-omics is an area of food science that deals with the interaction of food components with genes, proteins, and metabolites. The omics assess profiles of genes, proteins, or metabolites and explore the effects of food components on cells, tissues, or organs. The information obtained from nutrigenomic and proteomic studies can lead to the development of specialized foods and preparation of personalized nutrient guidelines for a person or a group of people to maintain health and prevent disease. Though genomic profiles as evaluated by microarray or mRNA expression analysis have the core components in determining modulated cell function, protein profile analysis may bear the real value linking the true effect of diet on an experimental subject. Due to limitations in protein profile analysis, mainly due to the inadequacy of affordable technology, most nutrigenomic studies focus on gene expression analysis.

In a number of studies, tea polyphenols have been shown to possess regulatory effects on the biochemical processes involved in the development of physical and neurological disorders, as discussed in the earlier sections, and a great number of studies have been devoted to gene expression analysis in both cancer and noncancer subjects. Anticancer efficacy of green tea polyphenols is mostly related to its inhibitory effect on the growth of cancer cells and preventing carcinogenicity on normal cells. Apart from the anticancer properties, the antiatherogenic, antiobesity, antidiabetic, anti-inflammatory, antioxidant, or antineurodegenerative effects of tea polyphenols have been studied in various models, and the modulation of numerous genes that control cell proliferation, cell death, cell cycle, angiogenesis, transcription regulation, signal transduction, cell differentiation, and fat or bone metabolism has been examined. Starting with a basic introduction on analytical techniques currently in use or under development for nutrigenomic/proteomic studies, this chapter covers the current knowledge on tea polyphenol-modulated gene or protein functions as evidenced by nutrigenomic and proteomic studies.

15.2 Advances in Analytical Techniques in Nutri-Omics Studies

Evaluating changes in mRNA expression has remained the first step to study the flow of information from genome to proteome and metabolome. mRNA expression can be quantitatively determined by Northern blotting. This technique involves the use of electrophoresis to separate RNA samples by size, followed by detection with a hybridization probe complementary to a part of the gene or of the entire target sequence. This method has gradually been replaced by real-time reverse transcription polymerase chain reaction (RT-PCR), which involves extraction of mRNA followed by cDNA construction and finally real-time PCR amplification using sequence-specific primers for each of the genes being amplified. Both Northern blotting and RT-PCR can only analyze gene expression for a limited number of genes at a time. This limitation is entirely overcome by two analytical approaches, based on either microarray technology or DNA sequencing. Microarray technology also includes extraction of RNA from cells/tissues, and labeling with a detectable marker (fluorescent/chemiluminescent), followed by hybridization of oligonucleotides from a target DNA sequence covalently arrayed on a solid surface of a glass/silicone bed. Once hybridization is complete, the nonspecific bonding sequences are washed off, and the remaining strongly hybridized oligonucleotide pairs are detected using a confocal laser scanner. On the other hand, in sequencing-based techniques the abundance of a particular mRNA is estimated from the count of tags derived from one particular section of its sequence. First, restriction enzymes are applied to obtain short tags of 14 to 21 bp, derived from one end of mRNA, which are then cloned and sequenced to determine the expression profile of their corresponding mRNAs.

In nutri-omics studies, the proteome analysis offers a clear picture of the effect of functional nutrients and diets in an organism, tissue, or cells at a particular moment. Conventionally, two-dimensional gel electrophoresis (2DE) followed by mass spectrometry (MS) analysis is used to study large-scale proteomes. The limitations persist in separating high molecular weight or extremely hydrophobic proteins. Gel-to-gel variation, one of the major problems of using 2DE, can be reduced by using difference gel electrophoresis (DIGE), wherein up to three protein samples can be applied in the same gel by labeling with highly sensitive fluorescent dyes (e.g., Cy3, Cy5, and Cy2) prior to separation on the gel. After image analysis, the protein of interest is spotted and subjected to proteolytic cleavage. Utilizing a peptide mass fingerprint (PMF) approach or tandem MS, the digested peptides are identified by queuing databases of protein sequences. The new proteomic approaches utilizing array technology are also gaining popularity in identifying specific protein biomarkers. One approach may be utilizing antibody arrays, wherein a great number of specific antibodies can be immobilized on the surface of specific substrate. In the development of antibody arrays, the first approach may be the creation of a 96-well antibody array plate designed for standard enzyme-linked immunosorbent assays (ELISAs).

15.3 Green Tea Polyphenol Interacts with Multiple Receptors to Modulate Cell Function

The physicochemical nature of tea polyphenols or its metabolites, as discussed in earlier chapters, has enabled these molecules, particularly epigallocatechin gallate (EGCG), to interact with various intra/extracellular components, including carbohydrates, proteins, lipids, nucleic acids, and other molecules. This interaction is believed to be closely related to the physiological functions of tea polyphenol (Ninomiya et al., 1997). Since the identification of the laminin receptor, several reports have indicated that green tea polyphenol, particularly EGCG, may interact with this target to promote antimetastasis (Suzuki and Isemura, 2001) and antiproliferation (Tachibana et al., 2004) in cancer cells, antidegranulation effects on the cytoskeleton (Fujimura et al., 2006), antiallergic effects on human basophilic KU812 cells (Fujimura et al., 2008), and inhibition of adipocyte (3T3) proliferation (Ku et al., 2009). Tea polyphenols have also been shown to interact with low-density lipoprotein (LDL) receptors (Bursill et al., 2001) on HepG2 cells, and the membrane type 1 matrix metalloproteinase (MT1-MMP) receptor (Dell'Aica et al., 2002), exert anti-inflammatory effects via Toll-like receptor 4 (Lee et al., 2004), and inhibit smooth muscle adhesion and migration via integrin β1 receptor (Lo et al., 2007). In addition, EGCG has also been shown to modulate the expression and activity of a variety of growth factor receptors, including epidermal growth factor receptor (Adachi et al., 2008; Kim, C.Y., et al., 2009), platelet-derived growth factor receptor (PDGFR) (Chen and Zhang, 2003; Sachinidis et al., 2000), and vascular endothelial growth factor receptors-1 and -2 in human endothelial cells (Lamy et al., 2002; Neuhaus et al., 2004). Catechins were also found to inhibit expression of peroxisome proliferator-activated receptor (PPAR) gamma2 in 3T3 cells (Furuyashiki et al., 2004). Moreover, direct association between EGCG and DNA/RNA is believed to modulate cell function (Kuzuhara et al., 2006).

15.4 Green Tea Polyphenol Alters Gene and Protein Expression Profile in Cancer Cells

Cancer is a disease that is widely linked with abnormal gene and protein activity. Numerous investigations, including laboratory tests, clinical trials, and epidemiologic studies, have indicated that dietary nutrition has profound effects on the incidence of cancer. Increased consumption of green vegetables, colored fruits, fermented milk, and phytochemicals is widely shown to lower the incidence of cancer in the oral cavity, esophagus, gastrointestinal tract, lung, liver, kidney, and pancreas. Gene and protein profile analysis on cancer cells/subjects exposed to tea polyphenols indicated that polyphenols modulate the expression of various genes and proteins that are critical for control of cell proliferation, apoptosis, angiogenesis, invasion, and metastasis (Table 15.1). Since occurrence of cancer is mostly associated with irregular gene functioning, protein performance data obtained from the profile-based analysis may have significant impact in designing prototypes of future chemoprevention drugs containing green tea catechin.

15.4.1 Regulation of Genes or Proteins Related to Cell Proliferation and Apoptosis

Genetic abnormalities such as the activation of oncogenes or inactivation of tumor suppressor genes results in hyperactive growth and division of cancer cells, and protection against apoptosis. Thereby, research focusing on cancer prevention includes investigating effectiveness of a potential component in suppressing cell proliferation and inducing apoptosis in cancer cells via increased expression of tumor suppressor/death genes or inactivation of tumor promoter genes or gene products (Berletch et al., 2008). Until recently, a great number of studies have focused on the ability of green tea polyphenols,

TABLE 15.1
Effect of Green Tea Polyphenols on Cancer-Related Cell Signaling

Type of Cancer	Genes	Pathway/Gene Function	Effect
Bladder cancer	Tmepai	Androgen/cell growth	Down
(TCCSUP)	Wnt2	Hedgehog/cell growth	Down
(Philips et al., 2009)	Wisp1	Wnt/cell survival	Down
Breast cancer	MMP-9	AKT/cell growth, cell invasion	Down
(Sen et al., 2010; Thangapazham et al., 2007)			
Breast cancer	Tissue inhibitor of MMP 1 (TIMP-1)	Inhibit tissue–MMP-1	Up
(Sen et al., 2010)			
Breast cancer	hTERT	Cell proliferation	Down
(Berletch et al., 2008)			
Oral carcinoma	p21^{WAF1}	CDK inhibitor/promote growth arrest and apoptosis	Up
(OSC2)			
(Hsu et al., 2005)			
Bronchial epithelial	BMPR2, SMAD7	BMP signaling/cell survival	Down
21BES cells	MAP3K8, MAPK-APK3	MAPK pathway	Down
(Vittal et al., 2004)	CDH3, DUSP5, FKBP5, FOXP1, ZNF184	BMP signaling	Up
Prostate cancer	Tyrosine receptor kinase type E mRNAb	Inhibit proliferation	Up
(LNCaP)	Phosphoglycerate kinase	Metabolic biosynthesis	Up
(Wang and Mukhtar, 2002)	Adenylate kinase 2A		Up
	CDK8 protein kinase	Regulates cell cycle	Up
	Putative serine/threonine protein kinase	Cell shrinkage	Up
	Ribosomal protein kinase B	Metabolic biosynthesis	Up
	Mevalonate kinase	Metabolic biosynthesis	Up
	Protein tyrosine phosphatase	Regulates phosphorylation in signaling molecules	Up
	Prostatic acid phosphatase	Deactivates erbB-2 and p38 MAP kinase	Up
	Receptor type protein tyrosine phosphatase g	Tumor suppressor	Up
	Protein tyrosine phosphatase 1C	Tumor suppressor	Up
	STE-20-related kinase SPAK	Cell differentiation	Up
	IAR/receptor-like protein tyrosine phosphatase		Up
	Pyrroline 5-carboxylate synthase	Inhibit cell proliferation/ survival	Up
	Glomerular epithelial protein 1 (GLEPP1)		Up
	Platelet-derived growth factor A type receptor		Up
	Protein kinase C-a b	Promote intracellular cyclic-nucleotides signaling cascade	Down

TABLE 15.1 (Continued)
Effect of Green Tea Polyphenols on Cancer-Related Cell Signaling

Type of Cancer	Genes	Pathway/Gene Function	Effect
	41 kDa protein kinase related to rat ERK2b	Promote cell growth by the MAPK and PI-3 kinase pathways	Down
	Type I b cGMP-dependent protein kinase b	Promote intracellular cyclic-nucleotides signaling cascade	Down
	Adenosine kinase short form	Promote intracellular cyclic-nucleotides signaling cascade	Down
	Phosphatidylinositol 3-kinase homolog	Promote cell growth by the MAPK and PI-3 kinase pathways	Down
	Protein tyrosine phosphatase PIR1		Down
	Protein tyrosine phosphatase zeta	Promote cell migration	Down
	KIAA0369 gene	Promote cell proliferation	Down
	Leukocyte common antigen T200	Cell surface component	Down
Lung cancer cell (PC-9)	Retinoic acid receptor alpha 1 gene		Up
(Fujiki et al., 2001)	NF-kappaB-inducing kinase (NIK)	Promote NFkappaB signaling	Down
	Death-associated protein kinase (DAPK 1)	Tumor suppressor	Down
	Rho B		Down
	Tyrosine-protein kinase (SKY)		Down
Colorectal cancer	VEGF	Promote angiogenesis	Down
(SW837)	HIF-1 alpha		Down
(Shimizu et al., 2010)	IGF-1	Promote cell growth and proliferation	Down
	IGF-2	Promote cell growth and proliferation	Down
	EGF		Down
	Heregulin		Down
Prostate cancer	Inhibitor of DNA binding 2 (ID2)		Down
Lung cancer cell line	NF-kB-inducing kinase (NIK)		Down
(PC-9)	Death-associated protein kinase 1 (DAPK1)		Down
(Okabe et al., 2001)			
	CDC42 GTPase-activating protein		Down
	Envoplakin		Down
	MAP kinase p38g		Down
	CDC25B/M phase inducer phosphatase 2		Down
	Tyrosine-protein kinase (SKY)		Down
	Rho B		Down
	T-Lymphoma invasion and metastasis		Down
	Inducing TIAM1		Down

Continued

TABLE 15.1 (Continued)
Effect of Green Tea Polyphenols on Cancer-Related Cell Signaling

Type of Cancer	Genes	Pathway/Gene Function	Effect
	Disheveled 1 (DVL1)		Down
	Synapse-associated protein 102 (SAP102)		Down
	Epidermal growth factor receptor (EGFR)		Down
	Retinoblastoma binding protein (RBQ1)		Up
	Vascular endothelial growth factor (VEGF)		Up
	Retinoic acid receptor a1 (RAR-a1)		Up
	Insulin-like growth factor-binding protein 3 (IGFBP 3)		Up
CaSki cells	Cyclin G-associated kinase		Down
(Ahn et al., 2003)	ATPase		Down
	NADH dehydrogenase (ubiquinone) 1 beta		Down
	DKFZP586O0120 protein		Down
	CD83 antigen (activated B lymphocytes)		Down
	HSPC135 protein		Down
	KIAA0250 gene product		Down
	DKFZP586J0917 protein		Down
	Membrane protein, palmitoylated 1 (55 kDa)		Down
	Ras homolog gene family, member G (rho G)		Down
	KIAA0210 gene product		Down
	KIAA0256 gene product		Down
	KIAA0015 gene product		Down
	DKFZP564I1922 protein		Down
	Expressed in activated T/LAK lymphocytes		Down
	Dual-specificity phosphatase 1		Down
	Solute carrier family 25 (mitochondrial carrier)		Up
	Vimentin		Up
	Ribosomal protein L19		Up
	Mitogen-activated protein kinase kinase kinase kinase 3		Up

particularly EGCG, to inhibit proliferation and induce apoptosis in *in vitro* and *in vivo* models of various cancers, including prostate cancer (Gupta et al., 2000), liver cancer (Huang et al., 2009), bladder tumor (Philips et al., 2009), colon cancer (Hwang et al., 2007), leukemia (Harakeh et al., 2008), lung cancer (Yamauchi et al., 2009), breast cancer (Thangapazham et al., 2007), ovarian cancer (Huh et al., 2004), and oral carcinoma (Hsu et al., 2005a), wherein increased expression of pro-apoptotic or death genes/proteins such as Bax, Bak, Bad, Mtd, or MH3, or inhibition of cell death suppressor genes such as Bcl-2, Bcl-xL, Bcl-w, Mcl-1, or A1 were emphasized. Upregulation of pro-apoptotic genes such as Bax, caspase-6, Fas ligand, and cell cycle inhibitor *GADD45*, and downregulation of Bcl-2 and Bcl-xL were observed in neuroblastoma cells exposed to a cytotoxic dose of EGCG. Tumor suppressor p53, which requires activation to transmit signals to downstream effector proteins such as p21, was shown to be upregulated in a variety of cancer cells exposed to tea polyphenol. Qin et al. (2008) demonstrated that EGCG activated tumor suppressor p53 and promoted apoptosis in JB-6 cell lines. The investigators found that EGCG-induced serine/threonine phosphatase-2A (PP-2A) inhibition enhanced p53 phosphorylation at Ser15 and upregulated Bak expression to promote EGCG-induced apoptosis. EGCG's ability to induce apoptosis in the absence of p53 was also proven in MEF p53−/− cells, wherein EGCG caused activation and translocation of Bax protein into the mitochondria (Shankar et al., 2007). Recent research focusing on the ability of tea polyphenols to induce proliferation inhibition and apoptosis in cancer cells includes genome-wide analysis of cancer cells. In one study, Luo and colleagues (2010) utilized Affymetrix Hu133 2.0 gene chip and found that the top 20 downregulated genes were involved in transcription regulation, proteolysis, and cellular defense systems, whereas the top 20 upregulated genes were involved in nutrient/ion transfer, phosphorylation, cell motility, and inflammatory response. Among the downregulated genes, ID2, which inactivates Rb protein and allows the cell cycle to progress, was downregulated fourfold. Similarly, utilizing an RT2 profiler PCR array constituting 84 genes, including 18 genes specific for signal transduction pathways, Philips et al. (2009) have shown that EGCG treatment of tumorigenic bladder cancer cell lines downregulated several genes, including *Tmepai* and *Wnt2*, which promotes cell survival and growth, indicating EGCG's ability to control cell growth at the genome level. Viral oncogenes E6 and E7 play critical roles for promotion of carcinogenesis, progression, invasion, and metastasis of cervical cancer; thereby, inhibition of these genes may be important for prevention and treatment of cervical cancer. In a recent study, Qiao et al. (2009) have shown that EGCG suppressed E6 and E7 mRNA expression in addition to aromatase and ERalpha in cervical cancer cells. The expression of E6 protein was also confirmed by protein analysis. Fluorescence-activated cell sorting (FACS) and microarray analysis by Huh et al. (2004) demonstrated that SKOV-3 cells treated with ECGC results in upregulation of p21, Bax, cyclin G genes and p21, Bax proteins accompanied with downregulation of cyclin-dependent kinase (CDK) 6, E2F-4, cyclin A genes and Rb, cyclin D1, CDK2, E2F-1, E2F-4, Bcl-xL, PCNA, and CD4 proteins. *NUDT6* gene is known to have a role in modulating proliferation in response to growth signals via the protein itself. Sukhthankar et al. (2010) have analyzed the *NUDT6* gene using array technology and found that ECG downregulated gene expression by modulating RNA stability of the NUDT6 transcript through the p38 mitogen-activated protein kinase (MAPK) and extracellular signal-regulated kinase (ERK) pathways. It is also widely acknowledged that microRNA (miRNA), which is composed of 20 to 25 nucleotides, regulates numerous fundamental cellular processes and disease development (Sun and Tsao, 2008). Since 2% of all known human genes encode miRNAs, it is speculated that miRNAs might be deregulated in proliferative diseases, such as cancer. In recent studies, miRNA has become a new layer of gene regulation to mediate the anticancer effects of green tea polyphenols. Using microarray technology, Tsang and Kwok (2010) have investigated miRNA expression in EGCG-treated human hepatocellular carcinoma HepG2 cells, and found miR-16, one of the 13 upregulated miRNAs, was confirmed to be targeting the Bcl-2 gene, since EGCG's effect on Bcl-2 downregulation and apoptosis induction was prevented in the cells transfected with anti-miR-16 inhibitor, which also suppressed miR-16 expression.

15.4.2 Regulation of Cell Cycle Checkpoints

Cell cycle events comprised of G1, S, and G2/M phases are tightly regulated by several cyclins and their catalytic counterparts, CDKs. Cyclin D1, the catalytic component of CDK1 and CDK6, plays key roles in traversing cells from the G1 to the S phase. A wealth of studies have proven that EGCG, the main counterpart of tea polyphenol, suppresses the growth of a variety of cancer cells, including colon cancer, breast cancer, lung cancer, prostate cancer, and ovarian cancer, by arresting cell cycle at the G1 phase (Shimizu et al., 2008), S phase (Huh et al., 2004), or G2/M phase (Qiao et al., 2009). Utilizing Western blotting, RT-PCR, microarray or transient transfection, and luciferase reporter technology, several studies have identified molecular targets of tea polyphenols involved in the activation of cell cycle regulator genes or in modulating kinase activity, stopping one or more steps of cell cycle checkpoints to suppress the growth and accelerate death of cancer cells. Ahn and colleagues (2003) found that EGCG at 35 μM arrests cell cycles at the G1 phase that precedes apoptosis, with altered gene expression in the EGCG-exposed human cervical carcinoma cell line, CaSki. When tested using 384 cDNA microarray, four genes were found to be upregulated

(>2-fold), and 16 genes were found to be downregulated (>2-fold), in particular cyclin G-associated kinase gene expression, which plays a key role in arresting cells at the G1 stage by interrupting transition to the S phase by activated cyclin G. FACS and microarray analysis by Huh et al. (2004) demonstrated that the cell cycle arrest at the G1 phase observed when SKOV-3 cells were treated with EGCG was associated with upregulation of p21, Bax, and cyclin G genes and p21 and Bax proteins, accompanied by a downregulation of CDK6, E2F-4, cyclin A genes and Rb, cyclin D1, CDK2, E2F-1, E2F-4, Bcl-xL, PCNA, and CD4 proteins. Shankar et al. (2007) have shown that the EGCG-induced growth arrest at the G1 stage in pancreatic cells is associated with induction of cycle inhibitors p21/WAF1/CIP1 and p27/KIP1, and inhibition of the expression of cyclin D1, CDK4, and CDK6. In p53-positive Hep G2 cells, EGCG blocked the progression of cell cycle at the G1 phase by inducing p53 expression and further upregulating p21 expression (Huang et al., 2009). EGCG-induced cell cycle arrest at the G1 phase in human head and neck squamous cell carcinoma (HNSCC), breast cancer, and colon cancer cell lines appeared to be related to the inhibition of ERK activation and TGF-α-stimulated c-fos and cyclin D1 promoter activity, resulting in reduced levels of cyclin D1 and Bcl-xL, whereas activation of EGFR, HER2, and HER3 and their multiple downstream signaling molecules were shown downregulated (Shimizu et al., 2008). Qiao et al. (2009) have observed that EGCG induced cell cycle arrest at the G2/M phase in CaSki (HPV16-positive) cells, wherein downregulation of E6/E7 oncogenes was demonstrated. Green tea extract-induced G2/M phase cell cycle arrest and apoptosis induction in U937 cells was found to be associated with upregulation of p21/Waf1 mRNA/protein expression (Ohata et al., 2005).

15.4.3 Regulation of Genes/Proteins Related to Cell Signaling

Since the development of cDNA microarray techniques it has become a useful tool to monitor expression levels of numerous cell signaling molecules simultaneously. In 2001, Okabe et al. from Saitama Cancer Research Center, Saitama, Japan, used a cDNA expression array constituting 588 genes involved in signal transduction pathways to assess the effect of EGCG on the human lung cancer cell line PC-9. In the study, PC-9 cells were treated with 200 μM of EGCG for 7 h, and the gene expression profile was analyzed using an Atlas Human Cancer cDNA Expression Array. Out of 588 genes, nontreated control PC-9 cells expressed 163 genes, of which 12 genes, including NF-kappaB-inducing kinase and MAPK p38gamma, were downregulated by EGCG, supporting involvement of p38-mediated NF-kappaB cell signaling pathway in EGCG-induced lung cancer prevention. Based on this evidence, it was concluded that downregulation of NIK gene expression involved in inhibition of NF-kappaB activation is significant for cancer prevention (Okabe et al., 2001). Utilizing cDNA microarray technology, Wang and Mukhtar (2002) investigated the expression of 250 kinase and phosphatase genes involved in various cell signaling pathways regulating proliferation, cell cycle, and apoptosis. The results indicated that treatment of LNCaP cells with 12 μM EGCG for 12 h induced proliferation inhibition and was related to modulation of G-protein signaling networks (Wang and Mukhtar, 2002). Also, Phillips et al. (2009) have used a Human Signal Transduction RT[2] Profiler array consisting of 84 key genes from 18 distinct signaling pathways to study the effect of EGCG on cell signaling pathways, including Wnt, TGB-beta, NF-kappaB, p53, protein kinase C, and Hedgehog. The results indicated that EGCG-mediated anticancer effects may be due to the modulation of multiple cell signaling pathways involved in cell proliferation and tumorogenesis. For example, gene expression of *Tmepai*, a gene of the androgen signaling pathway, was downregulated fourfold. Also, the *Wnt2* gene, which is known to be involved in Hedgehog signaling and development of some cancers, was also downregulated. Similarly, *Wisp-1*, a key regulator of WNR cell survival signaling, was found downregulated in the EGCG-treated cancer cells. On the other hand, EGCG was shown to upregulate *Ccl20* and *Il-8*, the two key components of the NF-kappaB signaling pathway, an observation that may be considered as a pro-oxidant response of EGCG on cancer cell lines. Utilizing a cDNA microarray, Vittal et al. (2004) have established the involvement of the bone morphogenetic protein (BMP) signaling pathway in EGCG-mediated anticancer effects. In the study, EGCG treatment resulted in downregulation of BMP receptors (BMPR2) and upregulation of BMP's negative modulators, such as CDH3, DUSP5, FKBP5, and FOXP1. Shim et al. (2010) conducted proteomic analysis using 2D gel electrophoresis and matrix-assisted laser desorption/ionization-time-of-flight (MALDI-TOF) mass spectrometry, and have shown that EGCG interacted with the Ras-GTPase-activating protein SH3 domain-binding protein 1 (G3BP1) and suppressed the activation of Ras. The study found that EGCG effectively attenuated G3BP1 downstream signaling, including extracellular signal-regulated kinase and mitogen-activated protein kinase/extracellular signal-regulated kinase in wild-type H1299 and shMock H1299 cells. In addition, a wealth of studies have utilized RT-PCR, Western blot, and transduction/transfection technology to monitor the modulation of cell signaling pathways in a variety of cancer cells. Such studies also provide evidence supporting modulation of numerous other signaling pathways, including NF-kappaB, AP-1, MAPKs, PKB/Akt, COX, p53, TNF-alpha, or STAT, in EGCG-mediated anticancer mechanisms (Roy et al., 2009).

15.4.4 Regulation of Genes/Proteins Related to Angiogenesis and Metastasis

Angiogenesis involves the growth and development of new blood vessels from the preexisting ones to supply nutrition, which is unavoidable for the growth of a tumor. A growing body of evidence supports that upregulation of several growth factors, such as vascular endothelial growth factor (VEGF), transforming growth factor alpha (TGF-alpha), basic fibroblast growth factor (bFGF), proteins Ang1 and Ang2, interleukin 8 (IL-8), and Dll4, and downregulation of thrombospondin-1 are features of tumors undergoing angiogenic changes. Thus, antiangiogenic studies mostly include investigating mechanisms involved in opposing the activities of the above factors or related pathways. Shankar and colleagues (2008) have shown that EGCG inhibition of capillary tube formation in HUVEC cells is associated with increased activation of ERK. The study further revealed that EGCG treatment resulted in reduced expression of VEGF-positive tumor cells in nude mice (Shimizu et al., 2010). Poly E green tea extract, of which the predominant component is EGCG, inhibited angiogenesis in MDA-MB231 breast cancer and human dermal microvascular endothelial (HMVEC) cells, wherein disruption of STAT3-mediated transcription of genes, including VEGF and MMP9, was observed (Leong et al., 2009). EGCG inhibition on tubular structure formation of endothelial cells was shown to be mediated via VEGF signaling consisting of a phosphoactivation of VEGF receptor and vascular endothelial (VE)-cadherin, inhibition of Akt phosphorylation and IL-8 production, and interference with VEGF-induced VE-cadherin/b-catenin complex (Liu et al., 2008). To better understand the molecular mechanisms involved in EGCG's inhibitory effect on tumor cell migration, invasion, and angiogenesis, Liu et al. (2008) tested the effect of EGCG on gene expression in endothelial cells using an Affymetrix microarray system. Out of 14,500 human genes, 420 genes were identified as unregulated, whereas 72 genes were downregulated, and involvement of Wnt and Id signaling pathways was demonstrated in the EGCG-mediated antiangiogenesis efficacy. Mast cells are involved in various physiological processes, including tumor development and angiogenesis, and act by releasing many kinds of chemical modulators, including various growth factors, cytokines, histamine, heparin, proteoglycans, tryptase, and chymase, several of which are involved in the progression of angiogenesis. Using cDNA microarray technology,

Melgarejo et al. (2007) have studied the effect of EGCG on the expression of angiogenesis-related genes in human mast cell line-1 (HMC-1) cells. The study found that EGCG treatment decreased the expression of nine genes by more than 1.3-fold, five of which (EFNA2, FGF6, GRO1, PDGFA, and PF4) encode for growth factors, two (ITGA5 and ITGB3) for adhesion molecules, one (MCP1) for a chemokine, and one (ERBB2) for a transcription factor, which are believed to decrease adhesion of mast cells with reduced activity to generate signals eliciting monocyte recruitment.

15.5 Regulation of Genes Related to Antioxidants and Antiaging

Mechanisms involved in tea polyphenol-mediated antioxidant effects include direct scavenging of ROS, chelation of redox-sensitive transition metals, inhibition of redox-sensitive transcription factors such as NF-kappaB, AP-1, and their downstream effectors, including lipoxygenase, xanthine oxidase, COX-2, and iNOS, and upregulation of antioxidant/phase II enzymes via the Nrf2 pathway. Meng and colleagues (2008a) have provided substantial evidence showing EGCG treatment at 25 to 50 µM for 24 h increases catalase, superoxide dismutase (SOD) 1, SOD2, and glutathione peroxidase gene expression and their enzyme activities, which also enhances the mitochondrial integrity in old PDL (>45) human diploid fibroblasts (HDFs), suggesting that EGCG-mediated antioxidant gene modulation may contribute to delay the aging process. To further examine EGCG-mediated antioxidant/antiaging effects, the same research group utilized a cDNA array consisting of 22,227 genes to determine the gene expression profile in the liver, kidney, heart, and brain of middle-aged male Fischer 344 rats receiving EGCG (Meng et al., 2008b). The microarray data revealed that EGCG modulated 76 genes, which were mostly involved in the metabolism of lipid, glutathione, hexose biosynthesis, and response to stress, indicating that EGCG may alleviate aging-related oxidative stress and oxidative damage. Hibernating mechanisms in certain species of animals are thought to hinder their aging process. The EGCG-mediated hibernating effect in human keratinocytes is associated with high cell survival featuring phosphorylation of Ser112 and Ser136 of Bad protein via Erk and Akt pathways and an increase in the Bcl-2-Bax ratio (Chung et al., 2003), protecting against the lethal effect of UV radiation. EGCG-mediated hibernation effects in normal cells not exposed to any radiation or extracellular stress may utilize a different mechanism. Bae et al. (2009)

examined 54,675 human genes to study the EGCG-mediated gene expression profile in neonatal human tarsal fibroblasts (nHFTs). The investigators observed that EGCG induced a temporal cell cycle arrest at the G0/G1 phase, which was accompanied with modulation of numerous genes associated with multiple cell functions, including metabolism, physiology, regulatory, cell, and signaling, in particular to the genes related with the G1, G1/S, and G2/M phases (Table 15.2). Therefore, it is speculated that EGCG-mediated temporal cell cycle arrest or halt in a metabolic process may have a profound effect on cellular life, slowing down the aging process and increasing the life span.

15.6 Regulation of Genes Related to Drug Metabolism

Drug metabolism is a biochemical process that leads to modification of pharmaceutical agents into more readily excreted forms. It is considered that tea polyphenol's *in vivo* antioxidant efficacy is mostly based on its ability to induce phase II and antioxidant enzymes, involving upregulation of a variety of metabolic genes, including human NAD(P)H: quinone oxidoreductase-1 (NQO1), glutathione S-transferase (GST), heme oxygenase (HO), and quinine reductase (QR) (Yang et al., 2006). The molecular events that lead to induction of these genes appear to involve activation of transcription factor Nfr2 and its subsequent binding to and activation of electrophile/antioxidant response elements (EpRE/ARE). To investigate the in-depth mechanisms of activation of potential response elements of the induced genes, Yang et al. have utilized a GEArray Q series human drug metabolism gene array: HS-011 containing 96 genes of drug metabolism and stress response. They found that treatment of cultured human hepatoma (HepG2) cells with green tea extract resulted in dramatically increased expression of at least 15 genes, of which microsomal glutathione S-transferase gene (MGST1), histone acetyltransferase 1 (HAT1), acetyl-coenzyme A acetyltransferase 1 (ACAT1), choline acetyltransferase (CHAT), and histamine N-methyltransferase (HNMT) genes were noted.

15.7 Regulation of Genes Related to Parkinson's Disease

The biological effects of tea polyphenols, including radical-scavenging, iron-chelating, anticarcinogenic, and anti-inflammatory actions, have been

TABLE 15.2
Effect of EGCG on Genes Related to Aging

Gene	Function	
Cyclin A2 (CCNA2)	Essential for G1/S phase and G2/M phase transition	Down
Cyclin B1 (CCNB1)	Essential for mitosis, associate with CDC2	Down
Cyclin B2 (CCNB2)	Essential for mitosis, associate with CDC2	Down
Cyclin D1 (CCND1)	Essential for G1 phase, with CDK4 and CDK6	Down
Cyclin D3 (CCND3)	Essential for G1 phase transition, with CDK4 and CDK6	Down
Cyclin E2 (CCNE2)	Essential for G1/S phase transition, with CDK2	Down
Cyclin F (CCNF)	Essential for cell cycle transitions	Down
Cyclin-dependent kinase 1 (CDK1)	Complex with CCNA and B, essential for G2/M phase transition	Down
Cyclin-dependent kinase 2 (CDK2)	Complex with CCNA and E, essential for G1/S phase transition	Down
Cyclin-dependent kinase 6 (CDK6)	Complex with CCND1, D2, and D3, essential for early G1 phase	Down
CDK inhibitor 1A (p21, Cip1)	Inhibitor of CCN-CDK2 or -CDK4 complexes	Up
CDK inhibitor 1C (p57, Kip2)	Inhibitor of several CCN-CDK complexes	Up
CDK inhibitor 2B (p15)	Inhibitor of CDK4 and CDK6	Up
CDK inhibitor 2D (p19)	Inhibitor of CDK4 and CDK6	Up
Cyclin-dependent kinase 3 (CDK3)	Complex with CCNA and B, essential for G2/M phase transition	Up
Cyclin-dependent kinase 4 (CDK4)	Complex with CCND1, D2, and D3, essential for G1/S phase transition	Down
Cyclin-dependent kinase 7 (CDK7)	Complex with CCNH and MAT1	Down
Cyclin-dependent kinase 8 (CDK8)	Complex with CCNC	Down
Cyclin-dependent kinase 9 (CDK9)	Complex with CCNT and CCNK	Down
Cyclin-dependent kinase 10 (CDK10)	Regulator of G2/M phase	Down
CDK inhibitor 1B (p27, Kip1)	Inhibitor of CCNE-CDK2 or CCND-CDK4 complexes	Down
CDK inhibitor 2A (p16)	Inhibitor of CDK4	Up
CDK inhibitor 2C (p18)	Inhibitor of CDK4 and CDK6	Down
CDK inhibitor 3	Inhibitor of CDK2	Down

Source. Adapted from Bae, J.Y., et al., *Cell Transplant*, 18: 459–469, 2009.

widely investigated. Recent studies have also demonstrated EGCG's capacity to improve spatial cognition learning ability in rats (Haque et al., 2008) and to lessen cerebral amyloidosis in AD transgenic mice (Rezai-Zadeh et al., 2005). In accordance with *in vivo* findings, several levels of evidence have also proven that EGCG prevented parkinsonism-inducing neurotoxin 6-hydroxydopamine (6-OHDA) and 1-methyl-4-phenylpyridinium (MPP+)-induced cytotoxicity in neuronal cell lines, such as PC12, SH-SY5Y, or in primary hippocampal neurons. In a previous contribution we have accumulated evidence showing that the protective role of EGCG is linked to prevention of toxin-induced cell death via upregulation of Bax, Bad, and Mdm2, and downregulation of Bcl-2, Bcl-w, and Bcl-x(L). We have also emphasized that the antiapoptotic effect of EGCG is linked to the upregulation of the PI3k/Akt pathway in neuronal cells (Roy et al., 2009). Although green tea catechins have been shown to be potent antioxidants *in vitro*, there exist a large number of publications suggesting that green tea catechin's neuroprotective efficacy is mostly related to modulation of signal transduction pathways. In a recent study, Ma and colleagues (2010) utilized genome-wide microarray technology to interrogate 25,000 oligonucleotide coded microarrays to characterize genomic pathways that respond to the neurotoxin 6-OHDA and MMP+ in A53T alpha-Syn expressing SH-SY5Y cells to determine how the genes were affected by EGCG. The investigators found that EGCG inhibited transcript changes associated with 6-OHDA treatment, of which GCLM (glutamate-cysteine ligase, modifier subunit), HMOX1 (heme oxygenase decycling 1), MAFF (v-maf musculoaponeurotic fibrosarcoma oncogene homolog F), NQO1 (NAD(P)H dehydrogenase, quinone 1), NQO2 (NAD(P)H dehydrogenase, quinone 2), TXNRD1 (thioredoxin reductase 1), and the Nrf2/keap target genes are noted. In various neurological disorders, including Alzheimer's and Parkinson's disease, protein misfolding and formation of amyloid fibrils contribute to cellular toxicity. Recently, Bieschke et al. (2010) have shown that EGCG has the capacity to alter the size and activity of the large, grown alpha-synuclein and amyloid-beta fibrils into lesser, amorphous protein aggregates that are harmless to mammalian cells, and this action was found to be via direct binding to beta-sheet-rich amyloid aggregates that mediate the conformational change of the protein. Deposition of cerebral beta-amyloid (A*beta*) that originates from the proteolytic (by alpha-, beta-, and gamma-secretases) processing of an amyloid precursor protein (APP) is the primary cause of Alzheimer's disease (AD). Lin et al. (2009) demonstrated that EGCG reduced A*beta* levels by enhancing endogenous nonamyloidogenic proteolytic processing of APP in MC65 cells stably transfected with the APP-C99 gene. EGCG also decreased nuclear translocation of c-Abl and inhibited APP-C99-dependent GSK3-beta activation by interrupting of c-Abl/Fe65 interaction. It is known that lipopolysaccharide (LPS)-induced memory disorder occurs through the accumulation of A*beta* via the increase of beta- and gamma-secretase activity. Lee et al. (2009) suggested that EGCG may prevent elevation of A*beta* via the inhibition of beta- and gamma-secretases and suppression of inducible nitric oxide synthetase (iNOS)- and cyclooxygenase-2 (COX-2)-mediated neuronal cell death associated with inhibition of ERK and NF-kappaB activation. H.J. Kim et al. (2009) have also shown that EGCG prevented A*beta*-induced oxidative or nitrosative cell death in BV2 microglia by suppressing pro-apoptotic signals, increasing expression of inducible nitric oxide synthase (iNOS) and subsequent production of nitric oxide (NO) and peroxynitrite by increasing cellular GSH levels and mRNA expression of gamma-glutamylcysteine ligase (GCL), the rate-limiting enzyme in the glutathione biosynthesis. Maintenance of mitochondrial membrane potential, inhibition of caspase-3 activity, and downregulation of the expression of pro-apoptotic protein Smac in cytosol were observed in EGCG-mediated neuroprotection in PC-3 cells exposed to neurotoxin paraquat (PQ) (Hou et al., 2008). Proteomic technology has been used to analyze EGCG-mediated protective mechanisms against long-term serum starvation-induced oxidative insult/cytotoxicity on model SH-SY5Y cells. Three-day serum-starved SH-SY5Y cells were treated with 0.1 and 1 µmol EGCG, and the proteins from the EGCG-treated and untreated control cells were subjected to two-dimensional polyacrylamide gel electrophoresis (2D-PAGE) and mass spectrometry. In the study, EGCG was found to increase the level of cell signaling binding protein 14-3-3 gamma that regulates various cellular events, including expression of cytoskeleton protein and prevention of apoptosis. In fact, EGCG was found to induce beta-tubulin IV and tropomyosin 3, which take part in cell assembly. Additionally, EGCG decreased the expression of both mRNA and protein prolyl 4-hydroxylase (beta subunit) that negatively regulates several cell survival/differentiation proteins, supporting the involvement of EGCG in other mechanisms of neuroprotection (Weinreb et al., 2007).

15.8 Regulation of Genes Related to Metabolic Diseases

The antidiabetic, antiobesity, and antiatherosclerosis efficacy of green tea polyphenols is widely established. It is widely acknowledged that EGCG mimics various actions of insulin, including inhibition of hepatic glucose production and suppression of PEFCK and G6Pase mRNA in several *in vitro* and *in vivo* models. Recently, Abe et al. (2009) have utilized microarray technology to investigate the effect of chronic consumption of green tea beverage on gene profiles consisting of several metabolic genes responsible for hepatic gluconeogenesis and fatty acid biosynthesis. In the study, rats given tea beverage for 28 days were examined for hepatic gene expression profile changes. The results indicated that tea beverage consumption downregulated expression of genes for glucose-6-phosphatase (G6Pase) and fatty acid synthase, and upregulated expression of peroxisome proliferator-activated receptor alpha (Abe et al., 2009). Wolfram et al. (2006) have explored antidiabetic effects of EGCG *in vivo* and have assessed the effects of EGCG on the expression of genes involved in lipid and glucose metabolism in H4IIE rat hepatoma cells, as well as in liver and adipose tissue of db/db mice by utilizing microarray and RT-PCR technology. The study revealed that EGCG downregulated several genes involved in gluconeogenesis and the synthesis of fatty acids, triacylgycerol, and cholesterol. EGCG also suppressed the mRNA expression of phosphoenolpyruvate carboxykinase in H4IIE cells as well as in liver and adipose tissue of db/db mice, wherein glucokinase mRNA expression was upregulated in a dose-dependent manner. Utilizing real-time PCR and DNA microarray, Isemura et al. (2007) have also shown that EGCG suppressed gene expression of hepatic gluconeogenic enzymes in mice, providing additional evidence of EGCG's ability to modulate gene function protecting health against diabetes. In addition, rats given a diet supplemented with tea polyphenol had increased mRNA expression of a number of genes involved in glucose transport, including Glut4, Gsk3b, Irs2, and Pik3cb, that are believed to facilitate glucose transport into liver and muscle tissues (Cao et al., 2007). Moreover, tea polyphenol suppressed SGLT1, which transports glucose across intestinal epithelial cells, and enhanced adiponectin expression, which increases cellular glucose uptake, providing further evidence linking genomics in the antidiabetic effects of green tea polyphenol (Roy et al., 2009). Regarding antiobesity effects, green tea polyphenols were found to inhibit expression of ACC1, FAS, G6PDH, ME, SCD-1 mRNA, and enzymes that are involved in fatty acid and triglyceride biosynthesis. In addition, tea polyphenols enhanced beta-oxidation by increasing mRNA and protein expression of ACO and MCAD in mice given a tea polyphenol-supplemented diet. Moreover, EGCG suppression of Rstn mRNA and leptin gene expression and induction of UCP-2 gene expression in either cell or animal models have been shown to be associated with increased lipid catabolism. On the other hand, microarray data from obese mice or 3T3-L1 cells treated with specialized green tea "Benifuuki" have shown downregulation of enzyme expression associated with both the synthesis and beta-oxidation of fatty acids in the obese model, providing additional information to support that tea polyphenol-mediated antiobesity effects may be linked to modulation of gene function (Oritani et al., 2009) (Table 15.3).

Tea polyphenol-mediated antiatherosclerosis mechanisms can be attributed to its antioxidant, hypolipemic, and antifibrinolytic activities (Vinson et al., 2004). However, other mechanisms involving proliferation/migration and

TABLE 15.3
Effect of EGCG on Genes Related to Diabetes and Obesity

Model	Gene	Function	Effect
Rat (EGCG 1%)			
(Wolfram et al., 2006)			
Liver			
	Glucokinase (GK)	Phosphorylation of glucose to glucose-6-phosphate	UP
	Phosphoenolpyruvate carboxykinase (PEPCK)	Gluconeogenesis	Down
	Acyl-CoA oxidase-1 (ACO-1)	Carboxylation of acetyl-CoA to produce malonyl-CoA	Up
	Carnitine palmitoyl transferase-1beta (CPT-1)	Transport long-chain fatty acids across the membrane	Up
Adipose tissue			
	Phosphoenolpyruvate carboxykinase (PEPCK)	Phosphorylation of glucose to glucose-6-phosphate	Down
	Acyl-CoA oxidase-1 (ACO-1)	Carboxylation of acetyl-CoA to produce malonyl-CoA	Down
	Carnitine palmitoyl transferase-1beta (CPT-1)	Transport long-chain fatty acids across the membrane	Up
Rat (Liver)			
(Abe et al., 2009)			
	PEFCK	Phosphorylation of glucose to glucose-6-phosphate	Up
	Glucose-6-phosphatase (G6Pase)	Gluconeogenesis	Down
	Fatty acid synthase (FASN)	Fatty acid synthesis	Down
	Peroxisome proliferator-activated receptor (PPARalpha)		Up
	PPARgamma		Up
	Forkhead box O1 (FOXO1a)	Transcription factor	Down
	HNF4alpha		Up

regulation of endothelial cells, vascular smooth muscle cells, or fibroblasts in the three layers of major blood vessels have been emphasized in advanced antiatherogenic studies of tea polyphenols. The vascular endothelium plays a critical role in vascular function and development of atherosclerosis. Green tea polyphenols alleviate angiotensin II (Ang II)-induced hyperpermeability of vascular endothelium by decreasing expression of nicotinamide adenine dinucleotide phosphate (NADPH) oxidase that generates ROS in Ang II-exposed cells (Ying et al., 2003). EGCG prevents Ang II-induced endothelial dysfunction by preventing the expression of HO-1, p22 (phox), and SOD-1 mRNA in the aorta (Antonello et al., 2007). In endothelial cells, EGCG stimulates vasodilatation by enhancing nitric oxide production via signaling pathways requiring activation of Fyn/PI3K/Akt/endothelial nitric oxide synthase (eNOS) (Kim et al., 2007). EGCG also contributes to vasodilatation by downregulating gene expression of caveolin-1 (Cav-1), a negative modulator or eNOS, by activating ERK1/2 and inhibiting p38 MAPK signaling (Li et al., 2009). Reduction in endothelin-1 synthesis can increase bioavailability of nitric oxide. Reiter et al. (2010) have reported that EGCG treatment (10 μM, 8 h) reduced the expression of ET-1 mRNA and protein, and ET-1 secretion in human aortic endothelial cells, wherein involvement of Akt- and AMPK-stimulated FOXO1 regulation of the ET-1 promoter was emphasized. To investigate the effect of EGCG on gene expression profiles in human umbilical vein endothelial cells (HUVECs), Liu et al. (2008) utilized cDNA microarray and found that EGCG treatment upregulated 4 genes and downregulated 14 genes. Among the genes, the TGF-β-inducible early growth responses gene (TIEG), which contributes to cell proliferation, was found to be downregulated, and tryptophanyl-tRNA synthetase (WARS), which inhibits angiogenesis, was upregulated, suggesting that EGCG-mediated proliferation inhibition and antiangiogenic activity may in part contribute to endothelial function. Tissue factor (TF) is an important mediator of blood coagulation. Thrombin-stimulated TF release from endothelial cells contributes to the pathogenesis of cardiovascular diseases. Wang et al. (2010a) have shown that a 25 μmol EGCG may decrease thrombin-stimulated TF mRNA expression via inhibiting phosphorylation of c-Jun terminal NH2 kinase (JNK). The role of oxidized low-density lipoprotein (oxLDL) in the pathogenesis of vascular diseases is widely acknowledged. EGCG inhibits the oxLDL-induced LOX-1-mediated signaling pathway in HUVECs by inhibiting NADPH oxidase and consequently ROS-enhanced LOX-1 expression, which contributes to further ROS generation and the subsequent activation of NF-kappaB via the p38 MAPK pathway (Ou et al., 2010).

Abnormal proliferation, invasion, and migration of vascular smooth muscle cell (VSMC) plays a critical role in the pathogenesis of cardiovascular diseases. Tea polyphenol-mediated proliferation inhibition of VSMC is partly linked with G1 phase cell cycle arrest mediated by increased expression of p21[WAFI] gene, and induction of apoptosis via p53-mediated NF-kappaB activation. Green tea extract inhibits thrombin-induced VSMC invasion by preventing MMP-2 expression and its activation by inhibiting MT1-MMP (El Bedoui et al., 2005), wherein C-reactive protein (CRP), an inflammatory marker, injures endothelial cells and stimulates proliferation of VSMCs. In a recent study, Wang et al. (2010b) have shown that EGCG inhibits ET-1-induced CRP production by inhibiting mRNA expression in aortic VSMCs. EGCG also suppressed pro-inflammatory marker IL-6 and atherogenic factor Ang II-induced CRP production at the mRNA level (Peng et al., 2010).

15.9 Regulation of Genes Related to Autoimmune Disorders

Autoimmune disorders are considered significant clinical problems in various countries, including the United States. The pathogenesis of the diseases includes systemic lupus erythematosus (SLE) and Sjogren's syndrome (SS), characterized by production of autoantibodies against normal cellular components that may be overexpressed. There are very few treatment options for these diseases, but in several animal models green tea polyphenols were shown to reduce autoimmune disorder-related pathological changes. When given to a SLE mouse model, a diet containing 2% green tea extract prolonged survival and reduced node hyperplasia and anti-DNA antibody levels, indicating tea polyphenol's possible role in inhibiting the expression of autoantigens (Sayama et al., 2003). Utilizing cDNA microarray, RT-PCR, and Western blotting, Hsu et al. (2005b) have shown that EGCG inhibited the transcription and translation of major autoantigens, including SS-B/La, SS-A/Ro, coilin, DNA topoisomerase I, and alpha-fodrin in normal human epidermal keratinocyte (NHEK) cells.

15.10 Regulation of Genes Related to Inflammation

Inflammation, a biological process activated in response to detrimental stimuli from pathogens, damaged cells, or irritants, has been implicated in the pathogenesis of several disease states, including cancer, cardiovascular diseases, and neurological disorders. The major causes that contribute to inflammation include smoking, poor dietary habits, obesity, infectious diseases, and diabetes, via the production of several biochemical mediators, including cytokines (TNF-alpha, IL-1beta, IL-6, IL-2, IFN-gamma), chemokines (Il-8, CXCL10, IP-10, MCP-1), pro-inflammatory enzymes (COX-2, MMP-2, granules), immunoreceptors, or other factors, like prostaglandin, histamine, or nitric oxide. Nuclear factor (NF)-kappaB and AP-1 play crucial roles in coordinating the transcriptional regulation of a variety of genes that encode most of the above-mentioned pro-inflammatory factors. In a variety of model studies, EGCG was shown to suppress activation of NF-kappaB or AP-1 with subsequent suppression of its downstream regulatory molecules by inhibiting inflammatory factor-induced phosphorylation of ERK1/2, JNK, and p38 MAPKs, resulting in suppression of PI3K and AKT. EGCG inhibits TNF-alpha-mediated IkappaB activation with subsequent inactivation of NF-kappaB to suppress IL-8 mRNA expression. EGCG suppressed pathogenic activation of TLR4 by inhibiting glycosylation-mediated activation associated with reduced binding of NF-kappaB to DNA, with subsequent suppression in IL-8, IL-1beta, IFN-gamma, TNF-alpha, COX-2, and LOX gene expression involving MyD88- and TRIF-dependent TLR signaling pathways.

15.11 Conclusion

Technology evolved in nutrigenomics and proteomics may have profound effects on the preparation of future dietary guidelines and the development of customized disease-specific nutraceuticals. Emerging evidence resulting from hybridization-based technology accompanied with 2DE and MS methodologies has demonstrated that tea polyphenols not only interact with receptor molecules to modulate cell signaling in the pathogenesis of cancer, neurodegenerative disease, metabolic syndromes, or aging, but also alter the function of DNA and protein activity to accelerate their effect in the prevention of the health disorders. Accordingly, epidemiological studies and clinical trial data have provided satisfactory results showing that tea polyphenol consumption is beneficial to improve cardiac health, weight loss, gut health, cancer prevention, or extending life span. Nevertheless, additional studies based on community-based human trials with subsequent analysis of genomic and proproteomic profiles are required to elucidate dose-response molecular mechanisms of tea polyphenols. Also, tea polyphenol-mediated genomic and proteomic modulation can be listed in a database, so that interested individuals, scientists, or industrial professionals can gain access to the current updated information on the many health-promoting benefits of these compounds.

References

Abe, K., Okada, N., Tanabe, H., Fukutomi, R., Yasui, K., Isemura, M., and Kinae, N. 2009. Effects of chronic ingestion of catechin-rich green tea on hepatic gene expression of gluconeogenic enzymes in rats. *Biomed Res*, 30: 25–29.

Adachi, S., Nagao, T., To, S., Joe, A.K., Shimizu, M., Matsushima-Nishiwaki, R., Kozawa, O., Moriwaki, H., Maxfield, F.R., and Weinstein, I.B. 2008. (–)-Epigallocatechin gallate causes internalization of the epidermal growth factor receptor in human colon cancer cells. *Carcinogenesis*, 29: 1986–1993.

Ahn, W.S., Huh, S.W., Bae, S.M., Lee, I.P., Lee, J.M., Namkoong, S.E., Kim, C.K., and Sin, J.I. 2003. A major constituent of green tea, EGCG, inhibits the growth of a human cervical cancer cell line, CaSki cells, through apoptosis, G(1) arrest, and regulation of gene expression. *DNA Cell Biol*, 22: 217–224.

Antonello, M., Montemurro, D., Bolognesi, M., Di Pascoli, M., Piva, A., Grego, F., Sticchi, D., Giuliani, L., Garbisa, S., and Rossi, G.P. 2007. Prevention of hypertension, cardiovascular damage and endothelial dysfunction with green tea extracts. *Am J Hypertens*, 20: 1321–1328.

Bae, J.Y., Kanamune, J., Han, D.W., Matsumura, K., and Hyon, S.H. 2009. Reversible regulation of cell cycle-related genes by epigallocatechin gallate for hibernation of neonatal human tarsal fibroblasts. *Cell Transplant*, 18: 459–469.

Berletch, J.B., Liu, C., Love, W.K., Andrews, L.G., Katiyar, S.K., and Tollefsbol, T.O. 2008. Epigenetic and genetic mechanisms contribute to telomerase inhibition by EGCG. *J Cell Biochem*, 103: 509–519.

Bieschke, J., Russ, J., Friedrich, R.P., Ehrnhoefer, D.E., Wobst, H., Neugebauer, K., and Wanker, E.E. 2010. EGCG remodels mature alpha-synuclein and amyloid-beta fibrils and reduces cellular toxicity. *Proc Natl Acad Sci USA*, 107: 7710–7715.

Bursill, C., Roach, P.D., Bottema, C.D.K., and Pal, S. 2001. Green tea upregulates the low-density lipoprotein receptor through the sterol-regulated element binding protein in HepG2 liver cells. *J Agric Food Chem*, 49: 5639–5645.

Cao, H., Hininger-Favier, I., Kelly, M.A., Benaraba, R., Dawson, H.D., Coves, S., Roussel, A.M., and Anderson, R.A. 2007. Green tea polyphenol extract regulates the expression of genes involved in glucose uptake and insulin signaling in rats fed a high fructose diet. *J Agric Food Chem*, 55: 6372–6378.

Chen, A., and Zhang, L. 2003. The antioxidant (–)-epigallocatechin-3-gallate inhibits rat hepatic stellate cell proliferation *in vitro* by blocking the tyrosine phosphorylation and reducing the gene expression of platelet-derived growth factor-beta receptor. *J Biol Chem*, 278: 23381–23389.

Chung, J.H., Han, J.H., Hwang, E.J., Seo, J.Y., Cho, K.H., Kim, K.H., Youn, J.I., and Eun, H.C. 2003. Dual mechanisms of green tea extract (EGCG)-induced cell survival in human epidermal keratinocytes. *FASEB J*, 17: 1913–1915.

Dell'Aica, I., Dona, M., Sartor, L., Pezzato, E., and Garbisa, S. 2002. (–)-Epigallocatechin-3-gallate directly inhibits MT1-MMP activity, leading to accumulation of nonactivated MMP-2 at the cell surface. *Lab Invest*, 82: 1685–1693.

El Bedoui, J., Oak, M.H., Anglard, P., and Schini-Kerth, V.B. 2005. Catechins prevent vascular smooth muscle cell invasion by inhibiting MT1-MMP activity and MMP-2 expression. *Cardiovasc Res*, 67: 317–325.

Fujimura, Y., Umeda, D., Kiyohara, Y., Sunada, Y., Yamada, K., and Tachibana, H. 2006. The involvement of the 67 kDa laminin receptor-mediated modulation of cytoskeleton in the degranulation inhibition induced by epigallocatechin-3-O-gallate. *Biochem Biophys Res Commun*, 348: 524–531.

Fujimura, Y., Umeda, D., Yamada, K., and Tachibana, H. 2008. The impact of the 67 kDa laminin receptor on both cell-surface binding and anti-allergic action of tea catechins. *Arch Biochem Biophys*, 476: 133–138.

Furuyashiki, T., Nagayasu, H., Aoki, Y., Bessho, H., Hashimoto, T., Kanazawa, K., and Ashida, H. 2004. Tea catechin suppresses adipocyte differentiation accompanied by down-regulation of PPARgamma2 and C/EBPalpha in 3T3-L1 cells. *Biosci Biotechnol Biochem*, 68: 2353–2359.

Gupta, S., Ahmad, N., Nieminen, A.-L., and Mukhtar, H. 2000. growth inhibition, cell-cycle dysregulation, and induction of apoptosis by green tea constituent (−)-epigallocatechin-3-gallate in androgen-sensitive and androgen-insensitive human prostate carcinoma cells. *Toxicol Appl Pharmacol*, 164: 82–90.

Haque, A.M., Hashimoto, M., Katakura, M., Hara, Y., and Shido, O. 2008. Green tea catechins prevent cognitive deficits caused by A[beta]1–40 in rats. *J Nutr Biochem*, 19: 619–626.

Harakeh, S., Abu-El-Ardat, K., Diab-Assaf, M., Niedzwiecki, A., El-Sabban, M., and Rath, M. 2008. Epigallocatechin-3-gallate induces apoptosis and cell cycle arrest in HTLV-1-positive and -negative leukemia cells. *Med Oncol*, 25: 30–39.

Hou, R.-R., Chen, J.-Z., Chen, H., Kang, X.-G., Li, M.-G., and Wang, B.-R. 2008. Neuroprotective effects of (−)-epigallocatechin-3-gallate (EGCG) on paraquat-induced apoptosis in PC12 cells. *Cell Biol Int*, 32: 22–30.

Hsu, S., Dickinson, D.P., Qin, H., Lapp, C., Lapp, D., Borke, J., Walsh, D.S., Bollag, W.B, Stoppler, H., Yamamoto, T., Osaki, T., and Schuster, G. 2005a. Inhibition of autoantigen expression by (−)-epigallocatechin-3-gallate (the major constituent of green tea) in normal human cells. *J Pharmacol Exp Ther*, 315: 805–811.

Hsu, S., Farrey, K., Wataha, J., Lewis, J., Borke, J., Singh, B., Qin, H., Lapp, C., Lapp, D., Nguyen, T., and Schuster, G. 2005b. Role of p21WAF1 in green tea polyphenol-induced growth arrest and apoptosis of oral carcinoma cells. *Anticancer Res*, 25: 63–67.

Huang, C.H., Tsai, S.J., Wang, Y.J., Pan, M.H., Kao, J.Y., and Way, T.D. 2009. EGCG inhibits protein synthesis, lipogenesis, and cell cycle progression through activation of AMPK in p53 positive and negative human hepatoma cells. *Mol Nutr Food Res*, 53: 1156–1165.

Huh, S.W., Bae, S.M., Kim, Y.-W., Lee, J.M., Namkoong, S.E., Lee, I.P., Kim, S.H., Kim, C.K., and Ahn, W.S. 2004. Anticancer effects of (−)-epigallocatechin-3-gallate on ovarian carcinoma cell lines. *Gynecol Oncol*, 94: 760–768.

Hwang, J.T., Ha, J., Park, I.J., Lee, S.K., Baik, H.W., Kim, Y.M, and Park, O.J. 2007. Apoptotic effect of EGCG in HT-29 colon cancer cells via AMPK signal pathway. *Cancer Lett*, 247: 115–121.

Isemura, O., Abe, I.K., Kinae, N., Yamamoto-Maeda, M., and Koyama, Y. 2007. Modulation of gene expression by green tea in relation to its beneficial health effects. *Yakugaku Zasshi—J Pharm Soc Jpn*, 127: 1–3.

Kim, C.Y., Lee, C., Park, G.H., and Jang, J.H. 2009. Neuroprotective effect of epigallocatechin-3-gallate against beta-amyloid-induced oxidative and nitrosative cell death via augmentation of antioxidant defense capacity. *Arch Pharm Res*, 32: 869–881.

Kim, H.J., Ryu, J.H., Kim, C.H., Lim, J.W., Moon, U.Y., Lee, G.H., Lee, J.G., Baek, S.J., Yoon, J.H. 2010. ECG suppresses oxidative stress-induced MUC5AC overexpression by interaction with EGFR. *Am J Respir Cell Mol Biol*, 43: 349–357.

Kim, J.A., Formoso, G., Li, Y., Potenza, M.A., Marasciulo, F.L., Montagnani, M., Quon, M.J. 2007. Epigallocatechin gallate, a green tea polyphenol, mediates NO-dependent vasodilation using signaling pathways in vascular endothelium requiring reactive oxygen species and Fyn. *J Biol Chem*, 282: 13736–13745.

Ku, H.C., Chang, H.H., Liu, H.C., Hsiao, C.H., Lee, M.J., Hu, Y.J., Hung, P.F., Liu, C.W., and Kao, Y.H. 2009. Green tea (–)-epigallocatechin gallate inhibits insulin stimulation of 3T3-L1 preadipocyte mitogenesis via the 67-kilodalton laminin receptor pathway. *Am J Physiol Cell Physiol*, 297: C121–C132.

Kuzuhara, T., Sei, Y., Yamaguchi, K., Suganuma, M., and Fujiki, H. 2006. DNA and RNA as new binding targets of green tea catechins. *J Biol Chem*, 281: 17446–17456.

Lamy, S., Gingras, D., and Beliveau, R. 2002. Green tea catechins inhibit vascular endothelial growth factor receptor phosphorylation. *Cancer Res*, 62: 381–385.

Lee, K.M., Yeo, M., Choue, J.S., Jin, J.H., Park, S.J., Cheong, J.Y., Lee, K.J., Kim, J.H., and Hahm, K.B. 2004. Protective mechanism of epigallocatechin-3-gallate against *Helicobacter pylori*-induced gastric epithelial cytotoxicity via the blockage of TLR-4 signaling. *Helicobacter*, 9: 632–642.

Lee, Y.K., Yuk, D.Y., Lee, J.W., Lee, S.Y., Ha, T.Y., Oh, K.W., Yun, Y.P., and Hong, J.T. 2009. (–)-Epigallocatechin-3-gallate prevents lipopolysaccharide-induced elevation of beta-amyloid generation and memory deficiency. *Brain Res*, 1250: 164–174.

Leong, H., Mathur, P.S., and Greene, G.L. 2009. Green tea catechins inhibit angiogenesis through suppression of STAT3 activation. *Breast Cancer Res Treat*, 117: 505–515.

Li, Y., Ying, C., Zuo, X., Yi, H., Yi, W., Meng, Y., Ikeda, K., Ye, X., Yamori, Y., and Sun, X. 2009. Green tea polyphenols down-regulate caveolin-1 expression via ERK1/2 and p38MAPK in endothelial cells. *J Nutr Biochem*, 20: 1021–1027.

Lin, C.-L., Chen, T.-F., Chiu, M.-J., Way, T.-D., and Lin, J.-K. 2009. Epigallocatechin gallate (EGCG) suppresses beta-amyloid-induced neurotoxicity through inhibiting c-Abl/FE65 nuclear translocation and GSK3 beta activation. *Neurobiol Aging*, 30: 81–92.

Liu, L., Lai, C.Q., Nie, L., Ordovas, J., Band, M., Moser, L., and Meydani, M. 2008. The modulation of endothelial cell gene expression by green tea polyphenol-EGCG. *Mol Nutr Food Res*, 52: 1182–1192.

Lo, H.-M., Hung, C.-F., Huang, Y.-Y., and Wu, W.-B. 2007. Tea polyphenols inhibit rat vascular smooth muscle cell adhesion and migration on collagen and laminin via interference with cell-ECM interaction. *J Biomed Sci*, 14: 637–645.

Luo, K.L., Luo, J.H., and Yu, Y.P. 2010. (–)-Epigallocatechin-3-gallate induces Du145 prostate cancer cell death via downregulation of inhibitor of DNA binding 2, a dominant negative helix-loop-helix protein. *Cancer Sci*, 101: 707–712.

Ma, L., Cao, T.T., Kandpal, G., Warren, L., Fred Hess, J., Seabrook, G.R., and Ray, W.J. 2010. Genome-wide microarray analysis of the differential neuroprotective effects of antioxidants in neuroblastoma cells overexpressing the familial Parkinson's disease alpha-synuclein A53T mutation. *Neurochem Res*, 35: 130–142.

Melgarejo, E., Medina, M.A., Sanchez-Jimenez, F., Botana, L.M., Dominguez, M., Escribano, L., Orfao, A., and Urdiales, J.L. 2007. (–)-Epigallocatechin-3-gallate interferes with mast cell adhesiveness, migration and its potential to recruit monocytes. *Cell Mol Life Sci*, 64: 2690–2701.

Meng, Q., Velalar, C.N., and Ruan, R. 2008a. Effects of epigallocatechin-3-gallate on mitochondrial integrity and antioxidative enzyme activity in the aging process of human fibroblast. *Free Radic Biol Med*, 44: 1032–1041.

Meng, Q.Y., Velalar, C.N., and Ruan, R.S. 2008b. Regulating the age-related oxidative damage, mitochondrial integrity, and antioxidative enzyme activity in Fischer 344 rats by supplementation of the antioxidant epigallocatechin-3-gallate. *Rejuvenation Res*, 11: 649–660.

Neuhaus, T., Pabst, S., Stier, S., Weber, A.-A., Schror, K., Sachinidis, A., Vetter, H., and Ko, Y.D. 2004. Inhibition of the vascular-endothelial growth factor-induced intracellular signaling and mitogenesis of human endothelial cells by epigallocatechin-3 gallate. *Eur J Pharmacol*, 483: 223–227.

Ninomiya, M., Unten, L., and Kim, M. 1997. Chemical and physicochemical properties of green tea polyphenols. In *Chemistry and applications of green tea*, ed. T. Yamamoto, L.R. Juneja, D.C. Chu, and M. Kim, 23–35. Boca Raton, FL: CRC Press.

Ohata, M., Koyama, Y., Suzuki, T., Hayakawa, S., Saeki, K., Nakamura, Y., and Isemura, M. 2005. Effects of tea constituents on cell cycle progression of human leukemia U937 cells. *Biomed Res*, 26: 1–7.

Okabe, S., Fujimoto, N., Sueoka, N., Suganuma, M., and Fujiki, H. 2001. Modulation of gene expression by (–)-epigallocatechin gallate in PC-9 cells using a cDNA expression array. *Biol Pharm Bull*, 24: 883–886.

Oritani, Y., Matsui, Y., Kurita, I., Kinoshita, Y., Kawakami, S., Yanae, K., Nishimura, E., Kato, M., Sai, M., Matsumoto, I., Abe, K., Maeda-Yamamoto, M., and Kamei, M. 2009. Mechanism of anti-obese effects of "Benifuuki" green tea. *J Jpn Soc Food Sci Technol—Nippon Shokuhin Kagaku Kogaku Kaishi*, 56: 412–418.

Ou, H.C., Song, T.Y., Yeh, Y.C., Huang, C.Y., Yang, S.F., Chiu, T.H., Tsai, K.L., Chen, K.L., Wu, Y.J., Tsai, C.S., Chang, L.Y., Kuo, W.W., and Lee, S.D. 2010. EGCG protects against oxidized LDL-induced endothelial dysfunction by inhibiting LOX-1-mediated signaling. *J Appl Physiol*, 108: 1745–1756.

Peng, N., Liu, J.T., Guo, F., and Li, R. 2010. Epigallocatechin-3-gallate inhibits interleukin-6- and angiotensin II-induced production of C-reactive protein in vascular smooth muscle cells. *Life Sci*, 86: 410–415.

Philips, B.J., Coyle, C.H., Morrisroe, S.N., Chancellor, M.B., and Yoshimura, N. 2009. Induction of apoptosis in human bladder cancer cells by green tea catechins. *Biomed Res*, 30: 207–215.

Qiao, Y., Cao, J., Xie, L., and Shi, X. 2009. Cell growth inhibition and gene expression regulation by (–)-epigallocatechin-3-gallate in human cervical cancer cells. *Arch Pharm Res*, 32: 1309–1315.

Qin, J., Chen, H.G., Yan, Q., Deng, M., Liu, J., Doerge, S., Ma, W., Dong, Z., and Li, D.W. 2008. Protein phosphatase-2A is a target of epigallocatechin-3-gallate and modulates p53-Bak apoptotic pathway. *Cancer Res*, 68: 4150–4162.

Reiter, C.E., Kim, J.A., and Quon, M.J. 2010. Green tea polyphenol epigallocatechin gallate reduces endothelin-1 expression and secretion in vascular endothelial cells: Roles for AMP-activated protein kinase, Akt, and FOXO1. *Endocrinology*, 151: 103–114.

Rezai-Zadeh, K., Shytle, D., Sun, N., Mori, T., Hou, H., Jeanniton, D., Ehrhart, J., Townsend, K., Zeng, J., Morgan, D., Hardy, J., Town, T., and Tan, J. 2005. Green tea epigallocatechin-3-gallate (EGCG) modulates amyloid precursor protein cleavage and reduces cerebral amyloidosis in Alzheimer transgenic mice. *J Neurosci*, 25: 8807–8814.

Roy, M.K., Kapoor, M.P., and Juneja, L.R. 2009. Green tea polyphenol-modulated genome functions for protective health benefits. In *Nutrigenomics and proteomics in health and disease*, ed. Y. Mine, K. Miyashita, and F. Shahidi, 201–237. Ames, IA, Wiley-Blackwell.

Sachinidis, A., Seul, C., Seewald, S., Ahn, H.-Y., Ko, Y., and Vetter, H. 2000. Green tea compounds inhibit tyrosine phosphorylation of PDGF [beta]-receptor and transformation of A172 human glioblastoma. *FASEB J*, 471: 51–55.

Sayama, K., Oguni, I., Tsubura, A., Tanaka, S., and Matsuzawa, A. 2003. Inhibitory effects of autoimmune disease by green tea in MRL-Faslprcg/Faslprcg mice. *In Vivo*, 17: 545–552.

Sen, T., Dutta, A., and Chatterjee, A. 2010. Epigallocatechin-3-gallate (EGCG) down-regulates gelatinase-B (MMP-9) by involvement of FAK/ERK/NFjB and AP-1 in the human breast cancer cell line MDA-MB-231. *Anticancer Drugs*, 21: 632–644.

Shankar, S., Ganapathy, S., Hingorani, S.R., and Srivastava, R.K. 2008. EGCG inhibits growth, invasion, angiogenesis and metastasis of pancreatic cancer. *Front Biosci*, 13: 440–452.

Shankar, S., Suthakar, G., and Srivastava, R.K. 2007. Epigallocatechin-3-gallate inhibits cell cycle and induces apoptosis in pancreatic cancer. *Front Biosci*, 12: 5039–5051.

Shim, J.H., Su, Z.Y., Chae, J.I., Kim, D.J., Zhu, F., Ma, W.Y., Bode, A.M., Yang, C.S., and Dong, Z. 2010. Epigallocatechin gallate suppresses lung cancer cell growth through Ras-GTPase-activating protein SH3 domain-binding protein 1. *Cancer Prev Res (Phila)*, 3: 670–679.

Shimizu, M., Shirakami, Y., and Moriwaki, H. 2008. Targeting receptor tyrosine kinases for chemoprevention by green tea catechin, EGCG. *Int J Mol Sci*, 9: 1034–1049.

Shimizu, M., Shirakami, Y., Sakai, H., Yasuda, Y., Kubota, M., Adachi, S., Tsurumi, H., Hara, Y., and Moriwaki, H. 2010. (–)-Epigallocatechin gallate inhibits growth and activation of the VEGF/VEGFR axis in human colorectal cancer cells. *Chemicobiol Interact*, 185: 247–252.

Sukhthankar, M., Choi, C.K., English, A., Kim, J.-S., and Baek, S.J. 2010. A potential proliferative gene, NUDT6, is down-regulated by green tea catechins at the posttranscriptional level. *J Nutr Biochem*, 21: 98–106.

Sun, B.K., and Tsao, H. 2008. Small RNAs in development and disease. *J Am Acad Dermatol*, 59: 725–737.

Suzuki, Y., and Isemura, M. 2001. Inhibitory effect of epigallocatechin gallate on adhesion of murine melanoma cells to laminin. *Cancer Lett*, 173: 15–20.

Tachibana, H., Koga, K., Fujimura, Y., and Yamada, K. 2004. A receptor for green tea polyphenol EGCG. *Nat Struct Mol Biol*, 11: 380–381.

Thangapazham, R.L., Passi, N., and Maheshwari, R.K. 2007. Green tea polyphenol and epigallocatechin gallate induce apoptosis and inhibit invasion in human breast cancer cells. *Cancer Biol Ther*, 6: 1938–1943.

Tsang, W.P., and Kwok, T.T. 2010. Epigallocatechin gallate up-regulation of miR-16 and induction of apoptosis in human cancer cells. *J Nutr Biochem*, 21: 1409–1446.

Vinson, J.A., Teufel, K., and Wu, N. 2004. Green and black teas inhibit atherosclerosis by lipid, antioxidant, and fibrinolytic mechanisms. *J Agric Food Chem*, 52: 3661–3665.

Vittal, R., Selvanayagam, Z.E., Sun, Y., Hong, J., Liu, F., Chin, K.V., and Yang, C.S. 2004. Gene expression changes induced by green tea polyphenol (–)-epigallocatechin-3-gallate in human bronchial epithelial 21BES cells analyzed by DNA microarray. *Mol Cancer Ther*, 3: 1091–1099.

Wang, C.J., Liu, J.T., and Guo, F. 2010a. (–)-Epigallocatechin gallate inhibits endothelin-1-induced C-reactive protein production in vascular smooth muscle cells. *Basic Clin Pharmacol Toxicol*, 107: 669–675.

Wang, H.J., Lo, W.Y., Lu, T.L., and Huang, H. 2010b. (–)-Epigallocatechin-3-gallate decreases thrombin/paclitaxel-induced endothelial tissue factor expression via the inhibition of c-Jun terminal NH2 kinase phosphorylation. *Biochem Biophys Res Commun*, 391: 716–721.

Wang, S.I., and Mukhtar, H. 2002. Gene expression profile in human prostate LNCaP cancer cells by (–)-epigallocatechin-3-gallate. *Cancer Lett*, 182: 43–51.

Weinreb, O., Amit, T., and Youdim, M.B.H. 2007. A novel approach of proteomics and transcriptomics to study the mechanism of action of the antioxidant-iron chelator green tea polyphenol (–)-epigallocatechin-3-gallate. *Free Radical Biol Med*, 43: 546–556.

Wolfram, S., Raederstorff, D., Preller, M., Wang, Y., Teixeira, S.R., Riegger, C., and Weber, P. 2006. Epigallocatechin gallate supplementation alleviates diabetes in rodents. *J Nutr*, 136: 2512–2518.

Yamauchi, R., Sasaki, K., and Yoshida, K. 2009. Identification of epigallocatechin-3-gallate in green tea polyphenols as a potent inducer of p53-dependent apoptosis in the human lung cancer cell line A549. *Toxicol In Vitro*, 23: 834–839.

Yang, S.-P., Wilson, K., Kawa, A., and Raner, G.M. 2006. Effects of green tea extracts on gene expression in HepG2 and Cal-27 cells. *Food Chem Toxicol*, 44: 1075–1081.

Ying, C.J., Xu, J.W., Ikeda, K., Takahashi, K., Nara, Y., and Yamori, Y. 2003. Tea polyphenols regulate nicotinamide adenine dinucleotide phosphate oxidase subunit expression and ameliorate angiotensin II-induced hyperpermeability in endothelial cells. *Hypertens Res*, 26: 823–828.

16

Green Tea Polyphenols in Food and Nonfood Applications

Mahendra P. Kapoor, Tsutomu Okubo, Theertham P. Rao, and Lekh R. Juneja

Contents

16.1 Introduction

The demand of botanical-derived dietary supplements is continuously growing through their traditional channels of natural and health food stores, multinational marketing companies, and e-commerce marketing of mail order and Internet sales. Although many people are aware of the health benefits associated with drinking green tea, not that many are aware of the properties when dealing with the topical use of green tea, without going into the health benefits of consuming green tea. The benefits of green tea range from a role in the prevention of chronic diseases to the ability to increase metabolism and help burn fat. Although sipping a freshly brewed cup of green tea is one way to enjoy its delicious taste and health properties, there are a variety of other ways to appreciate this popular drink. In this chapter, recent developments in the application of green tea polyphenols in foods, cosmetics, nutraceuticals/supplements, and so on, are summarized. Primarily the processed tea leaves are used as such or in minced form for household consumption as well as in tea bags. Apart from the above, processed green tea leaves are being used to manufacture green tea extracts with consolidated certain properties, including catechin/tea polyphenols as a majority. Green tea leaves can be powdered with a variety of well-known procedures, including grinding using a stone mill, impulsive fracturing with a power mill, and frost shattering. Depending on the appropriate method used, different particle sizes and flavors of green tea result. As an example, grinding with a stone mill is unique in Japan, to produce matcha powder with strong flavor and vibrant color of leaves, and is used not only in tea ceremonies but also for a number of popular applications, such as ice cream flavors, in confectionaries/bakeries, and so on.

Among the green tea catechin products currently available in the global market, most of the purified extracts are made by extracting tea leaves with hot water and then with a solvent, followed by column chromatography to elute concentrated or purified polyphenols. With subsequent concentration or pulverization of filtrate, highly concentrated or purified tea polyphenols can be made. Thus, refined polyphenols such as 90% epigallocatechin gallate (EGCG), which is the major catechin, can be manufactured. Currently, a large fraction of commercial green tea extracts sold worldwide is produced in the developing world, where federal regulations on pesticide use are seldom followed. Therefore, most of the green tea extract products available in the worldwide marketplace are not pesticide-free and probably contaminated with some poisonous pesticides, such as *endosulfan* and *dicofol*, that are well known for their endocrine disruptive features and related deleterious effects. In the United States, actually a limited fraction of imported green tea extracts are randomly analyzed by the Food and Drug Administration (FDA), leaving tons of defective and tainted green tea extract in the market for diverse applications yielding the worst scenarios. On the other hand, although sometimes the green tea extract is labeled as clean from pesticides, the analysis showed contamination with *benzopyrene*, a hazardous and carcinogenic hydrocarbon that might be used during the extraction process or caused by polluted/contaminated soil used for the tea plantations. Worldwide, the use of various pesticides in the cultivation of green tea is internationally regulated, and their residues are allowed in final botanical extracts (USFDA, 2011).

16.2 Sunphenon®: Safe, Functional, and Versatile Range of Green Tea Extracts

Taiyo Kagaku, Japan, one of the pioneers in green tea research, with over 30 years of intensive research, excelled in the perfection of developing safe and innovative green tea extracts under the brand name of Sunphenon®. Various scientists and institutions around the globe have evaluated Sunphenon in various animal and clinical studies, and confirmed its health benefits, such as physiological antioxidant, oral care, prebiotic, anitimicrobial, immunological, allergy, satiety, weight management, including in the alleviation of diabetic, renal, and muscle dystrophy diseases. Taiyo, with the initiation of its Taiyo Green Power factory in Wuxi, China, has become the world's largest green tea extract manufacturer, with a capacity of more than 2000 tons per year. Sunphenon is produced from tea leaves that are fully compliant with stringent USFDA residual pesticide regulations for tea, as outlined in 40 CFR 180. The tea leaves are extracted via a water infusion process and decaffeinated using only approved food-grade solvents. No chloroform or other illegal solvents are used in the processing of any Sunphenon extracts. Further, Taiyo is the only one of its kind now implementing a traceability system to monitor the growth and procurement of tea leaves, which offers a guarantee on the above residues, as well as traceability to confirm its suitability for safe use in various food materials.

Taiyo also excels in the development of a versatile range of Sunphenon products, offering low- to high-concentrate catechin extracts, purified ECGC with more than 95% EGCG content (Sunphenon EGCG), and low bitter green tea extracts. Further, it specializes in offering decaffeinated green tea extracts (Sunphenon DCF), natural caffeine-rich extract (Sunphenon CF), natural L-theanine-rich green tea extract (Sunphenon THD), natural fluoride-rich green tea extract (Sunphenon MRF), and white tea extract (Sunphenon WT). Taiyo also offers the exotic taste and flavors of East Asia, namely, sencha, hojicha, and genmaicha, for instant beverage. The versatile range of Sunphenon products fits very well in most food, beverages, dairy products, supplements, cosmetics, and industrial applications. Figure 16.1

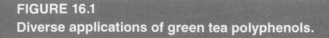

FIGURE 16.1
Diverse applications of green tea polyphenols.

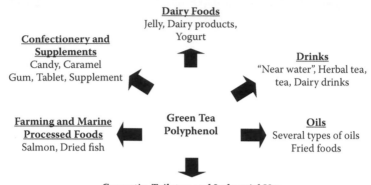

illustrates the schematic presentation of current commercial applications of green tea polyphenols in diverse areas. The Sunphenon brand of green tea catechins offers pure and tasty formulations that give refreshment, relaxation, health, and more energy in life.

16.3 Food and Beverage Applications of Green Tea Polyphenols

Green tea polyphenols, generally known as tea catechins, are frequently used in beverage applications, wherein tea prepared by extracting tea leaves with hot water may be directly bottled in a can or polyethylene terephthalate (PET) bottle. Alternatively, green tea extract powder may be diluted or catechins may be added to fortify the beverages. In a typical beverage currently available in the Japanese market, green tea extracts are further enriched with tea catechins or formulated ester-catechins and are sold as health beverages for specified claims, such as reduction of body fat or cholesterol when consumed in accordance to ascribed daily doses (Hase et al., 2001; Kajimoto et al., 2003; Kobayashi et al., 2005). Rapid growth of bottled tea drinks (e.g., black tea, green tea, oolong tea, and a combination of teas) in Japan is extraordinary among all types of canned/bottled drinks and is gaining popularity in the Western countries and North America. Figure 16.2 displays

FIGURE 16.2
Trend on consumption of different beverages in Japan.

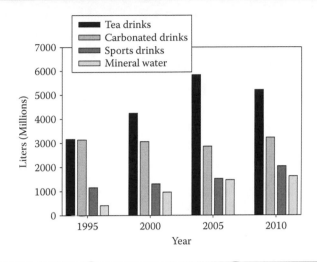

the incremental growth of different types of ready-to-drink (bottled/canned) beverages in the last 15 years in Japan, wherein the tea drinks dominate the overall growth rate. Such a growth curve of tea drinks can be attributed to an influence of the scientific record available, which evidently implies that tea catechins have definite strong health benefits to humans. An increasing interest and positive trend in green tea drinks is surfacing in Western countries, with improved flavor and an increased awareness of green tea health benefits through proper advertisements and new media enticement to consumers.

In the application of tea catechins for snacks, it has been successfully incorporated into candies, chocolates, chewing gums, caramels, jellybeans, and so on. A variety of snacks containing sucrose as a sweetener use green tea catechins as a dental caries prevention agent to foreclose cavity bacteria proliferation (Terajima et al., 1997) and plaque formations (Nishihara et al., 1993; Oiwa et al., 1993), and thus such commercial products are well recognized and established as oral care foods in the marketplace (Figure 16.3). Candies and chewing gums containing catechins are generally for the purpose of antiflu effects and protection from airborne contamination and prevention from viruses. However, the regulation does not allow claims on the product label concerning its effectiveness. Furthermore, tea fluorides in combination with tea catechins have gained reasonable interest for tooth and gum health (Fujii et al., 2004). It has been estimated that 1 L of tea prepared with fluoridated water can provide around 2.2 mg of fluoride per day (Gardner

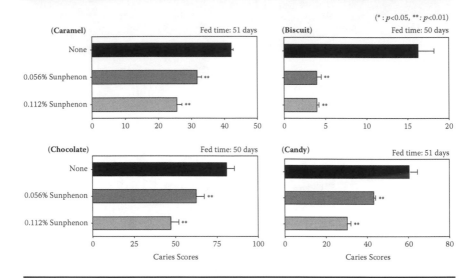

FIGURE 16.3
Inhibitory effect of Sunphenon on dental caries caused by various foods.

et al., 2007). Although a number of alternative products are available in the marketplace, tea fluorides are considered to strengthen teeth and protect them from decay. Green tea fluoride is a suitable alternative to currently used synthetic fluorides in the formulation of chewing gums and other oral care products. There exist a few examples of the use of green tea fluorides in such commercial products in Europe as well as in Japan. Certainly, green tea fluorides are expected to be applied in diverse food applications in the near future beyond their use in chewing gums. Also, mouth-rinsing fluids containing green tea catechins are also commercially available for significant reduction of dental plaque formation. In addition, green tea catechin formulations in food and beverages work to inhibit the growth of putrefactive bacteria without interfering with bifidobacteria, and thus result in a reduction of putrefactive bacteria and an increase of bifidobacteria/*Lactobacillus* bacteria. Shah et al. (2010) recently published a report about the ability of green tea extracts in stabilizing probiotics such as *Lactobacillus* and bifidos in fruit juices. So it is possible that use of green tea extracts not only limits prebiotic effects but also extends protection to probiotics in beverages and dairy applications. In another unique use of green tea catechins as a feed for lying hens, the resultant patented catechin eggs are labeled with much lower total lipid, total cholesterol, and peroxide values in the yolk. Such products are marketed in

Japan by Nippon Formula Feed Mfg. Co. Ltd.; however, knowledge about their efficacy may be spread widely only by word of mouth. On the other hand, the vibrant green color green tea powders largely known as matcha can also be used in unusual ways, such as green tea ice cream, green tea milk, and yogurt.

16.3.1 Antioxidative and Antimicrobial Applications of Green Tea Catechins in Foods and Beverages

Another example of the use of catechins is in health drinks with combinations of vitamins (i.e., beta-carotene) and various herbal extracts to enhance the scavenging actions against oxygenative free radicals. In addition to the above, green tea-derived antioxidant polyphenols are also added to beverage formulations with energy-enhancing substances or other health-promoting constituents in energy drinks. Such potent health drinks are recommended for use daily for complete protection during flu season. Products high in antioxidants are important because they help protect the cells in the body from the damaging effects of free radicals, fight against various forms of diseases, and protect health (Awika et al., 2003; Valko et al., 2007). It is generally known that antioxidants are required to prevent diseases, sustain health, and improve longevity (Harman, 2006). Antioxidants neutralize free radicals that are highly reactive chemical substances, which can damage cells such as DNA, lipids, and proteins, leading to premature aging and diseases implicated in the development of chronic pathologies. A number of studies have proved that green tea catechins are more powerful than vitamin C and nearly two to six times more powerful than vitamin E as an antioxidant (Wiseman et al., 1997). Also, a beverage containing green tea extract acts as a powerful immune system booster, as it consists of great antioxidant strength. Although the antioxidant activity of green tea polyphenols has long been proposed beneficial for human health, scientific opinion is currently shifting from this belief because of a lack of evidence in humans. It is challenging to quantify antioxidant intakes; however, a roughly average intake of 60 to 270 mg per day is estimated with regular tea consumers (Song and Chun, 2008).

Apart from the living systems and their defense against oxidative stress, green tea extracts are also equally important for food preservation (Masuda et al., 2003), wherein such phenolic antioxidants interrupt the propagation of the free radical autoxidation chain by contributing a hydrogen atom from a phenolic hydroxyl group, with the formation of a relatively stable free radical that does not initiate or further propagate the oxidation process (Kaur and Kapoor, 2001). Fried foods, confectionary products, margarine, and processed seafood may contain high amounts of oil, so they may easily cause oxidative degradation with heat and light. Green tea catechins might effectively eliminate active oxygen produced through catalysis with light, oxygen, and metals. So green tea catechins are commercially used as a natural antioxidant for the above foods that easily cause oxidative degradation due to high contents of oil. In a recent report, an approved retention of green tea extracts in biscuits by a reduction in pH of the dough was reported to enhance the stability of green tea catechins in fortified baked products (Sharma and Zhou, 2011). Also, oil-soluble catechin formulas are being developed and applied to a variety of edible oils, margarines, and so forth. The antioxidant potency of green tea catechins in edible oils and their synergism with surrounding components are well-deserved topics for in-depth understanding of the related mechanism. It is also well known that antioxidation potency is directly proportional to the catechin content in the green tea extracts, which means that green tea extracts with higher catechin contents could have higher antioxidant efficacies; however, it depends on the nature and molecular structure of the catechins. Apart from the radical scavenging potency, the speed of scavenging actions and the scavenging potency per unit weight of the green tea extracts are also important factors. Additionally, the antioxidative efficacies of green tea extracts have been demonstrated on a variety of edible oils (Figure 16.4) and fried foods (Figure 16.5), which showed an increased tendency in preserving food qualities (Okubo, 2004). In addition to soy oil and rapeseed oil, fish oil preservation is also effectively attained using green tea extracts wherein the other antioxidant (vitamin E, etc.) showed the least efficacy. Furthermore, green tea extracts could be effectively used for the preservation of freshness of salted and dried fish as well as suppression of natural food colors.

The antimicrobial activities of green tea extract against bacterial pathogens have been inconsistent due to variability of the production parameters of such extracts. In a report, green tea extracts were evaluated as a preservative treatment for fresh-cut lettuce at varied concentrations and temperatures and compared to other oxidants. Green tea extracts have been found to be suitable for use as antioxidant and antimicrobial agents to prevent the microbial growth and spoilage of fresh/preserved vegetables and foods (Barry-Ryan et al., 2008). Since consumers have become more critical to the use of artificial additives as antioxidants in order to preserve food and enhance characteristics such as flavor, color, and nutritional value (Bruhn, 2000), natural antioxidants such as green tea extracts have emerged to meet the challenge of replacing traditional synthetic antioxidants (butylated hydroxytoluene [BHT],

FIGURE 16.4
Inhibitory effect of Sunkatol No. 1 on the peroxidation of animal fat (lard).

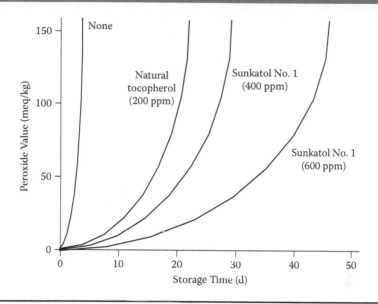

tert-butylhydroquinone [TBHQ], etc.) for enhanced prevention while retaining nutritional and sensory qualities. Therefore, in recent years a special renewed interest in the use of natural green tea products as decontaminates has been noticed, wherein commercial and research applications have evidently shown that green tea extracts as natural antimicrobials could replace traditional sanitizing agents (Cherry, 1999; Martin-Diana et al., 2006). Green tea extracts have been used to extend the shelf-life of dry fermented sausage (Bozkurt, 2006), enriched candy jellies (Gramza-Michalowska and Regula, 2007), and cooked pork patties (Nissen et al., 2004). Furthermore, green tea extracts are extremely effective against various strains of foodborne bacteria that can be harmful, poisonous, and even fatal. Antibacterial uses of green tea extracts are very important and directly related to soft drink manufacturing, storage, and distribution industries, wherein they are used to prevent propagation of pathogenic bacteria. Heat-tolerant bacteria (*Clostridium botulinum*), which can be deadly, are a major concern in the canning industry, wherein a low acidic environment germinates spores to grow into vegetative bacteria and produce toxins. PET bottle packaging of soft drinks with a neutral pH range is not perfectly free from bacterial contamination because PET cannot withstand postfilling temperatures. The risk of contamination by

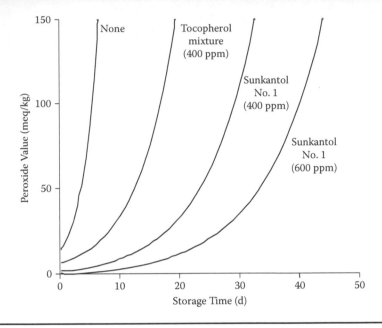

FIGURE 16.5
Antioxidative effect of Sunkatol No. 1 on oil-fried noodles.

heat-tolerated bacteria is always found in incomplete sterilized bottles. Thanks to the bacteriostatic uses of green tea extracts, however, a number of herbal soft drinks have been commercially fortified with green tea catechins.

Green tea extracts can be successfully added to ethylene vinyl copolymer films to produce antioxidant active packaging (de Dicastillo et al., 2011). It is claimed that green tea extract containing packaging films can be used for all type of foods, ranging from aqueous to fatty products, to reduce oxidation of sensitive products. Such a novel alternative is interesting, as a modified packaging system is used instead of addition of green tea extract antioxidants to foodstuffs for preserving antioxidant activity. Similarly, catechins are found effective in antiflu masks and recommended for use during the peak season of influenza for the aging population for enhanced protection from airborne seasonal contaminations. According to a recent market survey, about 60% of the U.S. respondents stated that they look for safe antioxidants when shopping for functional foods and beverages (Lal, 2007; Mintel, 2007, 2009). Therefore, considering the addition of appropriate contents of natural oxidants, such as green tea extracts in beverage formulations, could be a key driver to increased sales. A future trend that could prove to be very effective would be for beverage companies to increase the functionality of their beverages with the aid of antioxidants to target the growing number of health-conscious consumers.

16.3.2 Deodorizing Activities of Green Tea Catechins in Foods and Beverages

Green tea extracts are also known for their deodorizing properties, and their effect on halitosis has been well established. The deodorizing effect of green tea extract against methyl mercaptan, which is the main source of halitosis, is reported elsewhere. Among the number of green tea catechins, EGCG has shown a pronounced deodorizing effect against unpleasant smells. Additionally, the deodorizing effect of green tea extracts on meat products to suppress their offensive order is also known and is currently in use in hamburger and sauce manufacturing, as they are made from cheap quality meat and usually develop peculiar and unsavory meat odors when processed at cooking temperatures. They also have a deodorizing effect on stool, wherein green tea catechins show marked improvements in intestinal conditions and reduce fecal odors, particularly of elderly populations, due to inhibited action of α-amylase (Hara and Honda, 1990), helping to reduce the ammonia and sulfide contents in human feces and maintain a lower pH (nearly 3) while increasing the total organic acid concentration of the feces. Further, trapping of formaldehyde from homes and commercial buildings by tea catechins is another example of their deodorizing effect to help a serious air quality problem in close houses and installations, and to minimize sick house syndromes that could cause nausea, skin disorders, headaches, fatigue, and watery eyes due to an enhanced level of volatile organic compounds, such as formaldehyde.

16.4 Green Tea Polyphenols (Catechins) as Supplements

Green tea polyphenols (catechins) are mostly used as supplements, and most of them are claimed as "antioxidants" and are well established in the marketplace. In most food supplements tea polyphenol powdered extracts are in tablet or capsule form with suitable diluents. A commercial product, formulated with a combination of tea catechins plus galacto-oligosaccharide, is the existing example for improved health of bowel-modulating action with green tea polyphenols. Also, green tea polyphenol supplements are being promoted to help fight mental and physical fatigue, fight cancer, prevent blood clotting tendencies, lower blood cholesterol levels, regulate blood sugar levels, and assist in weight loss programs. Green tea polyphenol supplements are also known to confer protection against respiratory and digestive infections and food poisoning, while encouraging acidophilus growth and regularizing bowel habits. It is recommended that 500 mg of green tea polyphenols per day is enough to significantly lower blood pressure and possess antimutagenic activity. Furthermore, at very high levels (0.5 to 1% of daily diet) green tea polyphenols are capable of considerable lowering of high, total, and low-density lipoprotein (LDL) cholesterol levels, as claimed by several diet supplements with proven helpful studies. The caffeine content of green tea catechins also acts as a natural fat burner; thus, such supplements are available as tablets and capsules with varied caffeine contents. Further, green tea polyphenols containing nutritional supplements can very successfully control blood sugar in diabetic patients. Also, there are many other benefits documented from taking green tea polyphenol, as they are noted for their anticancer efficacies as well as their ability to assist in weight loss. Caffeine-free green tea polyphenol supplements available over the counter (OTC) are advisable for caffeine-sensitive individuals for safety reasons. Individuals who are diabetic also tend to be overweight; thus they can take advantage of caffeine-free green tea polyphenol supplements. Since no specific deficiencies result from not consuming green tea polyphenols, it cannot be categorized as an essential nutritional supplement. However, in therapeutic use its dosage is usually increased considerably, but the toxicity level is an important issue.

16.5 Green Tea Polyphenols in Skin Health and Oral Care Cosmetics

This section especially deals with antiaging (senescence), antibacterial, antiviral, skin care, beauty, and personal care applications of green tea polyphenols. Extensive research has shown that green tea extracts not only have amazing antioxidant and cell protective qualities, but also protect collagen by inhibiting collagenase, which is a collagen-reducing enzyme responsible for the breakdown of collagen. Since green tea extracts are well known as a potent anti-inflammatory agent, they are capable of reducing inflammation in the skin through an inhibitory action on collagenase. As free radicals are prone to attacking and damaging the DNA of cells, an antioxidant effect of green tea extracts plays a protective role to prevent free radical damage that could result in skin tumors and cancers (Chiu and Kimball, 2003). The methylxanthines in green tea extracts help stimulate skin microcirculation and thus influence skin health. Therefore, green tea polyphenols can be used as an antiaging ingredient in skin care formulations as well as to help skin fight inflammation that might induce premature aging. Green tea polyphenols are also capable of limiting cell death when vigorously exposed to UV radiation, and thus exhibit a cell-protecting function as well (Draelos, 2001; Ahmad and Mukhtar, 2001). EGCG, one of the polyphenols in green tea extract, is generally known to be 20 times more powerful than vitamin E (tocopherols) for neutralizing free radicals since it is effectively absorbed through the skin and acts as an astringent to protect the skin. It has been shown that skin has measurably more elastic tissue content after uninterrupted application of a product containing green tea extract, thereby maintaining firm and elastic skin. Green tea polyphenols also exhibit a photoprotective effect when applied to skin and reduce erythema formation. At certain concentrations the mixture of green tea polyphenols is effective in stimulating aged keratinocytes to generate biological energy, which helps to protect DNA and possibly renew cell division (Hsu et al., 2003).

Generally, it is recommended to use natural products for skin health since most of commercial skin care products available on the market are formulated with synthetic chemical ingredients that can lead to harmful effects on the skin. Sometimes commercial skin care products are composed of harsh chemicals such as surfactants, detergents, and preservatives that can quickly dry up the skin while taking away the natural protective oils from the skin, causing abrupt irritations and permanently drying out the skin, which can cause allergic reactions. Currently green tea extracts are used in a number of skin health restoration and cosmetic products, such as creams, toners, scrubs, lotions, treatments, cleansing solutions, and gels. These products boost microcirculation and help ensure healthy skin tone and function. The also provide desired astringent claims and help to maintain the elasticity and firmness of the skin while protecting the cells in the skin from premature cell death from radiation or extravagant free radicals. Furthermore, the benefit of green tea oil is widely accepted as a natural ingredient for skin health, as it can be derived from air-dried green teas only and is absolutely free from saponin as well as fatty acids. Its clear appearance makes it suitable for cosmetic formulations. Green tea polyphenol-formulated skin care products offer a number of amazing therapeutic benefits, as green tea polyphenols have been recognized as natural healing products for several centuries, caring for many types of nuisances and aches. Such formulations may contain several beneficial substances, such as natural theanine and natural caffeine, that can relax and rejuvenate the human skin. In addition, caffeine can cause the skin to look fresher, livelier, and younger by tightening up any slackness and removing small wrinkles, thus giving the skin a more tranquil and brisker appeal.

Moreover, it is well known that active oxygen species developed when exposed with UV rays can severely damage skin via attacking DNA in skin cells, oxidizing collagen, and stimulating tyrosinase activities (Vayalil et al., 2004). Studies have recently confirmed that application of green tea extracts protects against skin damage related to ultraviolet radiation from the sun, wherein it provides an extra level of protection for value-added markets. In particular, green tea polyphenols are known to absorb UVA and UVB rays and have emerged as a useful substance in skin care (sunscreen) applications (McCook et al., 1994). Specifically in Europe, an application of green tea polyphenols on skin has been developed to eliminate active oxygen species as well as to inhibit enzymes, and is increasingly being used for addressing problems of wrinkles and spots, including for lowering the possibilities of skin-related cancers (An et al., 2005a; Jeon et al., 2009). Further, products containing green tea polyphenols as an active ingredient are being developed and evaluated for promoting skin regeneration, and for the healing or treatment of certain epithelial conditions, such as aphthous ulcers, psoriasis, rosacea, and actinic keratosis.

Since green tea polyphenols have shown great potential for antiaging skin effects, natural antioxidative, enzyme-inhibitory, antibacterial, deodorant, astringent effects, and other cosmetic products (e.g., soap, toothpaste, and other miscellaneous goods) are appearing on the market that contain green tea extracts and are available OTC at drugstores and specialized cosmetics shopping counters. Green tea extract-formulated toothpastes and mouthwashes for oral care are now readily available in the marketplace. Oral care cosmetic products leverage a variety of multifunctional beneficial effects of green tea polyphenols, such as antibacterial and enzyme-inhibitory effects against oral bacteria, including cavity bacteria associated with tooth decay and gum disease, a deodorant effect against bad breath, and an astringent effect to tighten gums and promote a refreshing feeling in cavities (Hsu, 2005). Apart from the above, recently green tea polyphenol-formulated soaps, shampoos, and skin lotions have also attracted attention in the Japanese market, leading to increasing sales. Another beneficial use of green tea polyphenols could be as a natural perfume for a fresh and crisp scent, along with the relaxation effect for home and commercial interiors. In particular, oral cosmetic products for fragrance, for body odor reduction, and to diminish symptoms of atopic dermatitis are well accepted by consumers of all ages. Additional external uses of green tea extracts are for the treatment of sweaty and stinky feet. Compared to other OTC treatments, which usually have a scented odor, green tea extracts have a beneficial effect on odor elimination, but also get rid of odor-causing bacteria that thrive in sweat. Natural cures are delivered by green tea extract-formulated deodorants due to natural tannic acid that kills the bacteria that grow in the sweat of feet and also helps to prevent blisters.

16.6 Green Tea Polyphenols in Hair Care and Hair Cure Cosmetics

The popularity of green tea extracts is even increasing in hair care and hair cure applications. To date, some studies have reported that green tea polyphenols have the ability to cure hair loss, dandruff, baldness, and psoriasis. It seems that green tea polyphenols could be touted as a potential cure and provide a better alternative to many existing solutions, such as toupees that usually flap, minoxidil, hair plants, and surgery, which are either expensive or have minimum success in counteracting hair issues, especially hair loss and baldness. Much evidence has shown that green tea polyphenols are considered an anti-inflammatory and have stress-inhibitory characteristics, and also there is evidence that stress inhibits hair growth. Various research is focused on the antioxidant activity of green tea polyphenols to prove the plausible hair care/cure evidence of green tea extracts (Sueoka et al., 2001). It is believed that production of tumor necrosis factor-alpha (TNF-α), which has been implicated in cancer and related inflammatory diseases (arthritis, rheumatoid arthritis, etc.), could be suppressed by green tea polyphenols and prevent androgenetic hair loss/baldness. Actually, dihydrotestosterone (DHT) is the major hormone that regulates hair growth during pubescence. A sufficient amount of green tea polyphenols consumption every day increases the content of sex hormone binding globulin in the bloodstream, which helps to capture testosterone effectively before it is transformed to DHT. Thus, a reduced concentration of DHT in the bloodstream eventually protects hair follicles of the individual prone to have DHT-induced baldness. Other research has reported the role of green tea polyphenols on hair follicles and dermal papilla cells (found in hair follicles), which dictates the pattern of baldness as well as regulates hair growth (Kwon et al., 2007). Based on research evidence, some lotion-type treatments are available that contain green tea extract alcohol tincture for hair care/cure applications.

In addition to the above, green tea polyphenols containing hair treatments are also being examined for use against psoriasis and dandruff. Green tea extracts are widely known for soothing skin and inhibiting inflammation. Usually psoriasis causes unnecessary skin growth wherein fresh layers of skin develop before the old skin layers peel off, resulting in scaly and lesioned skin. Green tea polyphenols help to normalize the skin growth cycle by controlling the caspase-14 protein in skin cells, which is responsible for the normal growth of skin (Esfandiari and Kelly, 2005). Also, shampoos containing green tea polyphenols are available to eliminate affliction from dandruff and are free from carcinogenic substances. Hair conditioners formulated with green tea extracts can have multiple care/cures for hair, as they contain antioxidant properties and also can be a source of panthenol, which is a pro-vitamin B and a well-known component in hair conditioners, which strengthens as well as softens the hair without split end formation.

16.7 Other Industrial Applications

In addition to the aforementioned applications of green tea polyphenols, some researchers have also tried to develop cloth, paper, and filters containing green tea catechins that could be used for bed linens, handkerchiefs, socks, and air conditioners. With an aim to stabilize the catechins and improve their functions in particular applications, catechins are being formulated in synthetic resin, such as urethane and polypropylene, in combination with inorganic compounds such as silica. The resultant hybrid catechins are already being used in air conditioner filters (Inagaki et al., 2004). Such encapsulation or polymer complexes of catechins provides stable and long-lasting antibacterial and deodorant properties of green tea catechins, which can be characterized with minimal leaching of catechins from the filters when washed with water, compared to the filters containing free catechins. Also, paints formulated with catechins and their fluoride derivatives have been developed and indicate that there will be an increased use of functional nonfood applications wherein the catechins are combined with other substances.

Another recent development in the field is catechin-treated nanofibers, which are made by electrospinning technology (Nishio et al., 2008). Figure 16.6 displays an electron microscopic photograph of a representative nanofiber containing 5% catechins based on polylactic acid as an electrospun substrate. Such nanofibers are designed for growth inhibition of influenza viruses due to the intrinsic effects of catechins and are expected to be applied

FIGURE 16.6
Scanning electron micrograph of catechin nanofiber.

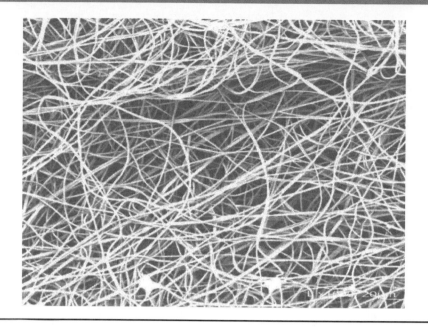

to high-function filters and proprietary face masks due to their good air permeability, despite their nanosized functional structure.

16.8 Conclusion

In summary, green tea polyphenols provide both therapeutic and functional health benefits in several food, beverage, cosmetics, and other industrial applications. The wavering fortunes of the processed green tea extract manufacturing industry are continuously being revived due to stringent worldwide regulatory and socioeconomic crises leading to a cost-price imbalance. This chapter particularly highlights the most recent and groundbreaking developments and applications of green tea polyphenols based on extraordinary research available in literature. The benefits of green tea polyphenols in the fortification of food and beverages, and their acceptance as supplements, are examples of consumer awareness and preference for selective natural antioxidants in their routine diets. Since there is fierce competition in the marketplace, green tea polyphenol extracts are a renewed focus of distinction to infiltrate the novel diligent business sectors with increasing efforts to search for innovative applications while protecting their originality and valid scientific credibility. Considering the scope of the marketing research and summary of the major findings translated to commercial applications, it is quite possible to learn market growth and future opportunities to design the product and application trends and competitive analysis. It would also bring new insightful for information on green tea industry dynamics, structure, acquisitions, legislation, and prolific challenges influencing the total tea polyphenols extract market. Although there is stringent European legislation hindering the growth of the green tea extract market, the major drivers, such as the growing number of positive research-driving market strategies and advancements due to increasing consumer demands, could be positively attributed to growth of the dietary supplements market and mounting demand for a functional foods and beverages market. This is clearly reflected in the market engineering measurement analysis, which is the basics of geographic, pricing, and market forecast analysis trends. Keeping in mind the competitive analysis, competitive market structure, competitive factors, product life cycle analysis, and competitive scenario analysis of market share analysis, which motivates market expansion, the strategic recommendation and evaluation could be designed to overcome challenges associated with the green tea polyphenol extracts industry.

References

Ahmad, N., and Mukhtar, H. 2001. Cutaneous photo chemoprotection by green tea: A brief review. *Skin Pharmacol Appl Skin Physiol*, 14: 69–76.

An, B.J., Kwak, J.H., Park, J.M., Lee, J.Y., Lee J.T., Son J.H., Jo, C., and Byan, Y.W. 2005. Inhibition of enzyme activities and the antiwrinkle effect of polyphenol isolated from the persimmon leaf (Diospyros kaki folium) on human skin. *Dermatol Surg*, 31(s1): 848–855.

An, B.J., Kwak, J.H., Park, J.M., Son, J.H., Lee, J.Y., Park, T.S., and Kim, S.O. 2005. Physiological activity of irradiated green tea polyphenol on the human skin. *Am J Chin Med*, 33: 535–546.

Awika, J.M., Rooney, L.W., Wu, X., Prior, R.L., and Cisneros-Zevallos, L. 2003. Screening methods to measure antioxidant activity of sorghum (*Sorghum bicolor*) and sorghum products. *J Agric Food Chem*, 51: 6657–6662.

Barry-Ryan, C., Martin-Diana, A.B., and Rico, D. 2008. Green tea extract as a natural antioxidant to extend the shelf-life of fresh-cut lettuce. *Inn Food Sci Emerg Technol*, 9: 593–603.

Bozkurt, H. 2006. Utilization of natural antioxidants: Green tea extract and Thymbra spicata oil in Turkish dry-fermented sausage. *Meat Sci*, 73: 442–450.

Bruhn, C. 2000. Food labelling: Consumer needs. In *Food labeling*, 5–18, ed. J. Ralph. Cambridge: Woodhead Publishing Ltd.

Cherry, J.P. 1999. Improving the safety of fresh produce with antimicrobials. *Food Technol*, 53: 54–59.

Chiu, A., and Kimball, A.B. 2003. Topical vitamins, minerals and botanical ingredients as modulators of environmental and chronological skin damage. *Br J Dermatol*, 149: 681–691.

de Dicastillo, C.L., Nerín, C., Alfaro, P., Catalá, R., Gavara, R., and Hernández-Muñoz, P. 2011. Development of new antioxidant active packaging films based on ethylene vinyl alcohol copolymer (EVOH) and green tea extract. *J Agric Food Chem*, 59: 7832–7840.

Draelos, Z.D. 2001. Botanicals as topical agents. *Clin Dermatol*, 19: 474–477.

Esfandiari, A., and Kelly, A.P. 2005. The effects of tea polyphenolic compounds on hair loss among rodents. *J Natl Med Assoc*, 97: 1165–1169.

Fujii, A., Okubo, T., Ishigaki, S., Aoi, N., Juneja, L.R., Ozaki, T., and Yoshida, S. 2004. Use of green tea extract on remineralization and acid resistance in bovine dental slabs. *J Jpn Council Adv Food Ingred Res*, 7: 33–39.

Gardner, E.J., Ruxton, C.H.S., and Leeds, A.R. 2007. Black tea—Helpful or harmful? A review of the evidence. *Eur J Clin Nutr*, 61: 3–18.

Gramza-Michalowska, A.G., and Regula, J. 2007. Use of tea extracts (*Camelia sinensis*) in jelly candies as polyphenols sources in human diet. *Asian Pac J Clin Nutr*, 16: 43–46.

Hara, Y., and Honda, M. 1990. The inhibition of α-amylase by tea polyphenols. *Agric Biol Chem*, 54: 1939–1945.

Harman, D. 2006. Free radical theory of aging: An update increasing the functional life span—Understanding and modulating aging. *Ann NY Acad Sci*, 1067: 10–20.

Hase, T., Komine, Y., Meguro, S., Takeda Y., Takahashi, H., Matsui, Y., Inaoka, S., Katsuragi, Y., Tokimitsu, I., Shimasaki, H., and Hiroshige, I. 2001. Anti-obesity effects of tea catechins in humans. *J Oleo Sci*, 50: 599–605.

Hsu, S. 2005. Green tea and the skin. *J Am Acad Dermatol*, 52: 1049–1059.

Hsu, S., Bollag, W.B., Lewis, J., Huang, Q., Singh, B., and Sharawy, M. 2003. Green tea polyphenols induce differentiation and proliferation in epidermal keratinocytes. *J Pharmacol Exp Ther*, 306: 29–34.

Inagaki, J., Katou, R., Suga, R., Nakajima, T., Mori, Y., Souma, N., Hashiguchi, K., and Gensui, K. 2006. Air cleaner, functional filter and method of manufacturing the filter, air cleaning filter and air cleaner device. U.S. Patent No. 2006/0728086 A1.

Jeon, H.Y., Kim, J.K., Kim, W.G., and Lee, S.J. 2009. Effects of oral epigallocatechin gallate supplementation on the minimal erythema dose and UV-induced skin damage. *Skin Pharmacol Physiol*, 22: 137–141.

Kajimoto, O., Kajimoto, Y., and Kakuda, T. 2003. Tea catechins reduce serum cholesterol levels in mild and borderline hypercholesterolemia patients. *J Clin Biochem Nutr*, 33: 101–111.

Kaur, C., and Kapoor, H.C. 2001. Antioxidants in fruits and vegetables—The millennium's health. *Int J Food Sci Technol*, 36: 703–725.

Kobayashi, M., Unno, T., Suzuki, Y., Nozawa, A., Sagesaka, Y., Kakuda, T., and Ikeda, I. 2005. Heat-epimerized tea catechins have the same cholesterol-lowering activity as green tea catechins in cholesterol-fed rats. *Biosci Biotechnol Biochem*, 69: 2455–2458.

Kwon, O.S., Han, J.H., Yoo, H.G., Chung, J.H., Cho, K.H., Eun, H.C., and Kim, K.H. 2007. Human hair growth enhancement *in vitro* by green tea epigallocatechin-3-gallate (EGCG). *Phytomedicine*, 14: 551–555.

Lal, G.G. 2007. Getting specific with functional beverages. *Food Technol*, 61: 25–31.

Martin-Diana, A., Rico, D., Frias, J., Mulcahy, J., Henehan, G., and Barry-Ryan, C. 2006. Whey permeate as a bio-preservative for shelf-life maintenance of fresh-cut vegetables. *Innov Food Sci & Emerg Technol*, 7(1–2): 112–123.

Masuda, T., Inaba, Y., Maekawa, T., Takeda, Y., Yamaguchi, H., Nakamoto, K., Kuninaga, H., Nishizato, S., and Nonaka, A. 2003. Simple detection method of powerful antiradical compounds in the raw extract of plants and its application for the identification of antiradical plant constituents. *J Agric Food Chem*, 51: 1831–1838.

McCook, J.P., Meyers, A.J., Dobkowski, B.J., and Burger, A.R. 1994. Cosmetic sunscreen composition containing green tea and a sunscreen. U.S. Patent No. 5306486.

Mintel Global New Products Database. 2007. Energy drinks. Chicago: Mintel International Group Ltd.

Mintel Global New Products Database. 2009. Energy drink ingredients continue down unhealthy path. Available from http://www.mintel.com/press-release/Energy-drink-ingredients-continue-down-unhealthy-path?id=386 (accessed December 21, 2011).

Nishihara, Y., Aoki, T., Ohkawa, T., Wada, Y., Makimura, M., Hirasawa, M., and Otake, S. 1993. Inhibitory effects of food containing sucrose added tea catechins on dental caries in rats. *Nihon Univ J Oral Sci*, 19: 217–223.

Nishio, T., Sugino, T., Okubo, T., Kaihatsu, K., and Kaihatsu, S. 2008. Preparation of PLLA sheet with tea catechin. *Proc Soc Fiber Sci Technol*, 76.

Nissen, L.R., Byrne, D.V., Bertelsen, G., and Skibsted, L.H. 2004. The antioxidative activity of plant extracts in cooked pork patties as evaluated by descriptive sensory profiling and chemical analysis. *Meat Sci*, 68: 485–495.

Oiwa, T., Sakanaka, S., Kim, M., Ozaki, T, Kashiwagi, M., Hasegawa, Y., Yoshihara, Y., and Yoshida, S. 1993. Inhibitory effect of tea polyphenols (Sunphenon) on human plaque formation. *Jpn J Prd Dent*, 31: 247–250.

Okubo, T. 2004. Research and development on green tea catechin as antioxidant. *Oleoscience*, 4: 401–407.

Shah, N.P., Ding, W.K., Fallourd, M.J., and Leyer, G. 2010. Improving the stability of probiotic bacteria in model fruit juices using vitamins and antioxidants. *J Food Sci*, 75: M278–M282.

Sharma, A., and Zhou, W. 2011. A stability study of green tea catechins during the biscuit making process. *Food Chem*, 126: 568–573.

Song, W.O., and Chun, O.K. 2008. Tea is major source of flavan-3-ol and flavanol in the US diet. *J Nutr*, 138: 1543S–1547S.

Sueoka, N., Suganuma, M., Sueoka, E., Okabe, S., Matsuyama, S., Imai, K., Nakachi, K., and Fujiki, H. 2001. A new function of green tea: Possible prevention of hair loss and other lifestyle-related diseases. *Ann NY Acad Sci*, 928: 274–280.

Terajima, T., Terajima, H., Togashi, K., Hasegawa, Y., Takahashi, R., Ozaki, T., Sakanaka, S., Kim, M., and Yoshida, S. 1997. Preventive effects of tea polyphenols (Sunphenon) on plaque formation in men. *Nihon Univ Dent J*, 71: 654–659.

U.S. Food and Drug Administration (USFDA). 2011. CFR—Code of Federal Regulations Title 21. Cite: 21 CFR 175.300: 21(3).

Valko, M., Leibfritz, D., Moncola, J., Cronin, M.T.D., Mazura, M., and Telser, J. 2007. Free radicals and antioxidants in normal physiological functions and human disease. *Int J Biochem Cell Biol*, 39: 44–84.

Vayalil, P.K., Mittal, A., Hara, Y., Elmets, C.A., and Katiyar, S.K. 2004. Green tea polyphenols prevent ultraviolet light-induced oxidative damage and matrix metalloproteinases expression in mouse skin. *J Invest Dermatol*, 122: 1480–1487.

Wiseman, S.A., Balentine, D.A., and Frei, B. 1997. Antioxidants in tea. *Crit Rev Food Sci Nutr*, 37: 705–718.

Index

Printed and bound by CPI Group (UK) Ltd, Croydon, CR0 4YY

21/10/2024

01777042-0015